TEXTS IN COMPUTER SCIENCE

Editors
David Gries
Fred B. Schneider

TEXTS IN COMPUTER SCIENCE

(continued after index)

Dexter C. Kozen

Theory of Computation

With 75 Illustrations

 Springer

Dexter C. Kozen
Department of Computer Science
Upson Hall
Cornell University
Ithaca, NY 14853-7501, USA
kozen@cs.cornell.edu

Series Editors:
David Gries
Department of Computer Science
Upson Hall
Cornell University
Ithaca, NY 14853-7501, USA

Fred B. Schneider
Department of Computer Science
Upson Hall
Cornell University
Ithaca, NY 14853-7501, USA

British Library Cataloguing in Publication Data
A catalogue record for this book is available from the British Library

Library of Congress Control Number: 2005937504

ISBN-10: 1-84628-297-7
ISBN-13: 978-1-84628-297-3

Printed on acid-free paper

Printed in the United States of America   (HAM)

9 8 7 6 5 4 3 2 1

Springer Science+Business Media

springer.com

To Frances

Preface

These are my lecture notes from CS682: Theory of Computation, a one-semester course for first-year graduate students in computer science at Cornell, which I have taught off and on for many years. I took the course myself as a PhD student at Cornell from Juris Hartmanis, and his influence can be seen in the presentation and selection of topics.

Overview and Goals

The course serves a dual purpose: to cover core material in the foundations of computing for graduate students in computer science preparing for their PhD qualifying exams, and to provide an introduction to some more advanced topics in the theory of computational complexity for those intending to pursue further study in the area. The course is thus a mixture of core and advanced material.

Most of the course is concerned with *computational complexity*, or the classification of computational problems in terms of their inherent complexity. This usually refers to time or space usage on a particular computational model, but may include other complexity measures as well, such as randomness, number of alternations, or circuit size or depth. We include a rigorous treatment of computational models, including deterministic, nondeterministic, and alternating Turing machines, circuits, probabilistic machines, interactive proof systems, automata on infinite objects, and various logical formalisms. Also included are various approximation and inapproximation results and some lower bounds. According to most treatments, the complexity universe stops at polynomial space, but we also look at higher levels of complexity all the way up through the primitive recursive functions, partial recursive functions, and the arithmetic and analytic hierarchies.

Despite the title of this book, there are many beautiful areas of theoretical computer science that I could not cover for lack of time and space.

Intended Audience

The course is aimed at an audience of advanced undergraduates and first-year graduate students in computer science or mathematics with an interest in the theory of computation and computational complexity. It may also be of interest to computer professionals and other scientists who would like to learn more about the classical foundations of computing and contemporary research trends.

Familiarity with the content of standard undergraduate courses in algorithms and the theory of computation are helpful prerequisites. In particular, we make free use of discrete mathematical structures, including graphs, trees, and dags; $O(\ )$ and $o(\ )$ notation; finite automata, regular expressions, pushdown automata, and context-free languages; and Turing machines, computability, undecidability, and diagonalization. There are many good undergraduate texts that cover this material, for example, [61, 76, 113].

Organization and Features

The course consists of 41 primary lectures and a handful of supplementary lectures covering more specialized or advanced topics. In my previous texts [75, 76], the basic unit is a *lecture*, which is a more or less self-contained chunk of 4–7 pages. I have received much positive feedback regarding this organization, so I have stuck with it here. In addition to the lectures, there are 12 homework sets and several miscellaneous homework exercises of varying levels of difficulty, many with hints and complete solutions.

Acknowledgments

Many people have contributed in many ways to this work, and there is not enough I can ever say in thanks. First and foremost, I owe an incalculable debt of gratitude to my wife, Frances, for her constant love, patience, and forbearance.

My administrative assistants, Rosemary Adessa, Beth Howard, and Kelly Patwell, deserve special thanks for making sure that the real world did not escape my attention for too long a period.

The professional staff at Springer, Catherine Brett, Valerie Greco, Natacha Menar, and Wayne Wheeler, have been most accommodating throughout the entire production process. It has been a pleasure working with them.

There are many students who over the years have contributed to these notes and kept me honest. In particular, I would like to thank Kamal Aboul-Hosn, Adam Arbree, Adam Barth, Yuri Berkovich, Chavdar Botev, Greg Bronevetsky, Walter Chen, Pablo Fierens, Suman Ganguli, Dan Grossman, Vincent Gu, Milos Hasan, Yannet Interian-Fernandez, Aaron Kaufman, Omar H. Khan, Łucja Kot, Brian Kulis, Milind Kulkarni, Ashwin Machanavajjhala, David Martin, Wojtek Moczydlowski, Antonio Montalban, Jonathan Moon, Bryan Renne, Andrew Scukanec, Alexa Sharp, Zoya Svitkina, and Ryan Williams.

I am especially indebted to my teaching assistants Suresh Chari, Reba Schuller, and Chaitanya Swamy for their help with proofreading, preparation of solution sets, and occasional lecturing.

I am grateful to my colleagues Jin-Yi Cai, Robert Constable, Uriel Feige, John Hopcroft, Jon Kleinberg, Bakhadyr Khoussainov, Stephen Mahaney, Anil Nerode, Christos Papadimitriou, Ronitt Rubinfeld, Erik Meineche Schmidt, Bart Selman, Richard Shore, Michael Sipser, and Éva Tardos for their interest in this project and for valuable comments and interesting exercises. Uriel's and Ronitt's notes on probabilistically checkable proofs were an especially useful source. Finally, I wish to express my sincerest gratitude to Juris Hartmanis, who has taught the course many times, and who is always an inspiration.

I would be most grateful for suggestions or criticism from readers.

Cornell University
Ithaca, New York

Dexter Kozen
October 2005

Contents

Hints and Solutions   **317**

References   **391**

Notation and Abbreviations   **399**

Index   **409**

Lectures

Lecture 1

The Complexity of Computations

In this course we are concerned mainly with two broad issues:

- The definition and study of various computational models and programming constructs, with an eye toward understanding their relative power and limitations;

- The classification of computational problems in terms of their inherent complexity. This usually means time or space complexity on a particular model, but may include other measures as well, such as randomness, number of alternations, or circuit size.

This area of study is generally known as *computational complexity theory*. It has deep roots in the theory of computability as developed by Church, Kleene, Post, Gödel, Turing, and others in the first half of the twentieth century.

It is widely acknowledged that the genesis of the theory as we know it today was the 1965 paper, "On the Computational Complexity of Algorithms," by Juris Hartmanis and Richard Stearns [55]. Although mathematicians had previously studied the complexity of algorithms, this paper showed that computational problems often have an inherent complexity, which can be quantified in terms of the number of time steps needed on a simple model of a computer, the multitape Turing machine, but is largely independent of the particular model of computation. In a subsequent paper with Philip Lewis [115], they showed that space complexity (number of tape cells used) can also be treated as a complexity measure in much the same way as time. Other pioneering work on computational complexity that appeared around the same time included papers of Cobham [30] and Edmonds [37], who are generally credited with the invention of the notion of polynomial time, and Hennie and Stearns [60]. Edmonds was apparently the first to conjecture that $P \neq NP$. These papers were immediately recognized as a fundamental advance. Indeed, it was the original Hartmanis and Stearns paper [55] that gave the name *computational complexity* to the discipline.

The fundamental contribution of Hartmanis and Stearns was not so much the specific results regarding the complexity of Turing machine computations, but the assimilation of concrete notions of resource usage into a general theory of computational complexity. Although they worked primarily with multitape Turing machines, they argued rightly that the concepts were universal and that the same behavior would emerge in any reasonable model. The fundamental notion of *complexity class* laid down in their original paper still pervades the field. The theory has been further generalized by Manuel Blum [16] using an abstract notion of complexity measure, and many results generalize to this more abstract setting (see Supplementary Lecture J). Other resources besides time and space, from area in VLSI layout problems to randomness in probabilistic computation, have been successfully treated in this framework.

Today the field also includes a wide variety of notions such as probabilistic complexity classes, interactive proof systems, approximation and inapproximation results, circuit complexity, and many others. The primary goal of this field is to understand what makes computational problems complex, with the hope that by doing so, we might better understand how to make them simpler.

Turing Machines

A convenient starting point for our study is Turing machine (TM) complexity. Turing machines were invented in 1936 by Alan M. Turing [123] at around the same time as several other formalisms purporting to capture the notion of *effective computability*: Post systems [94, 95], μ-recursive functions [47], λ-calculus [28, 71], and combinatory logic [109, 35].

All these models are computationally equivalent in the sense that they can simulate one another. This led Alonzo Church to formulate *Church's thesis* [29, 123], which states that these models exactly capture our intuitive notion of effective computability. But one aspect of computability that Church's thesis does not address is the notion of complexity. For example, it is true that a deterministic two-counter automaton can simulate an arbitrary nondeterministic multitape Turing machine, but only at great cost. We are thus left with the task of defining reasonable models in which these complexity questions can be formulated and studied.

The Turing machine is a good, albeit imperfect, model for defining basic time and space complexity, because at least for higher levels of complexity, the definitions are robust and reflect fairly accurately our expectations of real-life computation.

The One-Tape Turing Machine: A Quick Review

We review here the definition of deterministic, one-tape Turing machines that act as *acceptors*. Inputs to such a machine are finite-length strings over some fixed finite alphabet Σ. The length of a string x is denoted $|x|$. We also use the same notation $|A|$ for the size (cardinality) of a set A. The unique string of length 0 is called the *null string* and is denoted ε. The set of all finite-length strings over Σ is denoted Σ^*.

Informally, the machine has a finite set of states Q, a semi-infinite tape that is delimited on the left end by an endmarker $\vdash$ and is infinite to the right, and a head that can move left and right over the tape, reading and writing symbols from a finite alphabet Γ that contains Σ as a subset.

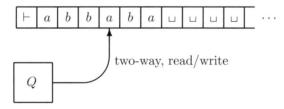

The input string is initially written on the tape in contiguous tape cells snug up against the left endmarker. The infinitely many cells to the right of the input all contain a special blank symbol $\sqcup$.

The machine starts in its start state s with its head scanning the left endmarker. In each step it reads the symbol on the tape under its head. Depending on that symbol and the current state, it writes a new symbol on that tape cell, moves its head either left or right one cell, and enters a new state. The action it takes in each situation is determined by a transition function δ. It *accepts* its input by entering a special accept state t and *rejects* by entering a special reject state r. If it either accepts or rejects its input, then it is said to *halt* on that input. On some inputs it may run infinitely without ever accepting or rejecting, in which case it is said to *loop* on that input. The subset of Σ^* consisting of all input strings accepted by the TM M is denoted $L(M)$.

Formally, a *deterministic one-tape Turing machine* is a 9-tuple

$$M \; = \; (Q, \Sigma, \Gamma, \vdash, \sqcup, \delta, s, t, r),$$

where

- Q is a finite set of *states*;

- Σ is a finite *input alphabet*;

- Γ is a finite *tape alphabet* containing Σ as a subset;

- $\sqcup \in \Gamma - \Sigma$ is the *blank symbol*;

- $\vdash \in \Gamma - \Sigma$ is the *left endmarker*;

- $\delta : Q \times \Gamma \to Q \times \Gamma \times \{L, R\}$ is the *transition function*;

- $s \in Q$ is the *start state*;

- $t \in Q$ is the *accept state*; and

- $r \in Q$ is the *reject state*, $r \neq t$.

Intuitively, $\delta(p, a) = (q, b, d)$ means, "When in state p scanning symbol a, write b on that tape cell, move the head in direction d, and enter state q." The symbols L and R stand for left and right, respectively.

We restrict TMs so that the left endmarker is never overwritten with another symbol and the machine never moves off the tape to the left of the endmarker; that is, we require that for all $p \in Q$ there exists $q \in Q$ such that

$$\delta(p, \vdash) \; = \; (q, \vdash, R). \tag{1.1}$$

We also require that once the machine enters its accept state, it never leaves it, and similarly for its reject state; that is, for all $b \in \Gamma$ there exist $c, c' \in \Gamma$ and $d, d' \in \{L, R\}$ such that

$$\delta(t, b) \; = \; (t, c, d), \qquad \delta(r, b) \; = \; (r, c', d'). \tag{1.2}$$

We refer to the state set and transition function collectively as the *finite control*.

There are many variations on Turing machines: two-way infinite tapes, multiple tapes, multiple accept and reject states, various forms of stacks and counters. Most of these variations produce machines that are computationally equivalent in that they are all capable of accepting the same sets. However, as mentioned, they are not necessarily equivalent from the standpoint of resource usage.

A TM as described above is an *acceptor*; that is, for each input x, it either accepts x, rejects x, or loops on x. This amounts to computing a partial $\{0, 1\}$-valued function. TMs can also be equipped with an output tape and specified output alphabet Δ to compute partial functions with range Δ^*.

An important variation is the *nondeterministic* Turing machine. A deterministic TM, as defined above, has a uniquely determined next configuration from any given configuration as specified by its transition function δ. A nondeterministic machine has a fixed finite choice of moves at each transition. Formally, δ is a relation, not a function. The computation of a deterministic machine can be described as a sequence of configurations beginning with the start configuration. A nondeterministic machine, on the other hand, determines a tree of possible computations, the root of which is the start configuration. The children of any node are the possible configurations that can be reached in one step. A nondeterministic machine is said to *accept* its input if there is some path in the computation tree leading to an accept configuration.

Because a Turing machine is a finite object, it is possible to encode it as a sequence of symbols over some alphabet in such a way that the resulting codes can be read and interpreted by another Turing machine and simulated. This leads to the notion of a *universal Turing machine*.

See [61, 76, 113] for a more complete treatment.

Crossing Sequences

Let us reacquaint ourselves with Turing machines by deriving a couple of results from [59, 115] on Turing machine time and space usage. These results illustrate the use of a counting technique to show lower complexity bounds.

Theorem 1.1   *Let* $\Sigma = \{0, 1, \#\}$. *The set of* palindromes

$$\text{PAL} \overset{\text{def}}{=} \{z \in \Sigma^* \mid z = \text{rev}\, z\}$$

requires $\Omega(n^2)$ *time on a one-tape TM.*

Here $\text{rev}\, x$ denotes x written backwards. Note that this result holds for one-tape TMs only; PAL can be accepted in linear time on a two-tape machine.

Proof. Let M be any machine accepting PAL. Assume without loss of generality that whenever M accepts, it first moves to the right end of the nonblank portion of the tape before entering its accept state. For each n that is a multiple of 4, consider the action of M on elements of PAL of the form

$$\text{PAL}_n \overset{\text{def}}{=} \{x \,\#^{n/2}\, \text{rev}\, x \mid x \in \{0, 1\}^{n/4}\}.$$

All elements of PAL_n are of length n and $\text{PAL}_n \subseteq \text{PAL}$. For each $x \in \text{PAL}_n$ and each position i, $0 \le i \le n$, let $c_i(x)$ denote the sequence $q_1, q_2, \ldots, q_k$ of states of the finite control Q of M that M is in as it passes over the line between the ith symbol and the $i + 1$st in either direction while scanning x. This is called the *crossing sequence* at i. Let

$$C(x) \stackrel{\text{def}}{=} \{c_i(x) \mid n/4 \le i \le 3n/4\}.$$

Lemma 1.2   *If $x, y \in \text{PAL}_n$ and $x \ne y$, then $C(x) \cap C(y) = \varnothing$.*

Proof. Suppose $c \in C(x) \cap C(y)$, say $c = c_i(x) = c_j(y)$. Let x' be the prefix of x consisting of the first i symbols, and let y' be the suffix of y consisting of the last $n - j$ symbols. Then $x'y'$ is accepted by M, because the machine behaves to the left of c as if it were scanning x and to the right of c as if it were scanning y; in particular, it accepts. But $x'y'$ is not in PAL, because it is not a palindrome. This is a contradiction.   □

Resume proof of Theorem 1.1. Let m_x be the length of the shortest crossing sequence in $C(x)$. We show that some m_x, $x \in \text{PAL}_n$, has to be long. Let $m = \max\{m_x \mid x \in \text{PAL}_n\}$. Then

$$\sum_{i=0}^{m} |Q|^i \quad = \quad \frac{|Q|^{m+1} - 1}{|Q| - 1} \quad \ge \quad 2^{n/4}.$$

The left hand side of the inequality gives the number of possible crossing sequences of length at most m. The right hand side is the number of elements of PAL_n. The inequality holds because all the shortest crossing sequences for elements of PAL_n must be different, by Lemma 1.2. By taking logs it follows that

$$m \quad \ge \quad \Omega(n).$$

Then there must be an $x \in \text{PAL}_n$ such that $m_x \ge \Omega(n)$. Because m_x is the length of the shortest crossing sequence in $C(x)$, all crossing sequences in $C(x)$ are of length $\ge \Omega(n)$, thus it takes at least $\frac{n}{2} \cdot \Omega(n) = \Omega(n^2)$ time to generate all the crossing sequences in $C(x)$.   □

The next result uses the same technique to show that $o(\log \log n)$ workspace is no better than no workspace at all.

Theorem 1.3   *If M runs in $o(\log \log n)$ space, then M accepts a regular set.*

There is a nonregular set accepted in $O(\log \log n)$ space:

$$\{\#b_k(0)\#b_k(1)\#b_k(2)\# \cdots \#b_k(2^k - 1)\# \mid k \ge 0\},$$

where $b_k(i)$ denotes the k-bit binary representation of i (Homework 2, Exercise 4).

Proof. Assume without loss of generality that M has a read-only input tape and one read/write worktape, and that M always moves its input head all the way to the right of the input string before accepting. If M is $S(n)$ space-bounded, then on inputs of length n there are at most

$$N \;=\; q \cdot S(n) \cdot d^{S(n)} \tag{1.3}$$

possible configurations of state, workhead position, and worktape contents, where q is the number of states and d is the size of the worktape alphabet of M (the position of the input head is not counted). These data constitute the total information that can be transferred across a vertical line drawn at some position i in the input string as the input head passes over that line. For this proof, the *crossing sequence* at i consists of the sequence of such configurations occurring at position i in the input string in either direction. There are

$$\sum_{i=0}^{m} N^i \;=\; \frac{N^{m+1} - 1}{N - 1}$$

possible crossing sequences of length at most m.

Lemma 1.4 *If there is a fixed finite bound k on the amount of space used by M on accepted inputs, then $L(M)$ is a regular set.*

Proof. If M uses at most k worktape cells on all accepted inputs, we can modify M to mark off k cells initially (k can be kept in the finite control) and reject if the computation ever tries to use more than k cells. That way we can make sure that M uses no more than k tape cells on any input. But then no worktape memory is needed at all; the contents of the worktape can be kept in the finite control. Thus M is equivalent to a two-way finite automaton. □

Resume proof of Theorem 1.3. Suppose $L(M)$ is not regular. By Lemma 1.4, there is no fixed finite bound on the amount of worktape used on inputs in $L(M)$. Thus for each k, there exists a string $x \in L(M)$ of minimal length for which at least k worktape cells are used. Here "of minimal length" means that for all shorter strings in $L(M)$, M uses fewer than k worktape cells. Let $n = |x|$ and let c be a crossing sequence containing an occurrence of a configuration using k worktape cells. If c occurs in the right half of x, then the first $n/2$ crossing sequences must all be distinct, otherwise a section of x could be cut out to obtain a shorter string accepted with crossing sequence c, contradicting the minimality assumption. If c occurs in the left half, then the last $n/2$ crossing sequences must all be distinct for the same reason. In either case, there are at least $n/2$ distinct crossing sequences.

In order to have $n/2$ distinct crossing sequences, there must be a crossing sequence of length at least m, where

$$\frac{n}{2} \ \leq \ \sum_{i=0}^{m} N^i \ = \ \frac{N^{m+1} - 1}{N - 1}. \tag{1.4}$$

Moreover, no crossing sequence can be of length greater than $2N$, otherwise a configuration would be repeated in the crossing sequence twice in the same direction, thus M would be looping, contradicting the fact that x is accepted. Thus

$$m \ \leq \ 2N. \tag{1.5}$$

Combining (1.3), (1.4), and (1.5) and taking logs, we get

$$S(n) \ \geq \ \Omega(\log \log n).$$

$\square$

Lecture 2

Time and Space Complexity Classes and Savitch's Theorem

Let $T : \mathbb{N} \to \mathbb{N}$ and $S : \mathbb{N} \to \mathbb{N}$ be numeric functions, which serve as asymptotic time and space bounds for Turing machine computations. These functions are usually written as functions of a single numeric variable n standing for the length of the input string; for example, $\log n$, $(\log n)^2$, n, $n \log n$, n^3, $2^{(\log n)^2}$, 2^n, $n!$, 2^{2^n}, and so on.

Definition 2.1   *We say that a nondeterministic TM runs in time $T(n)$ or is $T(n)$ time-bounded if on all but finitely many inputs x, no computation path takes more than $T(|x|)$ steps before halting, where $|x|$ denotes the length of x.*

We say that a nondeterministic TM runs in space $S(n)$ or is $S(n)$ space-bounded if on all but finitely many inputs x, no computation path uses more than $S(|x|)$ worktape cells, where $|x|$ is the length of x.

In Definition 2.1, the "but finitely many" is there mainly to avoid trivial counterexamples involving the null string, but it is also often technically convenient in asymptotic complexity to be able to ignore small inputs. We can always store a finite amount of data in the finite control and do table lookup for finitely many inputs.

Note that to run in a certain amount of time or space, the machine must not exceed the stated bounds even on nonaccepting computations.

The following are the basic time and space complexity classes.

Definition 2.2

$$DTIME(T(n)) \quad \stackrel{\text{def}}{=} \quad \{L(M) \mid M \text{ is a deterministic multitape TM running in time } T(n)\},$$

$$NTIME(T(n)) \quad \stackrel{\text{def}}{=} \quad \{L(M) \mid M \text{ is a nondeterministic multitape TM running in time } T(n)\},$$

$$DSPACE(S(n)) \quad \stackrel{\text{def}}{=} \quad \{L(M) \mid M \text{ is a deterministic TM with read-only input tape and read/write worktape running in space } S(n)\},$$

$$NSPACE(S(n)) \quad \stackrel{\text{def}}{=} \quad \{L(M) \mid M \text{ is a nondeterministic TM with read-only input tape and read/write worktape running in space } S(n)\}.$$

If $\mathcal{A}$ is a complexity class, the set of complements of sets in $\mathcal{A}$ is denoted co-$\mathcal{A}$. Note: this is not the complement of $\mathcal{A}$!

Linear Speedup

The next result says that for the TM model, it does not make sense to measure time and space complexity any more accurately than to within a constant factor.

Theorem 2.3   *Let $T(n) \geq n + 1$ and $S(n) \geq \Omega(\log n)$. For any constant $c \geq 1$,*

$$
\begin{aligned}
DTIME(cT(n)) &\subseteq DTIME(T(n)), \\
NTIME(cT(n)) &\subseteq NTIME(T(n)), \\
DSPACE(cS(n)) &\subseteq DSPACE(S(n)), \\
NSPACE(cS(n)) &\subseteq NSPACE(S(n)).
\end{aligned}
$$

Proof sketch. Expand the tape alphabet so that c tape cells of the old machine can be compressed into one tape cell of the new machine. This allows c steps of the old machine to be simulated in one step of the new machine. For the time bounds, we may need an extra tape to compress the input.   □

Some Common Complexity Classes

Definition 2.4

$$
\begin{aligned}
LOGSPACE &\stackrel{\text{def}}{=} DSPACE(\log n), \\
NLOGSPACE &\stackrel{\text{def}}{=} NSPACE(\log n),
\end{aligned}
$$

$$P \overset{\text{def}}{=} DTIME(n^{O(1)}) = \bigcup_{k>0} DTIME(n^k),$$

$$NP \overset{\text{def}}{=} NTIME(n^{O(1)}) = \bigcup_{k>0} NTIME(n^k),$$

$$PSPACE \overset{\text{def}}{=} DSPACE(n^{O(1)}) = \bigcup_{k>0} DSPACE(n^k),$$

$$NPSPACE \overset{\text{def}}{=} NSPACE(n^{O(1)}) = \bigcup_{k>0} NSPACE(n^k),$$

$$EXPTIME \overset{\text{def}}{=} DTIME(2^{n^{O(1)}}) = \bigcup_{k>0} DTIME(2^{n^k}),$$

$$NEXPTIME \overset{\text{def}}{=} NTIME(2^{n^{O(1)}}) = \bigcup_{k>0} NTIME(2^{n^k}),$$

$$EXPSPACE \overset{\text{def}}{=} DSPACE(2^{n^{O(1)}}) = \bigcup_{k>0} DSPACE(2^{n^k}),$$

$$NEXPSPACE \overset{\text{def}}{=} NSPACE(2^{n^{O(1)}}) = \bigcup_{k>0} NSPACE(2^{n^k}).$$

Basic Inclusions

The inclusions

$$
\begin{aligned}
DTIME(T(n)) &\subseteq NTIME(T(n)), \\
DSPACE(S(n)) &\subseteq NSPACE(S(n))
\end{aligned}
$$

are trivial, because by definition every deterministic TM is also a nondeterministic TM.

Theorem 2.5   *Let $S(n) \geq \log n$. Then*

$$
\begin{aligned}
DTIME(T(n)) &\subseteq DSPACE(T(n)), \\
NTIME(T(n)) &\subseteq NSPACE(T(n)), \\
DSPACE(S(n)) &\subseteq DTIME(2^{O(S(n))}), \\
NSPACE(S(n)) &\subseteq NTIME(2^{O(S(n))}).
\end{aligned}
$$

Proof. The first two inclusions follow from the fact that a machine can scan at most one new worktape cell in every step, thus the space usage can be no greater than the running time.

For the last two inclusions, we show how a machine running in space $S(n)$ can be modified to clock itself and shut down after $2^{O(S(n))}$ steps without affecting the set accepted. Assume without loss of generality that the machine has a single read-only input tape and a single read/write work-tape (multiple worktapes can be simulated on separate tracks of a single worktape without loss of space by increasing the size of the worktape alphabet). If the machine has d worktape symbols and q states, then there are at most $qnS(n)d^{S(n)}$ possible configurations on inputs of length n, because there are q possible states, n input head positions, $S(n)$ worktape head positions, and $d^{S(n)}$ worktape contents. This number is at most $c^{S(n)}$ for some sufficiently large constant c. Any computation path of length greater than this must have a repeated configuration, therefore there is a shorter computation path with the same outcome (accept or reject) obtained by deleting the segment between the two occurrences of the repeated configuration. This says that if there exists an accepting computation path, then there exists one of length at most $c^{S(n)}$. We can count up to $c^{S(n)}$ on a separate track of the worktape and reject if the simulated computation has not halted by then. This takes no extra space if the counting is done in c-ary, and the time overhead is $O(S(n))$ steps per simulated step of the old machine, or $O(S(n)c^{S(n)})$ in all, which is still $2^{O(S(n))}$.   □

We can show an even stronger result that subsumes Theorem 2.5.

Theorem 2.6   *Let $S(n) \geq \log n$. Then*

$$NTIME(T(n)) \subseteq DSPACE(T(n)),$$
$$NSPACE(S(n)) \subseteq DTIME(2^{O(S(n))}).$$

Proof. For the first inclusion, we do a depth-first search on the computation tree of the given nondeterministic $T(n)$-time-bounded machine, constructing the computation tree on the fly. We accept if an accept configuration is ever encountered. It may seem at first that we need $O(T(n)^2)$ space to keep a stack of configurations for the depth-first search, because each configuration requires up to $T(n)$ space and the depth of the tree is $T(n)$; but actually we only need to keep on the stack a k-ary string giving the path from the start configuration to the configuration currently being visited, assuming that all nondeterministic choices are at most k-ary. We can reconstruct the current configuration in space $T(n)$ at any time by starting at the start configuration and simulating the computation of the nondeterministic machine, using the k-ary string to resolve nondeterministic choices.

For the second inclusion, assume first that $S(n)$ is *space-constructible*. That means that there exists a Turing machine that, when started with any string of length n written on its input tape, marks off $S(n)$ worktape cells and halts, never using more than $S(n)$ space in the process. Not all functions are space-constructible, but the natural ones listed above are. We show later how to get rid of this assumption.

Using the assumption of space-constructibility, we first mark off $S(n)$ worktape cells. We then write down all the configurations of the nondeterministic machine that use no more than $S(n)$ space. As argued in the proof of Theorem 2.5, there are at most $c^{S(n)}$ of them, and we can write them all down in at most $S(n)c^{S(n)}$ time. Now we inductively mark all configurations that are reachable from the start configuration, accepting if we ever mark an accept configuration. A coarse analysis of this procedure still gives a time bound of $d^{S(n)}$ for sufficiently large constant d.

To get rid of the space-constructibility assumption, instead of marking off $S(n)$ space initially, we do the entire procedure above for $S = 0, 1, 2, \ldots$. For each S, if we ever encounter a configuration reachable from the start configuration that wants to use more than S space, we set $S := S + 1$ and restart. We eventually hit $S(n)$, at which point no reachable configuration will try to use more space. The time is at most

$$\sum_{S=0}^{S(n)} d^S \ \leq \ \frac{d^{S(n)+1} - 1}{d - 1},$$

which is still $2^{O(S(n))}$. $\qquad\qquad\qquad\qquad\qquad\qquad\qquad\qquad\square$

Savitch's Theorem

Probably the most important open question in theoretical computer science is the $P{=}NP$ question. The corresponding question for space was solved in 1970 by Walter Savitch [108]. This result is known as Savitch's theorem.

Theorem 2.7 (Savitch's Theorem) *Let $S(n) \geq \log n$. Then*

$$NSPACE(S(n)) \ \subseteq \ DSPACE(S(n)^2).$$

In particular, $PSPACE = NPSPACE$.

Proof. We prove the result under the assumption that $S(n)$ is space-constructible. We can get rid of this assumption the same way as in Theorem 2.6. By the construction of Theorem 2.5, we can also assume without loss of generality that the given nondeterministic $S(n)$-space-bounded machine M is also $d^{S(n)}$ time-bounded for some constant d.

A *configuration* of M consists of a state, head positions, and worktape contents. The input string need not be explicitly represented. Encode configurations of M as strings over a finite alphabet Δ in some reasonable way, and let $d = |\Delta|$. For inputs of length n, a configuration of M is represented by a string in $\Delta^{S(n)}$, and there are $d^{S(n)}$ such strings. If $\alpha, \beta \in \Delta^{S(n)}$, write $\alpha \xrightarrow{\leq k} \beta$ if α, β represent legal configurations of M and α goes to β in k

or fewer steps according to the transition relation of M without exceeding the space bound $S(n)$.

The deterministic machine will implement a recursive procedure SAV of three arguments α, β, k that determines whether $\alpha \xrightarrow{\leq k} \beta$. Note that by the pigeonhole argument of Theorem 2.5, if $\alpha \xrightarrow{\leq k} \beta$ for any k at all, then $\alpha \xrightarrow{\leq k} \beta$ for some $k \leq d^{S(n)}$, so we can restrict our attention to numbers in this range. Writing k in d-ary requires at most $S(n)$ space.

Thus to determine whether M accepts x, it suffices to check whether start $\xrightarrow{\leq d^{S(n)}}$ accept, where start and accept are the start and accept configurations, respectively, which we can assume without loss of generality are unique. The deterministic machine first constructs $S(n)$, then calls SAV(start, accept, $d^{S(n)}$).

The recursive procedure SAV(α, β, k) operates as follows. If $k = 0$ or 1, it checks directly whether $\alpha = \beta$ or, in the case $k = 1$, whether α goes to β in one step. It returns "yes" if so, otherwise it returns "no". If $k \geq 2$, it loops through all $\gamma \in \Delta^{S(n)}$ in some order, say lexicographic. For each such γ, it calls SAV$(\alpha, \gamma, \lceil k/2 \rceil)$ and SAV$(\gamma, \beta, \lfloor k/2 \rfloor)$ to determine whether $\alpha \xrightarrow{\leq \lceil k/2 \rceil} \gamma$ and $\gamma \xrightarrow{\leq \lfloor k/2 \rfloor} \beta$, respectively. If both recursive calls return "yes", it returns "yes". Otherwise, it goes on to the next γ. If it goes through all γ without success, then it returns "no".

This is a deterministic procedure. One can show easily by induction on k that SAV(α, β, k) returns "yes" iff $\alpha \xrightarrow{\leq k} \beta$. Each instantiation of SAV requires a stack frame of size $S(n)$ for preserving local data across recursive calls. The stack can be kept on the worktape. The depth of the recursion is $\log_2 d^{S(n)} = O(S(n))$, because k is halved with each recursive call. Thus the total storage required is $O(S(n)^2)$. $\square$

Lecture 3

Separation Results

Deterministic Separation Results

Theorem 3.1 *Let $S(n)$ be space-constructible. Then there exists a set in $DSPACE(S(n))$ that is not in $DSPACE(S'(n))$ for any $S'(n) = o(S(n))$.*

Proof. We prove this with a diagonalization argument. Let $M_0, M_1, \ldots$ be a list of all Turing machines with binary input alphabet. We assume that the binary representation of i gives an encoding of the machine M_i that allows universal simulation; that is, a universal TM, given the number i in binary and another binary string x, can read a description of M_i from i and simulate M_i step by step on input x. We assume that all binary strings code some machine; if i is a nonsense string not corresponding to any TM, we just take M_i to be some trivial machine that always halts immediately. (Such an encoding scheme is called a *Gödel numbering*. We study Gödel numberings from an axiomatic point of view in Lecture 33.)

We further assume that in our encoding scheme, leading zeros in the binary representation of i are ignored. If $\#(x)$ is the number represented by the binary numeral x, then for all sufficiently large n there is a binary string x of length n such that $\#(x) = i$, therefore any M_i has arbitrarily large codes obtained by padding on the left with extra zeros.

Now build a machine M that on a binary string x of length n does the following:

1. lays off $S(n)$ space on its worktape (this is possible by assumption of space-constructibility in the statement of the theorem); and

2. simulates M_i on input x, where $i = \#(x)$, never exceeding $S(n)$ space.

Exactly one of the following events has to occur, and in each case M takes the specified action.

(i)   If there is enough space to do the complete simulation and M_i halts, then M does the opposite—if M_i accepts, then M rejects, and vice versa.

(ii) If M_i loops infinitely, then so does M.

(iii) If the simulation requires more than $S(n)$ space, then M just halts and rejects.

Because of our assumption about padding in the encoding scheme, every Turing machine is simulated on some input of length n for every sufficiently large n. The meaning of "sufficiently large" depends on the TM being simulated.

Now if M_i is $o(S(n))$ space-bounded, then for all sufficiently large inputs x for which $\#(x) = i$, M will simulate M_i on input x and the simulation will have enough space to complete. This is true even if M_i's tape alphabet is much larger than M's—the simulation overhead is at most a constant factor c, and for sufficiently large n the space required by M_i is less than $S(n)/c$. Thus M and M_i will differ on input x, so $L(M_i) \neq L(M)$.

We also do not have to worry about the simulated machine not halting: recall from the proof of Theorem 2.5 that for every space-bounded machine, there is an equivalent self-clocking machine that always halts and uses no more space.

Thus M differs on some input from every machine running in $o(S(n))$ space. □

There is a slight subtlety in the above argument. The universal machine M may require logarithmic space in the length of the encoding of M_i to carry out the simulation of M_i. This is needed to compare states and tape symbols in the encoded description of M_i. However, it only requires logarithmic space in the length of the *unpadded* encoding of M_i, which is $\log \log i$, and this number is arbitrarily small compared to the length of the input x because of padding. Thus all we need is for $S(n)$ to be unbounded above. But if $S(n)$ is bounded, there is no function $S'(n) = o(S(n))$, so in that case the theorem is trivially true.

Theorem 3.2   *Let $T(n)$ be time-constructible, $T(n) \geq n$. Then there exists a set in $DTIME(T(n))$ that is not in $DTIME(T'(n))$ for any $T'(n)$ such that $T'(n) \log T'(n) = o(T(n))$.*

Here the extra log factor is overhead for the simulation. The proof is similar to the proof of Theorem 3.1, and we leave it as an exercise (Miscellaneous Exercise 1).

Nondeterministic Separation Results

Padding is also the basis of a technique for separating nondeterministic space and time classes. For space classes, stronger results can be obtained using the Immerman–Szelepcsényi theorem, which we cover next time; but our purpose here is to illustrate the padding technique, which predated the Immerman–Szelepcsényi theorem by 15 years.

We illustrate the technique by showing that $NSPACE(n^4)$ contains a set not in $NSPACE(n^3)$. Suppose for a contradiction that $NSPACE(n^4) \subseteq NSPACE(n^3)$. We show how we could then use padding to lift this into a proof that $NSPACE(n^5) \subseteq NSPACE(n^4)$.

Let M be an arbitrary nondeterministic machine running in space n^5, $A = L(M)$. Consider the set

$$A' = \{x \#^{|x|^{5/4} - |x|} \mid x \in A\},$$

where $\#$ is a new symbol not in M's input alphabet. In other words, we pad the input with a string of $\#$'s of length $|x|^{5/4} - |x|$. Build a machine M' which on an input of the form $x\#^m$:

(i)   checks whether $m = |x|^{5/4} - |x|$; and

(ii)   if so, runs M on x, ignoring the $\#$'s.

Because M runs in space $|x|^5$ on input x, M' runs in space

$$|x|^5 = (|x|^{5/4})^4 = |x\#^{|x|^{5/4} - |x|}|^4,$$

thus $A' = L(M') \in NSPACE(n^4)$.

By our soon-to-be-proven-erroneous assumption, $A' \in NSPACE(n^3)$, therefore there exists a machine M'' accepting A' running in space n^3.

Now build a new machine M''' for A, which does the following on input x:

(i)   appends a string of $\#$s of length $|x|^{5/4} - |x|$ to the end of x; and

(ii)   runs M'' on the resulting string.

Then on input x, M'''' runs in space

$$|x\#^{|x|^{5/4}-|x|}|^3 \;=\; (|x|^{5/4})^3 \;=\; |x|^{15/4} \;\leq\; |x|^4,$$

and $L(M''') = A$.

We have shown that under the assumption

$$NSPACE(n^4) \;\subseteq\; NSPACE(n^3),$$

we can conclude that

$$NSPACE(n^5) \;\subseteq\; NSPACE(n^4).$$

Repeating this process two more times, we get

$$NSPACE(n^6) \;\subseteq\; NSPACE(n^5),$$
$$NSPACE(n^7) \;\subseteq\; NSPACE(n^6).$$

Combining all these inclusions, we would get $NSPACE(n^7) \subseteq NSPACE(n^3)$. But then

$$
\begin{aligned}
NSPACE(n^7) \;&\subseteq\; NSPACE(n^3) \\
&\subseteq\; DSPACE(n^6) \quad \text{by Savitch's theorem} \\
&\subsetneq\; DSPACE(n^7) \quad \text{by Theorem 3.1} \\
&\subseteq\; NSPACE(n^7),
\end{aligned}
$$

which is a contradiction.

This theorem can be strengthened (Miscellaneous Exercises 4, 5).

For further containment and separation results, and a more detailed treatment, see [63].

A Space-Constructible Function $S(n) \leq O(\log\log n)$

For $n \geq 3$, let $\ell(n)$ be the least positive number not dividing n. The number $\ell(n)$ is always a prime power, because if km does not divide n, and if k and m are relatively prime (have no common factors), then one of k or m must not divide n. Also $\ell(n) = O(\log n)$, which is easily shown using the fact that for any ℓ,

$$\prod_{\substack{p \leq \ell \\ p \text{ prime}}} p \;\geq\; 2^{\Omega(\ell)}$$

(see [51, Theorem 414, p. 341]).

This gives an unbounded space-constructible function $S(n)$ that is $O(\log\log n)$, namely the space needed by a particular TM accepting the set

$$A \;=\; \{a^n \mid \ell(n) \text{ is prime}\}.$$

This TM checks the defining condition for A by writing down $k = 2, 3, 4, \ldots$ in binary and checking for each k whether k divides n until it finds one that does not; this is $\ell(n)$. It can test whether k divides n by counting the length of the input mod k, which can be done by repeatedly counting up to k in binary on a second track of its worktape, moving its read head one cell to the right in each step.

When it finds $\ell(n)$, it checks whether it is prime. It can do this by computing the remainder of $\ell(n)$ modulo $2, 3, 4, \ldots$ using integer division on the worktape.

The amount of space used is no more than $\log \ell(n) = O(\log \log n)$. This is space-constructible because only the length of the input string is used: the input alphabet is a single-letter alphabet.

Surprisingly, the function $\lceil \log \log n \rceil$ itself is not space constructible. In fact, it can be shown that for any space-constructible function $S(n) \leq o(\log n)$, there must exist a value k such that $S(n) = k$ for infinitely many n; in other words,

$$\liminf_{\substack{n \geq 0 \; m \geq n}} S(m) \; < \; \infty.$$

This can be proved by a crossing sequence argument (Homework 6, Exercise 1).

Lecture 4

The Immerman–Szelepcsényi Theorem

In 1987, Neil Immerman [65] and independently Róbert Szelepcsényi [119] showed that for space bounds $S(n) \geq \log n$, the nondeterministic space complexity class $NSPACE(S(n))$ is closed under complement. The case $S(n) = n$ gave an affirmative solution to a long-standing open problem of formal language theory: whether the complement of every context-sensitive language is context-sensitive.

Theorem 4.1   (Immerman–Szelepcsényi Theorem)   *For $S(n) \geq \log n$, $NSPACE(S(n)) = co\text{-}NSPACE(S(n))$.*

Proof. For simplicity we first prove the result for space-constructible $S(n)$. One can remove this condition in a way similar to the proof of Savitch's theorem (Theorem 2.7).

The proof is based on the following idea involving the concept of a *census function*. Suppose we have a finite set A of strings and a nondeterministic test for membership in A. Suppose further that we know in advance the size of the set A. Then there is a nondeterministic test for *nonmembership* in A: given y, successively guess $|A|$ distinct elements and verify that they are all in A and all different from y. If this test succeeds, then y cannot be in A.

Let M be a nondeterministic $S(n)$-space bounded Turing machine. We wish to build another such automaton N accepting the complement of

$L(M)$. Assume we have a standard encoding of configurations of M over a finite alphabet Δ, $|\Delta| = d$, such that every configuration on inputs of length n is represented as a string in $\Delta^{S(n)}$.

Assume without loss of generality that whenever M wishes to accept, it first erases its worktape, moves its heads all the way to the left, and enters a unique accept state. Thus there is a unique accept configuration $\mathtt{accept} \in \Delta^{S(n)}$ on inputs of length n. Let $\mathtt{start} \in \Delta^{S(n)}$ represent the start configuration on input x, $|x| = n$: in the start state, heads all the way to the left, worktape empty.

Because there are at most $d^{S(n)}$ configurations M can attain on input x, if x is accepted then there is an accepting computation path of length at most $d^{S(n)}$. Define A_m to be the set of configurations in $\Delta^{S(n)}$ that are reachable from the start configuration $\mathtt{start}$ in at most m steps; that is,

$$A_m = \{\alpha \in \Delta^{S(n)} \mid \mathtt{start} \xrightarrow{\leq m} \alpha\}.$$

Thus $A_0 = \{\mathtt{start}\}$ and

$$M \text{ accepts } x \quad \Leftrightarrow \quad \mathtt{accept} \in A_{d^{S(n)}}.$$

The machine N will start by laying off $S(n)$ space on its worktape. It will then proceed to compute the sizes $|A_0|$, $|A_1|$, $|A_2|$, $\ldots$, $|A_{d^{S(n)}}|$ inductively. First, $|A_0| = 1$. Now suppose $|A_m|$ has been computed and is written on a track of N's tape. Because $|A_m| \leq d^{S(n)}$, this takes up $S(n)$ space at most. To compute $|A_{m+1}|$, successively write down each $\beta \in \Delta^{S(n)}$ in lexicographical order; for each one, determine whether $\beta \in A_{m+1}$ (the algorithm for this is given below); if so, increment a counter by one. The final value of the counter is $|A_{m+1}|$. To test whether $\beta \in A_{m+1}$, nondeterministically guess the $|A_m|$ elements of A_m in lexicographic order, verify that each such α is in A_m by guessing the computation path $\mathtt{start} \xrightarrow{\leq m} \alpha$, and for each such α check whether $\alpha \xrightarrow{\leq 1} \beta$. If any such α yields β in one step, then $\beta \in A_{m+1}$; if no such α does, then $\beta \notin A_{m+1}$.

After $|A_{d^{S(n)}}|$ has been computed, in order to test $\mathtt{accept} \notin A_{d^{S(n)}}$ nondeterministically, guess the $|A_{d^{S(n)}}|$ elements of $A_{d^{S(n)}}$ in lexicographic order, verifying that each guessed α is in $A_{d^{S(n)}}$ by guessing the computation path $\mathtt{start} \xrightarrow{\leq d^{S(n)}} \alpha$, and verifying that each such α is different from $\mathtt{accept}$.

The nondeterministic machine N thus accepts the complement of $L(M)$ and can easily be programmed to run in space $S(n)$.

To remove the constructibility condition, we do the entire computation above for successive values $S = 1, 2, 3, \ldots$ approximating the true space bound $S(n)$. In the course of the computation for S, we eventually see all configurations of length S reachable from the start configuration, and can check whether M ever tries to use more than S space. If so, we know that S is too small and can restart the computation with $S + 1$.   □

Lecture 5

Logspace Computability

For either deterministic or nondeterministic machines, the following resources are equally powerful:

(i)   Logarithmic workspace;

(ii) Counting up to n, the length of the input; and

(iii) "Finite fingers".

By this we mean that the following classes of automata can simulate one another:

(i)   Logspace-bounded TMs;

(ii) Automata with a two-way read-only input head and a fixed finite number of integer counters that can hold an integer between 0 and n, the length of the input; and

(iii) Automata with a fixed finite number of two-way read-only input heads that may not move outside the input.

A machine of type (ii) with k counters is called, appropriately enough, a *k-counter automaton with linearly bounded counters*. In each step, such a machine may test each of its counters for zero. Based on this information and its current state, it may add one or subtract one from each of the counters, move its read head left or right, and enter a new state. Without the bound of n on the maximum value of the counters, a two-counter machine could simulate an arbitrary Turing machine. But with the bound, k-counter machines are no more powerful than logspace machines.

A machine of type (iii) with k heads is called a *k-headed two-way finite automaton* (k-FA).

Intuitively, logspace is enough to simulate a finite number of integer counters that may count only up to n, because the value of the counters can be kept on the worktape in binary. Conversely, the values in the counters can simulate the contents of the worktape, although this simulation is a little more complicated; it takes a bit of cleverness to figure out how to simulate reading and writing symbols on the worktape by manipulating the values in the counters (Homework 1, Exercise 2).

A two-way read head can simulate a linearly bounded counter by maintaining the count as the distance from the left endmarker; conversely, a counter can maintain the distance of a simulated read head from the left endmarker (Miscellaneous Exercise 9).

Logspace Transducers

A *logspace transducer* is a total deterministic logspace-bounded Turing machine with output. *Total* means it halts on all inputs. A logspace transducer has

- a two-way read-only input tape;

- a two-way read/write logspace-bounded worktape, initially blank; and

- a write-only left-to-right output tape, initially blank.

It has three finite alphabets Σ, Γ, and Δ, the input, worktape, and output alphabets, respectively.

The machine begins in its start state with all the heads positioned all the way to the left on the three tapes. The input string $x \in \Sigma^*$ is written on the input tape between endmarkers, and the input head may never go outside the endmarkers. It runs as a normal logspace TM, reading symbols on its input tape and reading and writing symbols on its worktape. Occasionally it may enter a special state that causes it to write a symbol in Δ on its output tape and advance the output tape head one cell. When the machine halts (which it must, by assumption), the string in Δ^* that is written on the output tape at that point is the value of the function computed by the machine on input x.

Only worktape space usage is counted in the space bound; the output tape is not counted.

A function $\sigma : \Sigma^* \to \Delta^*$ is called *logspace computable* if it is computed by some logspace transducer in this way.

The output of a logspace transducer is polynomially bounded in length. That is, for any logspace computable function $\sigma : \Sigma^* \to \Delta^*$, there is a constant d such that for all $x \in \Sigma^*$, $|\sigma(x)| \leq |x|^d$. This is because a logspace transducer can emit at most one output symbol per step, and it can run for at most polynomially many steps, because otherwise it would repeat a configuration, in which case it would be in an infinite loop.

Logspace Reducibility

To encode one problem in another, we use a reducibility relation known as *logspace reducibility*. This is a relation that is computed by a logspace transducer. Logspace reducibility was first studied by Savitch [108] and Jones [67].

If $A \subseteq \Sigma^*$ and $B \subseteq \Delta^*$, we write $A \leq_{\mathrm{m}}^{\log} B$ and say A is *logspace reducible to B* if there is a logspace-computable function $\sigma : \Sigma^* \to \Delta^*$ such that for all $x \in \Sigma^*$,

$$x \in A \quad \Leftrightarrow \quad \sigma(x) \in B.$$

The subscript m on $\leq_{\mathrm{m}}^{\log}$ stands for "many–one" and refers to the type of reducibility relation.

Lemma 5.1 *The relation $\leq_{\mathrm{m}}^{\log}$ is transitive. That is, if $A \leq_{\mathrm{m}}^{\log} B$ and $B \leq_{\mathrm{m}}^{\log} C$, then $A \leq_{\mathrm{m}}^{\log} C$.*

Proof. Homework 1, Exercise 1. This is nontrivial, because there is not enough space to write down an intermediate result in its entirety.   □

Lemma 5.2 *For $A \in \Sigma^*$, $A \in LOGSPACE$ iff $A \leq_{\mathrm{m}}^{\log} \{0, 1\}$.*

Proof. A logspace decision procedure for membership in A is essentially a reduction of A to a two-element set.   □

Lemma 5.3 *If $A \leq_{\mathrm{m}}^{\log} B$ and $B \in LOGSPACE$, then $A \in LOGSPACE$.*

Proof. This follows immediately from Lemmas 5.1 and 5.2.   □

In the theory of *NP*-completeness, you might have studied the polynomial-time many–one reducibility relation $\leq_{\mathrm{m}}^{\mathrm{P}}$, also known as *Karp reducibility*. The relations $\leq_{\mathrm{m}}^{\mathrm{P}}$ and $\leq_{\mathrm{m}}^{\log}$ are similar, except that $\leq_{\mathrm{m}}^{\mathrm{P}}$ is defined in terms of polynomial time transducers, which are the same as logspace transducers except that the bound is on time instead of space. Because logspace transducers can run for at most polynomial time, we have immediately that

Lemma 5.4   *If $A \leq_m^{\log} B$ then $A \leq_m^P B$.*

It is not known whether $\leq_m^{\log}$ is strictly stronger than $\leq_m^P$, however.

Although $\leq_m^P$ is adequate for studying the $P = NP$ question, we are interested in the stronger reducibility $\leq_m^{\log}$ because it has consequences for lower complexity classes such as *LOGSPACE* and *NLOGSPACE*. Most of the natural polynomial-time reductions appearing in the literature can actually be done in logspace.

Completeness

A set $A \in \Sigma^*$ is said to be $\leq_m^{\log}$*-hard* for a complexity class $\mathcal{C}$ if $B \leq_m^{\log} A$ for all $B \in \mathcal{C}$. Intuitively, A is as hard as any decision problem in $\mathcal{C}$, because it can encode any decision problem in $\mathcal{C}$. A set A is said to be *complete for $\mathcal{C}$ with respect to $\leq_m^{\log}$* if

(i) A is $\leq_m^{\log}$-hard for $\mathcal{C}$, and

(ii) $A \in \mathcal{C}$.

Sometimes we say that A is $\leq_m^{\log}$*-complete for $\mathcal{C}$* or just *$\mathcal{C}$-complete* if the reducibility relation $\leq_m^{\log}$ is understood. Intuitively, A is a hardest element of $\mathcal{C}$ in the sense that it encodes every other element of $\mathcal{C}$.

The Cook–Levin theorem says that Boolean satisfiability is *NP*-complete. The usual proofs of the Cook–Levin theorem only show $\leq_m^P$ completeness, but it is easily shown that the problem is complete with respect to the stronger reducibility $\leq_m^{\log}$. We derive this as a corollary next time.

Here is the canonical *NLOGSPACE*-complete problem that is to the *LOGSPACE = NLOGSPACE* question as Boolean satisfiability is to the $P = NP$ question.

Definition 5.5   *The problem* MAZE *(also known as* directed graph reachability*) is the problem of determining, given a directed graph $G = (V, E)$ and distinguished vertices $s, t \in V$, whether there exists a directed path from s to t.*

Theorem 5.6   (Jones, Lien, and Laaser [68])   MAZE *is $\leq_m^{\log}$-complete for NLOGSPACE.*

Proof. We must show

(i)   MAZE is $\leq_m^{\log}$-hard for *NLOGSPACE*, and

(ii)   MAZE $\in$ *NLOGSPACE*.

Part (ii) is easy, because we can trace our way through the given graph using finitely many fingers, guessing the path from s to t nondeterministically.

To show (i), let A be an arbitrary element of *NLOGSPACE*. We must show that $A \leq_{\mathrm{m}}^{\log}$ MAZE. Suppose M is a nondeterministic logspace-bounded TM accepting A. Let start and accept be the start and accept configurations of M, respectively. We can assume without loss of generality that accept is unique by making M erase its worktape and move its heads all the way to the left before accepting. We build a graph $G = (V, E)$, where V is a finite set of configurations of M containing all configurations of M on input x, and E is the next-configuration relation on input x. Then M accepts x iff there is a path from start to accept in G.

Recall that a configuration of M consists of the current state, the current head positions, and the current contents of the worktape. For an input x of length n, all this information takes only $\log n$ space to record, say as a string of length $\log n$ over an alphabet Δ of size d. We can thus take $V = \Delta^{\log n}$. This set is of size $d^{\log n} = n^{\log d}$. We take the edges E to be the pairs (α, β) such that α and β are encodings of configurations of M and β follows from α in one step on input x according to the transition rules of M.

It remains to argue that this graph can be produced by a logspace transducer on input x. The transducer first outputs the set of vertices $\Delta^{\log n}$. This is easy: it lays off $\log n$ space on its worktape, then cycles through all strings in $\Delta^{\log n}$ lexicographically, outputting them all. For the edges, it writes down all pairs α, β in lexicographic order. For each pair, it checks that they both encode configurations of M and that α goes to β in one step on input x. It has to read the ith symbol of its input x to determine this, where i is the position of the input head specified by α. If so, it outputs the edge (α, β). Finally, it outputs the two strings encoding the configurations start and accept.

Although the output $(V, E, \text{start}, \text{accept})$ is polynomial in length, only logarithmic workspace was used to produce it, because there were only at most two configurations of M written down on the transducer's worktape at any time. Thus the map $x \mapsto (V, E, \text{start}, \text{accept})$ is computable in logspace and constitutes a reduction from A to MAZE. $\square$

Corollary 5.7   MAZE $\in$ *LOGSPACE* iff *LOGSPACE* = *NLOGSPACE*.

Proof. Lemmas 5.1 and 5.2 and Theorem 5.6. $\square$

Omer Reingold [101] has very recently shown that the *undirected* graph reachability problem is solvable in deterministic *LOGSPACE*.

Lecture 6

The Circuit Value Problem

In the early 1970s, Stephen Cook [31] and independently Leonid Levin [79] showed that the Boolean satisfiability problem (SAT)—whether a given Boolean formula has a satisfying truth assignment—is NP-complete. Thus Boolean satisfiability is in P iff $P = NP$. This theorem has become known as the *Cook–Levin theorem*. Around the same time, Richard Karp [70] showed that a large number of optimization problems in the field of operations research such as the traveling salesperson problem, graph coloring, bin packing, and many others were all interreducible and therefore NP-complete. These two milestones established the study of NP-completeness as an important aspect of theoretical computer science. In fact, the question of whether $P = NP$ is today widely considered one of the most important open questions in all of mathematics.

There is a theorem of Ladner [78] that plays the same role for the $P = NLOGSPACE$ or $P = LOGSPACE$ question that the Cook–Levin theorem plays for the $P = NP$ question. The decision problem involved is the *circuit value problem* (CVP): given an acyclic Boolean circuit with several inputs and one output and a truth assignment to the inputs, what is the value of the output? The circuit can be evaluated in deterministic polynomial time; the theorem says that this problem is $\leq_{\mathrm{m}}^{\log}$-complete for P. It follows from the transitivity of $\leq_{\mathrm{m}}^{\log}$ that $P = NLOGSPACE$ iff CVP $\in NLOGSPACE$ and $P = LOGSPACE$ iff CVP $\in LOGSPACE$.

Formally, a *Boolean circuit* is a program consisting of finitely many assignments of the form

$$P_i \; := \; 0,$$
$$P_i \; := \; 1,$$
$$P_i \; := \; P_j \wedge P_k, \quad j, k < i,$$
$$P_i \; := \; P_j \vee P_k, \quad j, k < i, \quad \text{or}$$
$$P_i \; := \; \neg P_j, \quad j < i,$$

where each P_i in the program appears on the left-hand side of exactly one assignment. The conditions $j, k < i$ and $j < i$ ensure acyclicity. We want to compute the value of P_n, where n is the maximum index.

Theorem 6.1   *The circuit value problem is $\leq_{\mathrm{m}}^{\log}$-complete for P.*

Proof. We have already argued that CVP $\in P$. To show hardness, we reduce an arbitrary $A \in P$ to CVP. Let M be a deterministic single-tape polynomial-time-bounded TM accepting A, say with time bound n^c. Let Γ be the worktape alphabet of M and let Q be the set of states of M's finite control. The transition function δ of M is of type $\delta : Q \times \Gamma \to Q \times \Gamma \times \{L, R\}$. Intuitively, $\delta(p, a) = (q, b, d)$ says, "When in state p scanning symbol a on the tape, print b on that cell, move in direction d, and enter state q." We can encode configurations of M over a finite alphabet Δ as usual.

Now given x of length n, think of the successive configurations of M on input x as arranged in an $(n^c+1) \times (n^c+1)$ time/space matrix R with entries in Δ. The ith row of R is a string in Δ^{n^c+1} describing the configuration of the machine at time i. The jth column of R describes what is going on at tape cell j throughout the history of the computation.

For example, the ith row of the matrix might look like

⊢	a	b	a	a	b	a	b	a	a	b	a	b	⊔	⊔	⊔
						p									

and the $i + 1$st might look like

⊢	a	b	a	a	b	b	b	a	a	b	a	b	⊔	⊔	⊔
							q								

This would happen if $\delta(p, a) = (q, b, R)$. The elements of Δ are thus of the form

a		a
	or	q

for $a \in \Gamma$ and $q \in Q$.

We can specify the matrix R in terms of a set of *local consistency conditions* on $(n^c + 1) \times (n^c + 1)$ matrices over Δ. Each local consistency condition is a relation on the entries of the matrix in a small neighborhood of some location i, j. The conjunction of all these local consistency conditions is enough to determine R uniquely.

Our circuit will involve Boolean variables

$$P_{ij}^a, \ 0 \leq i, j \leq n^c, \ a \in \Gamma,$$

$$Q_{ij}^q, \ 0 \leq i, j \leq n^c, \ q \in Q.$$

The variable P_{ij}^a says, "The symbol occupying tape cell j at time i is a," and the variable Q_{ij}^q says, "The machine is in state q scanning tape cell j at time i."

Now we write down a set of conditions in terms of the P_{ij}^a and Q_{ij}^q describing all ways that the machine could be in state q scanning tape cell j at time i or that the symbol occupying tape cell j at time i is a.

For $1 \leq i \leq n^c$, $0 \leq j \leq n^c$, and $b \in \Gamma$, we include the assignment

$$P_{ij}^b \ := \bigvee_{\delta(p,a)=(q,b,d)} (Q_{i-1,j}^p \wedge P_{i-1,j}^a) \tag{6.1}$$

$$\vee \ (P_{i-1,j}^b \wedge \bigwedge_{p \in Q} \neg Q_{i-1,j}^p) \tag{6.2}$$

in our circuit. Intuitively, this says, "The symbol occupying tape cell j at time i is b if and only if either the machine was scanning tape cell j at time $i - 1$ and printed b (clause (6.1)), or the machine was not scanning tape cell j at time $i - 1$ and the symbol occupying that cell at time $i - 1$ was b (clause (6.2))." The join in (6.1) is over all states $p, q \in Q$, symbols $a \in \Gamma$, and directions $d \in \{L, R\}$ such that $\delta(p, a) = (q, b, d)$; that is, all situations that would cause b to be printed.

For $1 \leq i \leq n^c$, $1 \leq j \leq n^c - 1$ (that is, ignoring the left and right boundaries), and $q \in Q$, we include the assignment

$$Q_{ij}^q \ := \bigvee_{\delta(p,a)=(q,b,R)} (Q_{i-1,j-1}^p \wedge P_{i-1,j-1}^a) \tag{6.3}$$

$$\vee \bigvee_{\delta(p,a)=(q,b,L)} (Q_{i-1,j+1}^p \wedge P_{i-1,j+1}^a). \tag{6.4}$$

Intuitively, this says, "The machine is scanning tape cell j at time i if and only if either it was scanning tape cell $j - 1$ at time $i - 1$ and moved right (clause (6.3)), or it was scanning tape cell $j + 1$ at time $i - 1$ and moved left (clause (6.4))." The join in (6.3) is over all states $p \in Q$ and symbols

$a, b \in \Gamma$ such that $\delta(p, a) = (q, b, R)$; that is, all situations that would cause the machine to move right and enter state q.

For $j = 0$, that is, for the leftmost tape cell, we define Q_{ij}^q in terms of (6.3) only. Similarly, for $j = n^c$, we define Q_{ij}^q in terms of (6.4) only.

This takes care of everything except the first row of the matrix. The values P_{ij}^b and Q_{ij}^q are determined by the start configuration; these are the inputs to the circuit. If $x = a_1 \cdots a_n$, the start state is s, and the endmarker and blank symbol are $\vdash$ and $\sqcup$, respectively, we include

$$
\begin{aligned}
P_{0,0}^{\vdash} &:= 1, \\
P_{0,0}^b &:= 0, \ b \in \Gamma - \{\vdash\}, \\
P_{0,j}^{a_j} &:= 1, \ 1 \le j \le n, \\
P_{0,j}^b &:= 0, \ b \in \Gamma - \{a_j\}, \ 1 \le j \le n, \\
P_{0,j}^{\sqcup} &:= 1, \ n + 1 \le j \le n^c, \\
P_{0,j}^b &:= 0, \ b \in \Gamma - \{\sqcup\}, \ n + 1 \le j \le n^c, \\
Q_{0,0}^s &:= 1, \\
Q_{0,j}^s &:= 0, \ 1 \le j \le n^c, \\
Q_{0,j}^q &:= 0, \ 0 \le j \le n^c, q \in Q - \{s\}.
\end{aligned}
$$

This gives a circuit. Assuming that the machine moves its head all the way to the left before entering its accept state t, the Boolean value of

$$
Q_{n^c,0}^t \vee Q_{n^c,1}^t
$$

determines whether M accepts x.

The construction we have just given can be done in logspace. Even though the circuit is polynomial size, it is highly uniform in the sense that it is built of many identical pieces. The only differences are the indices i, j, which can be written down in logspace. $\qquad\square$

The Cook–Levin Theorem

Now we show how to derive the Cook–Levin theorem as a corollary of the previous construction. We would like to show that the Boolean satisfiability problem SAT—given a Boolean formula, does it have a satisfying truth assignment?—is *NP*-complete. The two main differences between SAT and CVP are:

(i) With SAT, the input values are not provided. The problem asks whether there exist values making the formula evaluate to true.

(ii)   CVP is defined in terms of circuits and SAT is defined in terms of formulas. The difference is that in circuits, Boolean values may be used more than once. A circuit is represented as a labeled directed acyclic graph (dag), whereas a formula is a labeled tree. Another way to look at it is that a circuit allows sharing of common subexpressions. The satisfiability problem is *NP*-complete, regardless of whether we use circuits or formulas; but the problem of evaluating a formula on a given truth assignment is apparently easier than CVP, because it can be done in logspace (Homework 2, Exercise 3).

We define a *circuit with unspecified inputs* exactly as above, except that we also include assignments

$$P_i \quad := \quad ?$$

denoting inputs whose value is unspecified.

Theorem 6.2   *Boolean satisfiability is $\leq_{\mathrm{m}}^{\log}$-complete for NP.*

Proof. Boolean satisfiability is in *NP*, because we can guess a truth assignment and verify that it satisfies the given formula or circuit in polynomial time.

To show that the problem is $\leq_{\mathrm{m}}^{\log}$-hard for *NP*, let A be an arbitrary set in *NP*, and let M be a nondeterministic machine accepting A and running in time n^c. Assume without loss of generality that the nondeterminism is binary branching. Then a computation path of M is specified by a string in $\{0,1\}^{n^c}$.

Let M' be a deterministic machine that takes as input $x\#y$, where $|y| = |x|^c$, and runs M on input x, using y to resolve the nondeterministic choices and accepting if the computation path of M specified by y leads to acceptance. By the construction of Theorem 6.1, there is a circuit that has value 1 iff M' accepts $x\#y$. Note from the construction that if $|z| = |y| = n^c$, the circuit constructed for $x\#z$ is identical to that for $x\#y$ except for the inputs corresponding to y; making these inputs unspecified, we obtain a circuit $C(P_1, \ldots, P_{n^c})$ with unspecified inputs $P_1, \ldots, P_{n^c}$ such that M accepts x if and only if there exist $y_1, \ldots, y_{n^c} \in \{0,1\}$ such that $C(y_1, \ldots, y_{n^c}) = 1$.

We can transform the circuit into a formula by replacing each assignment $P_i := E$ with the clause $P_i \leftrightarrow E$ and taking the conjunction of all clauses obtained in this way. The resulting formula is satisfiable iff M accepts x. $\square$

The Boolean satisfiability problem remains *NP*-hard even when restricted to formulas in conjunctive normal form (CNF with at most three literals per clause (3CNF) (Miscellaneous Exercise 10), whereas it is solvable in polynomial time for formulas in 2CNF (Homework 2, Exercise 2). The satisfiability problem for 3CNF formulas is known as 3SAT.

Supplementary Lecture A

The Knaster–Tarski Theorem

Transfinite Ordinals

Everyone is familiar with the set $\omega = \{0, 1, 2, \ldots\}$ of *finite ordinals*, also known as the *natural numbers*. An essential mathematical tool is the *induction principle* on this set, which states that if a property is true of zero and is preserved by the successor operation, then it is true of all elements of ω.

In theoretical computer science, we often run into inductive definitions that take longer than ω to close, and it is useful to have an induction principle that applies to these objects. Cantor recognized the value of such a principle in his theory of infinite sets. Any modern account of the foundations of mathematics will include a chapter on ordinals and transfinite induction.

Unfortunately, a complete understanding of ordinals and transfinite induction is impossible outside the context of set theory, because many issues impact the very foundations of the subject. Here we only give a cursory account of the basic facts, tools, and techniques we need.

Set-Theoretic Definition of Ordinals

Ordinals are defined as certain sets of sets. The key facts we need about ordinals, succinctly stated, are:

(i) There are two kinds: *successors* and *limits*.

(ii) They are well ordered.

(iii) There are a lot of them.

(iv) We can do induction on them.

We explain each of these statements in more detail below.

A set C of sets is said to be *transitive* if $A \in C$ whenever $A \in B$ and $B \in C$. Equivalently, C is transitive if every element of C is a subset of C; that is, $C \subseteq 2^C$. Formally, an *ordinal* is defined to be a set A such that

- A is transitive; and

- all elements of A are transitive.

It follows that any element of an ordinal is an ordinal. We use $\alpha, \beta, \gamma, \ldots$ to refer to ordinals. The class of all ordinals is denoted Ord. It is not a set, but a proper class.

This neat but rather obscure definition of ordinals has some far-reaching consequences that are not at all obvious. For ordinals α, β, define $\alpha < \beta$ if $\alpha \in \beta$. The relation $<$ is a strict partial order. As usual, there is an associated nonstrict partial order $\leq$ defined by $\alpha \leq \beta$ if $\alpha \in \beta$ or $\alpha = \beta$.

It follows from the axioms of set theory that the relation $<$ on ordinals is a linear order. That is, if α and β are any two ordinals, then either $\alpha < \beta$, $\alpha = \beta$, or $\alpha > \beta$. This is most easily proved by induction on the well-founded relation

$$(\alpha, \beta) \leq (\alpha', \beta') \quad \overset{\text{def}}{\Longleftrightarrow} \quad \alpha \leq \alpha' \text{ and } \beta \leq \beta'.$$

Then every ordinal is equal to the set of all smaller ordinals (in the sense of $<$). The class of ordinals is well-founded in the sense that any nonempty set of ordinals has a least element.

If α is an ordinal, then so is $\alpha \cup \{\alpha\}$. The latter is called the *successor* of α and is denoted $\alpha + 1$. Also, if A is any set of ordinals, then $\bigcup A$ is an ordinal, and is the supremum of the ordinals in A under the relation $\leq$.

The smallest few ordinals are

$$0 \overset{\text{def}}{=} \varnothing$$

$$1 \overset{\text{def}}{=} \{0\} = \{\varnothing\}$$

$$2 \overset{\text{def}}{=} \{0, 1\} = \{\varnothing, \{\varnothing\}\}$$

$$3 \overset{\text{def}}{=} \{0, 1, 2\} = \{\varnothing, \{\varnothing\}, \{\varnothing, \{\varnothing\}\}\}$$

Pictorially,

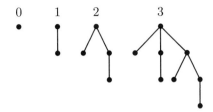

The first infinite ordinal is

$$\omega \;\overset{\text{def}}{=}\; \{0,1,2,3,\dots\}.$$

An ordinal is called a *successor ordinal* if it is of the form $\alpha+1$ for some ordinal α, otherwise it is called a *limit ordinal*. The smallest limit ordinal is 0 and the next smallest is ω. Of course, $\omega + 1 = \omega \cup \{\omega\}$ is an ordinal, so it does not stop there.

The ordinals form a proper class, thus there can be no one-to-one function Ord $\rightarrow A$ into a set A. This is what we mean above by, "There are a lot of ordinals." In practice, this comes up when we construct functions $f : \text{Ord} \rightarrow A$ from Ord into a set A by induction. Such an f, regarded as a collection of ordered pairs, is necessarily a class and not a set. We will always be able to conclude that there exist distinct ordinals α, β with $f(\alpha) = f(\beta)$.

Transfinite Induction

The *transfinite induction principle* is a method of establishing that a particular property is true of all ordinals (or of all elements of a class of objects indexed by ordinals). It states that in order to prove that the property is true of all ordinals, it suffices to show that the property is true of an arbitrary ordinal α whenever it is true of all ordinals $\beta < \alpha$. Proofs by transfinite induction typically contain two cases, one for successor ordinals and one for limit ordinals. The basis of the induction is often a special case of the case for limit ordinals, because $0 = \varnothing$ is a limit ordinal; here the premise that the property is true of all ordinals $\beta < \alpha$ is vacuously true.

The validity of this principle ultimately follows from the well-foundedness of the set containment relation $\in$. This is an axiom of set theory.

Zorn's Lemma and the Axiom of Choice

Related to the ordinals and transfinite induction are the axiom of choice and Zorn's lemma.

The *axiom of choice* states that for every set A of nonempty sets, there exists a function f with domain A that picks an element out of each set in A; that is, for every $B \in A$, $f(B) \in B$. Equivalently, any Cartesian product of nonempty sets is nonempty.

Zorn's lemma states that every set of sets closed under unions of chains contains a $\subseteq$-maximal element. Here a *chain* is a family of sets linearly ordered by the inclusion relation $\subseteq$, and to say that a set C of sets is closed under unions of chains means that if $B \subseteq C$ and B is a chain, then $\bigcup B \in C$. An element $B \in C$ is $\subseteq$-*maximal* if it is not properly included in any $B' \in C$.

The *well-ordering principle* states that every set is in one-to-one correspondence with some ordinal. A set is *countable* if it is either finite or in one-to-one correspondence with ω.

The axiom of choice, Zorn's lemma, and the well-ordering principle are equivalent to one another and independent of Zermelo–Fraenkel (ZF) set theory in the sense that if ZF is consistent, then neither they nor their negations can be proven from the axioms of ZF.

Complete Lattices

A *complete lattice* is a set U with a distinguished partial ordering relation $\leq$ defined on it (reflexive, antisymmetric, transitive) such that every subset of U has a *supremum* or *least upper bound* with respect to $\leq$. That is, for every subset $A \subseteq U$, there is an element $\sup A \in U$ such that

(i)   for all $x \in A$, $x \leq \sup A$ ($\sup A$ is an upper bound for A), and

(ii)   if $x \leq y$ for all $x \in A$, then $\sup A \leq y$ ($\sup A$ is the least upper bound).

It follows from (i) and (ii) that $\sup A$ is unique. We abbreviate $\sup\{x, y\}$ by $x \vee y$.

Any complete lattice U has a maximum element $\top \overset{\text{def}}{=} \sup U$ and a minimum element $\bot \overset{\text{def}}{=} \sup \varnothing$. Also, every subset $A \subseteq U$ has an *infimum* or *greatest lower bound* $\inf A \overset{\text{def}}{=} \sup\{y \mid \forall z \in A \; y \leq z\}$. One can show (Miscellaneous Exercise 19) that $\inf A$ is the unique element such that

(i)   for all $y \in A$, $\inf A \leq y$ ($\inf A$ is a lower bound for A), and

(ii)   if $x \leq y$ for all $y \in A$, then $x \leq \inf A$ ($\inf A$ is the greatest lower bound).

A common example of a complete lattice is the powerset 2^X of a set X, or set of all subsets of X, ordered by the subset relation $\subseteq$. The supremum of a set $\mathcal{C}$ of subsets of X is their union $\bigcup \mathcal{C}$ and the infimum of $\mathcal{C}$ is their intersection $\bigcap \mathcal{C}$.

Monotone, Continuous, and Finitary Operators

An *operator* on a complete lattice U is a function $\tau : U \to U$. Here we introduce some special properties of such operators such as monotonicity and closure and discuss some of their consequences. We culminate with a general theorem due to Knaster and Tarski concerning inductive definitions.

In the special case of set-theoretic complete lattices 2^X ordered by set inclusion $\subseteq$, we call such an operator a *set operator*.

An operator τ is said to be *monotone* if it preserves $\leq$:

$$x \leq y \quad \Rightarrow \quad \tau(x) \leq \tau(y).$$

A *chain* in U is a subset of U totally ordered by $\leq$; that is, for every x and y in the chain, either $x \leq y$ or $y \leq x$. An operator τ is said to be *chain-continuous* if for every chain A,

$$\tau(\sup A) \quad = \quad \sup_{x \in A} \tau(x).$$

For set operators $\tau : 2^X \to 2^X$, τ is said to be *finitary* if its action on $A \subseteq X$ depends only on finite subsets of A in the following sense:

$$\tau(A) \quad = \quad \bigcup_{\substack{B \subseteq A \\ B \text{ finite}}} \tau(B).$$

A set operator is finitary iff it is chain-continuous (Miscellaneous Exercise 20), and every chain-continuous operator on any complete lattice is monotone, but not necessarily vice versa (Miscellaneous Exercise 21). In many applications involving set operators, the operators are finitary.

Example A.1 For a binary relation R on a set V, define

$$\tau(R) \quad = \quad \{(a, c) \mid \exists b \ (a, b), \ (b, c) \in R\}.$$

The function τ is a set operator on V^2; that is,

$$\tau : 2^{V^2} \quad \to \quad 2^{V^2}.$$

The operator τ is finitary, because $\tau(R)$ is determined by the action of τ on two-element subsets of R. $\qquad \square$

Prefixpoints and Fixpoints

A *prefixpoint* of an operator τ on U is an element $x \in U$ such that $\tau(x) \leq x$. A *fixpoint* of τ is an element $x \in U$ such that $\tau(x) = x$. Every operator on U has at least one prefixpoint, namely $\sup U$. Monotone operators also have fixpoints, as we show below.

For set operators $\tau : 2^X \to 2^X$, we often say that a subset $A \subseteq X$ is *closed* under τ if A is a prefixpoint of τ, that is, if $\tau(A) \subseteq A$.

Example A.2   By definition, a binary relation R on a set V is *transitive* if $(a, c) \in R$ whenever $(a, b) \in R$ and $(b, c) \in R$. Equivalently, R is transitive iff it is closed under the finitary set operator τ defined in Example A.1.   □

Lemma A.3   *The infimum of any set of prefixpoints of a monotone operator τ is a prefixpoint of τ.*

Proof. Let A be any set of prefixpoints of τ. We wish to show that inf A is a prefixpoint of τ. For any $x \in A$, we have inf $A \leq x$, therefore

$$\tau(\inf A) \quad \leq \quad \tau(x) \quad \leq \quad x,$$

because τ is monotone and x is a prefixpoint. Because $x \in A$ was arbitrary, $\tau(\inf A) \leq \inf A$.   □

For $x \in U$, define

$$PF_\tau(x) \quad \overset{\text{def}}{=} \quad \{y \in U \mid \tau(y) \leq y, \ x \leq y\}, \tag{A.1}$$

the set of all prefixpoints of τ above x. Note that $PF_\tau(\bot)$ is the set of all prefixpoints of U, and all $PF_\tau(x)$ are nonempty, because $\top$ is in there at least.

It follows from Lemma A.3 that $PF_\tau(\bot)$ forms a complete lattice under the induced ordering $\leq$; however, whereas the infimum in $PF_\tau(\bot)$ of any set of prefixpoints A is inf A, the supremum is $\inf PF_\tau(\sup A)$, which is not the same as $\sup A$ in general, because $\sup A$ is not necessarily a prefixpoint (Miscellaneous Exercise 22). Thus we must be careful to say whether we are taking suprema in U or in $PF_\tau(\bot)$.

For $x \in U$, define

$$\tau^\dagger(x) \quad \overset{\text{def}}{=} \quad \inf PF_\tau(x). \tag{A.2}$$

By Lemma A.3, $\tau^\dagger(x)$ is the least prefixpoint of τ such that $x \leq \tau^\dagger(x)$.

Lemma A.4   *Any monotone operator τ has a $\leq$-least fixpoint.*

Proof. We show that $\tau^\dagger(\bot)$ is the least fixpoint of τ in U. By Lemma A.3, it is the least prefixpoint of τ. If it is a fixpoint, then it is the least one, because every fixpoint is a prefixpoint. But if it were not a fixpoint, then by monotonicity, $\tau(\tau^\dagger(\bot))$ would be a smaller prefixpoint, contradicting the fact that $\tau^\dagger(\bot)$ is the smallest.   □

Closure Operators

An operator σ on a complete lattice U is called a *closure operator* if it satisfies the following three properties.

(i)   The operator σ is monotone.

(ii)   For all x, $x \leq \sigma(x)$.

(iii)   For all x, $\sigma(\sigma(x)) = \sigma(x)$.

Because of clause (ii), fixpoints and prefixpoints coincide for closure operators. Thus an element is closed with respect to a closure operator σ iff it is a fixpoint of σ. As shown in Lemma A.3, the set of closed elements of a closure operator forms a complete lattice.

Lemma A.5   *For any monotone operator τ, the operator $\tau^\dagger$ defined in (A.2) is a closure operator.*

Proof. The operator $\tau^\dagger$ is monotone, because

$$x \leq y \;\;\Rightarrow\;\; PF_\tau(y) \subseteq PF_\tau(x) \;\Rightarrow\; \inf PF_\tau(x) \leq \inf PF_\tau(y),$$

where $PF_\tau(x)$ is the set defined in (A.1).

Property (ii) of closure operators follows directly from the definition of $\tau^\dagger$. Finally, to show property (iii), because $\tau^\dagger(x)$ is a prefixpoint of τ, it suffices to show that any prefixpoint of τ is a fixpoint of $\tau^\dagger$. But

$$\tau(y) \leq y \;\;\Leftrightarrow\;\; y \in PF_\tau(y) \;\;\Leftrightarrow\;\; y = \inf PF_\tau(y) = \tau^\dagger(y).$$

$\square$

Example A.6   The *transitive closure* of a binary relation R on a set V is the least transitive relation containing R; that is, it is the least relation containing R and closed under the finitary transitivity operator τ of Example A.1. The *transitive closure* of R is the relation $\tau^\dagger(R)$. Thus the closure operator $\tau^\dagger$ maps an arbitrary relation R to its transitive closure. $\square$

Example A.7   The *reflexive transitive closure* of a binary relation R on a set V is the least reflexive and transitive relation containing R; that is, it is the least relation that contains R, is closed under transitivity, and contains the identity relation $\iota = \{(a, a) \mid a \in V\}$. Note that "contains the identity relation" just means closed under the (constant valued) monotone operation $R \mapsto \iota$. Thus the reflexive transitive closure of R is $\sigma^\dagger(R)$, where σ denotes the finitary set operator $R \mapsto \tau(R) \cup \iota$. $\square$

The Knaster–Tarski Theorem

The *Knaster–Tarski theorem* is a useful theorem describing how least fixpoints of monotone operators can be obtained either "from above," as in the proof of Lemma A.4, or "from below," as a limit of a chain of elements defined by transfinite induction.

Let U be a complete lattice and let τ be a monotone operator on U. Let $\tau^\dagger$ be the associated closure operator defined in (A.2). We show how to attain $\tau^\dagger(x)$ starting from x and working up. The idea is to start with x and then apply τ repeatedly until achieving closure. In most applications, the operator τ is continuous, in which case this takes only ω iterations; but for monotone operators in general, it can take more.

Formally, we construct by transfinite induction a chain of elements $\tau^\alpha(x)$ indexed by ordinals α:

$$\tau^{\alpha+1}(x) \;\overset{\text{def}}{=}\; x \vee \tau(\tau^\alpha(x))$$

$$\tau^\lambda(x) \;\overset{\text{def}}{=}\; \sup_{\alpha<\lambda} \tau^\alpha(x), \quad \lambda \text{ a limit ordinal}$$

$$\tau^*(x) \;\overset{\text{def}}{=}\; \sup_{\alpha\in\text{Ord}} \tau^\alpha(x).$$

The base case is included in the case for limit ordinals:

$$\tau^0(x) \;=\; \bot.$$

Intuitively, $\tau^\alpha(x)$ is the set obtained by applying τ to x α times, reincluding x at successor stages.

Lemma A.8   *If $\alpha \leq \beta$, then $\tau^\alpha(x) \leq \tau^\beta(x)$.*

Proof. We proceed by transfinite induction on α. For two successor ordinals $\alpha + 1$ and $\beta + 1$,

$$\tau^{\alpha+1}(x) \;=\; x \vee \tau(\tau^\alpha(x)) \;\leq\; x \vee \tau(\tau^\beta(x)) \;=\; \tau^{\beta+1}(x),$$

where the inequality follows from the induction hypothesis and the monotonicity of τ. For a limit ordinal λ on the left-hand side and any ordinal β on the right-hand side,

$$\tau^\lambda(x) \;=\; \sup_{\alpha<\lambda} \tau^\alpha(x) \;\leq\; \tau^\beta(x),$$

where the inequality follows from the induction hypothesis. Finally, for a limit ordinal λ on the right-hand side, the result is immediate from the definition of $\tau^\lambda(x)$.   □

Lemma A.8 says that the $\tau^\alpha(x)$ form a chain in U. The element $\tau^*(x)$ is the supremum of this chain over all ordinals α.

Now there must exist an ordinal κ such that $\tau^{\kappa+1}(x) = \tau^\kappa(x)$, because there is no one-to-one function from the class of ordinals to the set U. The least such κ is called the *closure ordinal* of τ. If κ is the closure ordinal of τ, then $\tau^\beta(x) = \tau^\kappa(x)$ for all $\beta > \kappa$, therefore $\tau^*(x) = \tau^\kappa(x)$.

If τ is chain-continuous, then its closure ordinal is at most ω, but not for monotone operators in general (Miscellaneous Exercise 23).

Theorem A.9 (Knaster–Tarski) $\tau^\dagger(x) = \tau^*(x)$.

Proof. First we show the forward inclusion. Let κ be the closure ordinal of τ. Because $\tau^\dagger(x)$ is the least prefixpoint of τ above x, it suffices to show that $\tau^*(x) = \tau^\kappa(x)$ is a prefixpoint of τ. But

$$\tau(\tau^\kappa(x)) \;\leq\; x \vee \tau(\tau^\kappa(x)) \;=\; \tau^{\kappa+1}(x) \;=\; \tau^\kappa(x).$$

Conversely, we show by transfinite induction that for all ordinals α, $\tau^\alpha(x) \leq \tau^\dagger(x)$, therefore $\tau^*(x) \leq \tau^\dagger(x)$. For successor ordinals $\alpha + 1$,

$$
\begin{aligned}
\tau^{\alpha+1}(x) \;&=\; x \vee \tau(\tau^\alpha(x)) \\
&\leq\; x \vee \tau(\tau^\dagger(x)) \qquad \text{induction hypothesis and monotonicity} \\
&\leq\; \tau^\dagger(x) \qquad\qquad\ \text{definition of } \tau^\dagger.
\end{aligned}
$$

For limit ordinals λ, $\tau^\alpha(x) \leq \tau^\dagger(x)$ for all $\alpha < \lambda$ by the induction hypothesis; therefore

$$\tau^\lambda(x) \;=\; \sup_{\alpha<\lambda} \tau^\alpha(x) \;\leq\; \tau^\dagger(x).$$

$\square$

Lecture 7

Alternation

In this lecture we present a useful generalization of nondeterminism called *alternation*. The word "alternation" refers to the alternation of *and* and *or*. We introduce *alternating Turing machines* and present some simulation results relating alternating machines to deterministic machines. These results are useful for studying the complexity of problems with a natural alternating and/or structure, such as games or logical theories. Alternating Turing machines were introduced by Chandra, Kozen, and Stockmeyer [26].

We usually think of a nondeterministic computation as a single process that makes guesses at choice points, following a guessed computation path and accepting if that path causes the machine to enter an accept state.

Alternatively, we can think of a nondeterministic machine as a multiprocessor machine with a potentially unbounded number of processors available for allocation and assignment to tasks. In this view, the machine starts with a single root process in the start configuration. It computes as a normal Turing machine until it reaches a nondeterministic choice point. At that point, it spawns as many independent subprocesses as there are possible next configurations, then suspends execution, waiting for a report from one of the subprocesses it just spawned. Each of the subprocesses takes one of the possible next configurations and continues execution from there. This continues down the tree. If there are m configurations in the computation tree at depth i, then there will be m independent parallel processes running simultaneously at time i. When a process enters an accept state, it reports a Boolean 1 indicating acceptance to its parent and terminates. When a process enters a reject state, it reports a Boolean 0 indicating rejection to its parent and terminates. A suspended process, upon receiving a 1 from a subprocess, immediately reports the 1 up to its parent process and terminates. A suspended process, upon receiving a 0 from a subprocess, waits for a report from another subprocess. If and when all the subprocesses have reported 0, it reports 0 up to its parent and terminates. The input is accepted if a 1 is ever reported to the root process.

This description is of course just an intuitive device; there is no explicit mechanism for spawning subprocesses or reporting Boolean values back up the tree.

Now we wish to extend this idea by allowing "and" as well as "or" branching. In normal nondeterminism as described above, a suspended process waiting at a choice point checks whether any one of its subprocesses leads to acceptance. It essentially computes the "or" ($\vee$) of the Boolean values returned to it by its subprocesses. We might also allow a process to check whether all subprocesses lead to acceptance by computing the Boolean "and" ($\wedge$) of the Boolean values returned to it by its subprocesses. Whether a nondeterministic choice point is an $\wedge$-branch or an $\vee$-branch is determined by the state. The word "alternation" refers to the alternation of $\wedge$ and $\vee$ in the course of a computation.

We give a formal definition of alternating Turing machines below and prove a remarkable correspondence between alternation and determinism: *alternating time is the same as deterministic space, and alternating space is the same as exponentially more deterministic time.*

It will often be convenient to allow negating steps ($\neg$) as well as $\wedge$ and $\vee$ branches in alternating Turing machines. It turns out we can get rid of negations at no cost in either space or time.

Definition 7.1 *An* alternating Turing machine (ATM) *is exactly like a nondeterministic TM, except there is included in the specification of the machine a function*

$$\text{type} : Q \quad \rightarrow \quad \{\wedge, \vee, \neg\},$$

where Q is the set of states. The function type *tells whether a state is an and-state, an or-state, or a not-state. A configuration is called an* and-configuration, *an* or-configuration, *or a* not-configuration *according as the state in the configuration is an and-state, an or-state, or a not-state, respectively. We impose the restriction that not-configurations have exactly one successor.*

Accept and reject states do not need to be explicitly specified in the description of the machine. We can take an accept state *to be an and-state with no successors and a* reject state *to be an or-state with no successors.*

Acceptance for ATMs is defined in terms of an inductive labeling of the computation tree. For this definition, we consider two partial orders on the set $\{0, 1, \bot\}$:

- the natural order $0 \leq \bot \leq 1$ of three-valued logic, and

- the *information order* $\sqsubseteq$ in which $\bot \sqsubseteq 0$ and $\bot \sqsubseteq 1$.

The symbol $\bot$ stands for "don't know" and is used to handle infinite computations. The Boolean operations $\vee$, $\wedge$, and $\neg$ extend in a natural way to the three-element set $\{0, 1, \bot\}$ according to the following tables.

$\vee$	1	$\bot$	0
1	1	1	1
$\bot$	1	$\bot$	$\bot$
0	1	$\bot$	0

$\wedge$	1	$\bot$	0
1	1	$\bot$	0
$\bot$	$\bot$	$\bot$	0
0	0	0	0

$\neg$	
1	0
$\bot$	$\bot$
0	1

Thus $\vee$ gives supremum and $\wedge$ gives infimum in the natural order $0 \leq \bot \leq 1$.

Now we consider an inductive labeling of configurations with 1, 0, or $\bot$ that corresponds to the intuitive procedure of passing Boolean values back up the computation tree as outlined above. We do things this way in order to be completely precise about how the machine deals with infinite computation paths and negations.

Let $\mathcal{C}$ denote the set of configurations, and let $\xrightarrow[M]{1}$ denote the next configuration relation of M. Thus $\alpha \xrightarrow[M]{1} \beta$ if configuration β follows from configuration α in one step according to the transition rules of M.

A *labeling* is a map $\ell : \mathcal{C} \to \{0, 1, \bot\}$. The order $\sqsubseteq$ extends pointwise to labelings; that is, define

$$\ell \sqsubseteq \ell' \stackrel{\text{def}}{\iff} \forall \alpha \in \mathcal{C}\ \ell(\alpha) \sqsubseteq \ell'(\alpha).$$

The set of labelings forms a complete lattice under $\sqsubseteq$. Thus every set of labelings has a $\sqsubseteq$-least upper bound. There is a least labeling $\lambda\alpha.\bot$,[1] which is the least upper bound of the empty set of labelings.

Now define the labeling ℓ_* to be the $\sqsubseteq$-least solution of the recursive equation

$$\ell_*(\alpha) = \begin{cases} \bigwedge_{\alpha \to \beta} \ell_*(\beta) & \text{if } \alpha \text{ is an } \wedge\text{-configuration} \\ \bigvee_{\alpha \to \beta} \ell_*(\beta) & \text{if } \alpha \text{ is an } \vee\text{-configuration} \\ \neg\ell_*(\beta) & \text{if } \alpha \text{ is a } \neg\text{-configuration and } \alpha \to \beta. \end{cases}$$

---

[1] $\lambda x.E(x)$ is the function that on input x returns $E(x)$.

This is the least fixpoint of the $\sqsubseteq$-monotone map $\tau : \{\text{labelings}\} \rightarrow \{\text{labelings}\}$ defined by

$$\tau(\ell)(\alpha) \overset{\text{def}}{=} \begin{cases} \bigwedge_{\alpha \rightarrow \beta} \ell(\beta) & \text{if } \alpha \text{ is an } \wedge\text{-configuration} \\ \bigvee_{\alpha \rightarrow \beta} \ell(\beta) & \text{if } \alpha \text{ is an } \vee\text{-configuration} \\ \neg\ell(\beta) & \text{if } \alpha \text{ is a } \neg\text{-configuration and } \alpha \rightarrow \beta. \end{cases}$$

The labeling ℓ_* exists by the Knaster–Tarski theorem (Theorem A.9). In this case, the closure ordinal of the inductive definition is ω, thus the labeling ℓ_* is the supremum of the chain

$$\ell_0 \ \sqsubseteq \ \ell_1 \ \sqsubseteq \ \ell_2 \ \sqsubseteq \ \cdots, \tag{7.1}$$

where

$$\ell_0 \overset{\text{def}}{=} \lambda\alpha.\bot,$$
$$\ell_{i+1} \overset{\text{def}}{=} \tau(\ell_i).$$

Definition 7.2   *The machine* accepts *x if $\ell_*(\text{start}) = 1$, where* start *is the start configuration on input x. It* rejects *x if $\ell_*(\text{start}) = 0$.*

Of course, $\ell_*(\text{start}) = \bot$ is also possible, in which case the machine neither accepts nor rejects.

Intuitively, accept configurations ($\wedge$-configurations with no successors) are labeled 1 and reject configurations ($\vee$-configurations with no successors) are labeled 0 by ℓ_1, and the definition of ℓ_2, ℓ_3, ... models the computation of these Boolean values back up the tree. The supremum ℓ_* of these labelings labels a configuration 1 if it is ever labeled 1 by some ℓ_i and labels a configuration 0 if it is ever labeled 0 by some ℓ_i.

Lemma 7.3   *Every alternating Turing machine with negations can be simulated by an alternating Turing machine without negations at no extra cost in space or time.*

Proof. The *dual* of an alternating TM M is the alternating TM M' that looks exactly like M except that $\vee$- and $\wedge$-states are exchanged; that is, if state q is an $\vee$-state of M, then the corresponding state q' of M' is an $\wedge$-state, and if q is an $\wedge$-state, then q' is an $\vee$-state. One can show by induction that for every configuration α of M and its corresponding configuration α' of M', and for all i, $\ell_i(\alpha) = \neg\ell_i'(\alpha')$, thus $\ell_*(\alpha) = \neg\ell_*'(\alpha')$.

Now form a new machine M'' by taking the disjoint union of the finite controls of M and M' and altering the transition function such that if p is a $\neg$-state and $((p,a),(q,b,d))$ is a transition of M (hence p' is a $\neg$-state and $((p',a),(q',b,d))$ is a transition of M'), we make p an $\wedge$-state and p' an $\vee$-state of M'' and replace the transitions $((p,a),(q,b,d))$

and $((p', a), (q', b, d))$ with $((p, a), (q', b, d))$ and $((p', a), (q, b, d))$ in M''. Everything else in M'' is the same as in M and M'. Thus instead of negating, we jump to the dual machine. One can show inductively that $\ell''_i(\alpha) = \neg \ell''_i(\alpha') = \ell_i(\alpha)$, thus $\ell''_*(\alpha) = \neg \ell''_*(\alpha') = \ell_*(\alpha)$.   □

For an ATM without negations, one can show that acceptance of x is tantamount to the existence of a *finite accepting subtree* of the computation tree on input x. This is a finite subtree T of the computation tree containing the start configuration such that every $\lor$-configuration has at least one successor in T and every $\land$-configuration has all its successors in T (Miscellaneous Exercise 26).

Alternating Complexity Classes

An ATM is said to be $T(n)$-*time-bounded* if for any input of length n, all paths in the computation tree are of length at most $T(n)$. It is said to be $S(n)$-*space-bounded* if for any input of length n, no path in the computation tree uses more than $S(n)$ tape cells. The complexity class $ATIME(T(n))$ is the class of sets accepted by $T(n)$-time-bounded ATMs, and the class $ASPACE(S(n))$ is the class of sets accepted by $S(n)$-space-bounded ATMs. We also define

$$ALOGSPACE \stackrel{\text{def}}{=} ASPACE(\log n)$$
$$APTIME \stackrel{\text{def}}{=} ATIME(n^{O(1)})$$
$$APSPACE \stackrel{\text{def}}{=} ASPACE(n^{O(1)})$$
$$AEXPTIME \stackrel{\text{def}}{=} ATIME(2^{n^{O(1)}}),$$

and so on.

Complexity Results

The following four simulations show a strong correspondence between deterministic machines and alternating machines: to within a polynomial, alternating time is the same as deterministic space and alternating space is the same as exponentially more deterministic time.

Theorem 7.4   *Let $T(n) \geq n$ and $S(n) \geq \log n$. Then*

(i) $ATIME(T(n)) \subseteq DSPACE(T(n))$;

(ii) $DSPACE(S(n)) \subseteq ATIME(S(n)^2)$;

(iii) $ASPACE(S(n)) \subseteq DTIME(2^{O(S(n))})$;

(iv) $DTIME(T(n)) \subseteq ASPACE(\log T(n))$.

Proof. We assume for simplicity that the functions $S(n)$ and $T(n)$ are constructible. These assumptions can be removed.

(i) The proof of this result is similar to the proof of Theorem 2.6. We can perform a depth-first search on the computation tree of a $T(n)$-time-bounded alternating machine, computing the Boolean labels $\ell_*(\alpha)$. The position in the computation tree of the configuration currently being visited can be represented by a binary string of length at most $T(n)$.

(ii) We can actually show the stronger result

$$NSPACE(S(n)) \quad \subseteq \quad ATIME(S(n)^2)$$

using a parallel implementation of Savitch's theorem (Theorem 2.7). The main routine is a parallel recursive procedure $\text{PARSAV}(\alpha, \beta, k)$ that determines whether α goes to β in k or fewer steps. If $k = 0$ or 1, it checks directly whether $\alpha = \beta$ or, in the case $k = 1$, whether α goes to β in one step. If $k \geq 2$, it guesses $\gamma \in \Delta^{S(n)}$ using $\vee$-branching. This takes time $S(n)$ and results in $2^{S(n)}$ independent parallel processes, each with a different γ. The process handling γ checks in parallel whether $\text{PARSAV}(\alpha, \gamma, \lceil k/2 \rceil)$ and $\text{PARSAV}(\gamma, \beta, \lfloor k/2 \rfloor)$ using $\wedge$-branching. This alternating procedure can be implemented in alternating time $S(n)^2$; the analysis is essentially the same as in the proof of Savitch's theorem.

(iii) For this simulation, we can write down all configurations that fit in $S(n)$ space (there are $2^{O(S(n))}$ of them) and compute the labeling ℓ_* inductively. We start by labeling all configurations $\bot$. This is the labeling ℓ_0. Now suppose we have computed the labeling ℓ_i. We make a pass across the tape, computing $\ell_{i+1} = \tau(\ell_i)$. This takes time $2^{O(S(n))}$. We keep doing this until there are no more changes; we have found the least fixpoint. There are at most $2^{O(S(n))}$ passes in all, each taking time $2^{O(S(n))}$. Thus the entire algorithm runs in deterministic time $2^{O(S(n))}$.

(iv) Given a $T(n)$-time-bounded deterministic machine M and input x, an alternating machine can construct and evaluate on the fly a circuit describing the $T(n) \times T(n)$ computation matrix of M on input x like the one in the proof of the P-completeness of the circuit value problem (Theorem 6.1). Each process of the alternating machine tries to determine the value of some Boolean variable P_{ij}^b or Q_{ij}^q; it needs only $\log T(n)$ space to record the indices i, j of the variable. $\qquad \square$

Corollary 7.5 *For $T(n) \geq n$ and $S(n) \geq \log n$,*

$$
\begin{aligned}
ATIME(T(n)^{O(1)}) &= DSPACE(T(n)^{O(1)}), \\
ASPACE(S(n)) &= DTIME(2^{O(S(n))}).
\end{aligned}
$$

In other words, alternating time and deterministic space are the same to within a polynomial, and alternating space is the same as exponentially more deterministic time. Thus the hierarchy

$$LOGSPACE \subseteq P \subseteq PSPACE \subseteq EXPTIME \subseteq EXPSPACE \subseteq \cdots$$

shifts by exactly one level when alternation is introduced; in other words,

$$
\begin{aligned}
ALOGSPACE &= P \\
APTIME &= PSPACE \\
APSPACE &= EXPTIME \\
AEXPTIME &= EXPSPACE \\
&\vdots
\end{aligned}
$$

Lecture 8

Problems Complete for $PSPACE$

We have seen natural complete problems for $NLOGSPACE$, P, and NP, namely MAZE, CVP, and SAT, respectively. Here is a natural problem complete for $PSPACE$, the *quantified Boolean formula problem*. The problem has a natural alternating and/or structure, so we use ATMs liberally in proofs, but the theorem itself predated the invention of ATMs [118].

Definition 8.1 *The* quantified Boolean formula problem (QBF) *is the problem of determining the truth of quantified expressions*

$$Q_1 x_1 \ Q_2 x_2 \ \cdots \ Q_n x_n \ B(x_1, \ldots, x_n),$$

where $B(x_1, \ldots, x_n)$ is a Boolean formula with variables $x_1, \ldots, x_n$, each Q_i is either $\exists$ or $\forall$, and the quantification is over the two-element Boolean algebra $\{0,1\}$. This is essentially the decision problem for the first-order theory of the two-element Boolean algebra.

The Boolean satisfiability problem, which is NP-complete, is the restriction of QBF to *existential* formulas only, that is, formulas in which all Q_i are $\exists$. The Boolean validity problem, which is co-NP-complete, is the restriction of QBF to universal ($\forall$) formulas only.

Theorem 8.2 (Stockmeyer and Meyer [118])   QBF *is $\leq_{\mathrm{m}}^{\log}$-complete for $PSPACE$.*

Proof. We can determine the truth of a given quantified Boolean formula with an alternating TM by eliminating the quantifiers using $\vee$- and $\wedge$-branching, then evaluating the formula on the resulting truth assignment. Given a suitable encoding, this can be done in alternating linear time. By Theorem 7.4(i), QBF is in *PSPACE*.

To show *PSPACE*-hardness, we encode the computation of polynomial-time-bounded ATMs. The construction is very similar to the proof of the Cook–Levin theorem (Theorem 6.2).

Let A be an arbitrary set in *PSPACE*. We can assume without loss of generality that $A \subseteq \{0, 1\}^*$, because any set over a larger input alphabet is trivially $\leq_{\mathrm{m}}^{\mathrm{log}}$-reducible to a set over a binary alphabet. By Theorem 7.4(ii), there is a polynomial-time-bounded ATM M accepting A, say with time bound n^c. By Lemma 7.3, we can assume without loss of generality that M has no $\neg$-states. We can also assume without loss of generality, by adding dummy states if necessary, that the computation tree of M is binary branching and strictly alternates between $\vee$- and $\wedge$-configurations beginning with $\vee$.

Let x be an input to M of length n, say $x = x_1 x_2 \cdots x_n$. The computation tree of M on input x is of depth at most n^c, and each path is specified by a binary string y of length n^c. Exactly as in the proof of Theorem 6.2, construct a formula $B(X_1, \ldots, X_n, Y_1, \ldots, Y_{n^c})$ that has value 1 on a given instantiation x, y of the Boolean variables X, Y iff the computation path of M on input x specified by y leads to acceptance. Then M accepts x iff the quantified Boolean formula

$$\exists Y_1 \; \forall Y_2 \; \exists Y_3 \; \cdots \; Q_{n^c} Y_{n^c} \; B(x_1, \ldots, x_n, Y_1, \ldots, Y_{n^c})$$

is true; the alternation of quantifiers in the quantifier prefix of the formula exactly reflects the alternation of $\vee$- and $\wedge$-configurations in the computation tree. □

Complexity of Games

A *two-person perfect information game*, for our purposes, is a graph $G = (\text{BOARDS}, \text{MOVE})$ and a distinguished start node $s \in \text{BOARDS}$. The edge relation MOVE is a binary relation on BOARDS that specifies the legal moves of the game. The game starts at $s_0 = s$ and the players alternate with Player I moving first. Player I chooses $s_1 \in \text{BOARDS}$ such that $\text{MOVE}(s_0, s_1)$. Player II then chooses $s_2 \in \text{BOARDS}$ such that $\text{MOVE}(s_1, s_2)$, and so forth. A player wins by forcing the opponent into a position from which no move is possible (a "checkmate"), that is, a position t such that no u exists with $\text{MOVE}(t, u)$.

Most common two-person games—chess, checkers, go—are of this form. For chess, the set BOARDS consists of

{legal chessboards} $\times$ {white, black},

the second component telling whose move it is. The start board is

(the starting chessboard, white).

The relation MOVE encodes the legal chess moves.

In the game of *geography*, two players alternate thinking of names of countries. The first player may pick any country. Thereafter, each player must think of a country whose name begins with the same letter that the previously named country ends with; for example: Albania, Azerbaijan, Norway, Yemen, Nicaragua, and so on. A country may not be named more than once. Here the elements of the set BOARDS are pairs (A, B), where A is a set of countries and B is a set of letters of the alphabet. The starting board is ({all countries},{all letters}). A move $(A, B) \to (A', B')$ is a legal move if $A' = A - \{c\}$ for some country c whose name begins with a letter in B, and $B' = \{a\}$ where a is the last letter of c's name.

Definition 8.3   *Define*

$$\text{CHECKMATE}(y) \overset{\text{def}}{\iff} \forall z \; \neg\text{MOVE}(y, z).$$

A board position $x \in$ BOARDS is a forced win *for the player whose move it is if x satisfies the predicate* WIN(x)*, where the predicate* WIN *is the least solution of the recursive equation*

$$\text{WIN}(x) \quad \Leftrightarrow \quad \exists y \; \text{MOVE}(x, y) \land \tag{8.1}$$
$$(\text{CHECKMATE}(y) \lor \forall z \; (\text{MOVE}(y, z) \to \text{WIN}(z))); \tag{8.2}$$

equivalently, if WIN *is the least fixpoint of the monotone map τ on predicates defined by*

$$\tau(\varphi)(x) \quad \Leftrightarrow \quad \exists y \; \text{MOVE}(x, y) \land$$
$$(\text{CHECKMATE}(y) \lor \forall z \; (\text{MOVE}(y, z) \to \varphi(z))).$$

The least fixpoint exists by the Knaster–Tarski theorem (Theorem A.9).

Intuitively, if it is Player I's move, then the current board is a forced win for Player I if Player I can make a legal move that results in either (i) an immediate checkmate, or (ii) a board position from which all moves for Player II lead to a forced win for Player I.

A little elementary logic shows that (8.2) can be simplified to

$$\text{WIN}(x) \quad \Leftrightarrow \quad \exists y \; \text{MOVE}(x, y) \land \forall z \; (\text{MOVE}(y, z) \to \text{WIN}(z)).$$

This is because if CHECKMATE(y), then $\forall z \; (\text{MOVE}(y, z) \to \text{WIN}(z))$ is vacuously true.

We wish to study the complexity of deciding whether Player I has a forced win from a given board position. For this to make sense in terms of asymptotic complexity, we need to generalize the games in some reasonable way to allow arbitrarily large games. For example, we need to say explicitly what we mean by chess on an $n \times n$ board. This has been done for many common games, and the generalized versions have been shown to be complete for various complexity classes.

For example, we might generalize the game of geography as follows. An instance of the game is a tuple (C, E, s), where (C, E) is a directed graph and $s \in C$. The vertex set C corresponds to the cities. We start with a token on $s_0 = s$. In the even stages, Player I moves the token from s_{2i} to some vertex s_{2i+1} adjacent to s_{2i}, and in the odd stages, Player II moves the token from s_{2i+1} to some vertex s_{2i+2} adjacent to s_{2i+1}. No player may move to a vertex that has already been visited. A player wins by forcing the opponent to a position from which there is no legal next move.

The decision problem for generalized geography then becomes: given an instance (C, E, s), does Player I have a forced win?

Theorem 8.4   **(Stockmeyer and Chandra [116])**   *Generalized geography is $\leq_{\mathrm{m}}^{\log}$-complete for PSPACE.*

Proof. We show first that the problem is in *PSPACE* by giving an alternating polynomial time algorithm for it. Start by marking the vertex $s_0 = s$ as visited, then iterate the following procedure. Choose nondeterministically, using $\vee$-branching, a move s_1 for Player I. Move the token to that position and mark s_1 as visited. Then, using $\wedge$-branching, try all possible moves for Player II. Each new subprocess moves the token to some s_2 and marks it as visited. Then choose nondeterministically a move for Player I from s_2; and so on, alternating between the two players. Whenever it is Player I's move and there is no legal next move, halt and report failure (0). If it is Player II's move and there is no legal next move, halt and report success (1). Each computation path terminates after at most $|C|$ steps, because at least one new vertex gets marked as visited in each step.

Now we show that the problem is *PSPACE*-hard by reducing QBF to it. Given a quantified Boolean formula, we want to construct an instance of the game such that Player I has a forced win iff the formula is true. By applying elementary transformations of first-order logic and inserting dummy variables if necessary, we can assume without loss of generality that the given formula consists of a quantifier prefix followed by a quantifier-free part in conjunctive normal form, there are an even number of quantifiers, and the quantifiers alternate strictly beginning with $\exists$:

$$\exists x_1 \ \forall x_2 \ \exists x_3 \ \cdots \ \forall x_n \ c_1 \wedge \cdots \wedge c_m,$$

where each c_i is a clause of the form $\ell_1 \vee \cdots \vee \ell_k$ and each ℓ_i is a literal, either x or $\overline{x}$ for some variable x.

From such a formula, we construct an instance of generalized geography as follows.

(i)   For each i, $1 \leq i \leq n$, we create four vertices x_i, $\overline{x}_i$, u_i, and v_i. We have edges from u_i to x_i and to $\overline{x}_i$ and from x_i and $\overline{x}_i$ to v_i.

(ii)   We insert an edge from v_i to u_{i+1}, $1 \leq i \leq n - 1$.

(iii) We create a vertex for each clause c_j and an edge from v_n to each of $c_1, c_2, \ldots, c_m$.

(iv)   We insert an edge from c_j to the literal ℓ if ℓ appears in c_j.

The start vertex is u_1.

For example, for the formula

$$\exists x_1 \; \forall x_2 \; \underbrace{(x_1 \vee x_2)}_{c_1} \wedge \underbrace{(\overline{x}_1 \vee \overline{x}_2)}_{c_2},$$

the construction would produce the game

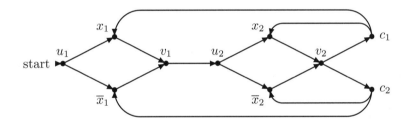

The first few moves of the game that take place on parts (i) and (ii) of the graph choose a truth assignment. The values of the odd-numbered variables are chosen by Player I and the values of the even-numbered variables are chosen by Player II. Each literal chosen is assigned the value 0 (false).

The last two steps are played on parts (iii) and (iv). Here Player II must choose a clause. Of course, Player II is trying to make the formula false, so if there is a clause that is false under the chosen truth assignment, then Player II chooses that clause. Player I has no move, because all the literals reachable from that clause are false, which means they have already been visited. On the other hand, if all clauses are true under that truth assignment, then each one must contain a true literal, so no matter what clause Player II picks, Player I will have a next move, after which Player II will be stuck. Thus Player I wins iff the chosen truth assignment satisfies the quantifier-free part of the formula.

In the example above, the formula is false, so Player II should always get the win by playing optimally. Indeed, if Player I chooses x in the first move, then Player II should choose y in the fourth move and c_1 in the sixth move (the other moves are forced); and if Player I chooses $\overline{x}$ in the first move, then Player II should choose $\overline{y}$ in the fourth move and c_2 in the sixth move. In either case Player II wins by checkmate.   □

See the exercises for more *PSPACE*-complete problems (Homework 6, Exercise 2 and Miscellaneous Exercises 31 and 29).

Lecture 9

The Polynomial-Time Hierarchy

The *polynomial-time hierarchy* (*PH*) is a hierarchy of complexity classes lying over P and inside *PSPACE*. It was first identified by Stockmeyer [117]. It is most easily defined in terms of alternating polynomial-time-bounded TMs, although it was originally defined in terms of oracle Turing machines. The hierarchy is analogous in many ways to the arithmetic hierarchy, which we introduce later in Lecture 35. However, unlike the arithmetic hierarchy, it is not known whether *PH* is strict.

In this lecture and the next, we define *PH* in two different ways, in terms of ATMs and oracle TMs, and prove the equivalence of the two definitions. We also give generic $\leq_{\mathrm{m}}^{\log}$-complete problems for each level of the hierarchy.

Definition of *PH* in Terms of ATMs

Informally, a Σ_k-machine (respectively, Π_k-machine) is an ATM without negations that on any input makes at most k alternations of $\vee$- and $\wedge$-configurations along any computation path, beginning with $\vee$ (respectively, $\wedge$).

Definition 9.1 *A Σ_k-machine is an ATM such that on any input, every computation path can be divided into contiguous intervals such that*

(i)   *in any interval, all configurations are either $\vee$-configurations or all are $\wedge$-configurations;*

(ii)   *there are at most k intervals; and*

(iii) *the first interval consists of $\vee$-configurations.*

A Π_k-machine is similar, except we change (iii) to:

(iii$'$) *the first interval consists of $\wedge$-configurations.*

A Σ_1-machine is just a nondeterministic TM. By convention, Σ_0- and Π_0-machines are deterministic TMs.

Definition 9.2   *The complexity classes Σ_k^{p} and Π_k^{p} are defined as follows.*

$$\Sigma_k^{\mathrm{p}} \ \overset{\mathrm{def}}{=} \ \{L(M) \mid M \text{ is a polynomial-time-bounded } \Sigma_k\text{-machine}\},$$

$$\Pi_k^{\mathrm{p}} \ \overset{\mathrm{def}}{=} \ \{L(M) \mid M \text{ is a polynomial-time-bounded } \Pi_k\text{-machine}\}.$$

Thus $\Sigma_1^{\mathrm{p}} = NP$, $\Pi_1^{\mathrm{p}} = \text{co-}NP$, and $\Sigma_0^{\mathrm{p}} = \Pi_0^{\mathrm{p}} = P$.

Lemma 9.3

$$
\begin{aligned}
\Pi_k^{\mathrm{p}} \ &= \ \text{co-}\Sigma_k^{\mathrm{p}} \ = \ \{\sim\!A \mid A \in \Sigma_k^{\mathrm{p}}\}, \\
\Sigma_k^{\mathrm{p}} \cup \Pi_k^{\mathrm{p}} \ &\subseteq \ \Sigma_{k+1}^{\mathrm{p}} \cap \Pi_{k+1}^{\mathrm{p}}, \\
\bigcup_k \Sigma_k^{\mathrm{p}} \ &= \ \bigcup_k \Pi_k^{\mathrm{p}} \ \subseteq \ PSPACE.
\end{aligned}
$$

Here $\sim\!A$ denotes the complement of A.

Proof. The first equation is obtained by interchanging $\vee$- and $\wedge$-states. This gives the dual machine, which accepts the complement of the set accepted by the original machine (provided the machine halts along all computation paths, which it does in this case because it is polynomial-time-bounded). The second inclusion is immediate from the definition. The third inclusion follows from the fact that $PSPACE = APTIME$, in which there are no restrictions on the number of alternations. $\square$

It is not known whether any of the inclusions in Lemma 9.3 are strict.

Generic Complete Problems

We can define generic problems that are complete for the various levels of the polynomial-time hierarchy. Define

$$H_k \ \overset{\mathrm{def}}{=} \ \{M\#x\#^m \mid M_k^m \text{ accepts } x\},$$

where M is any ATM and M_k^m is M modified so as to halt on any computation path that tries to alternate between $\vee$ and $\wedge$ states more than k

times, beginning with an $\vee$, or tries to take more than m steps. In other words, the computation tree of M_k^m on any input is essentially the same as that of M, except that it is artificially truncated to depth m and k alternations. Any leaves of the computation tree of M_k^m that are not leaves of the computation tree of M are either accept or reject configurations of M_k^m, according as the corresponding configuration of M is an $\wedge$-configuration or an $\vee$-configuration, respectively.

Theorem 9.4 *The set H_k is $\leq_{\mathrm{m}}^{\log}$-complete for Σ_k^{p}.*

Proof. To show that H_k is in Σ_k^{p}, we describe a polynomial-time-bounded Σ_k-machine N that accepts H_k. The machine N first checks that its input is of the form $M\#x\#^m$, where M is a valid ATM description and x is a valid encoding of a string over M's input alphabet. It then simulates M on input x, counting the number of simulated steps of M and the number of alternations. In the simulation, N makes $\vee$-branches whenever M would and $\wedge$-branches whenever M would. Because the branching degree of M may be arbitrarily large and that of N must be fixed, it may take several branches of N to simulate one branch of M; but this is ok, because the number of alternations is the same. If during the simulation, the number of steps of M exceeds m (the number provided in the input to N), or if M tries to make more than k alternations, N halts on that computation path. Otherwise, N accepts or rejects as M does.

It takes a small polynomial in the length of $M\#x\#^m$ to simulate one step of M, and there are at most m steps simulated, so N runs in polynomial time; moreover, the computation is Σ_k. Thus $H_k \in \Sigma_k^{\mathrm{p}}$.

To show that H_k is $\leq_{\mathrm{m}}^{\log}$-hard for Σ_k^{p}, let $A \in \Sigma_k^{\mathrm{p}}$ be arbitrary. Then there is a Σ_k-machine M accepting A with time bound n^c. Thus for $m = |x|^c$, the computation trees of M and M_k^m are essentially the same on input x; that is, no truncation takes place. Thus M accepts x iff M_k^m accepts x iff $M\#x\#^m \in H_k$. Therefore the map

$$x \;\mapsto\; M\#x\#^{|x|^c}$$

constitutes a reduction from A to H_k, and this map is easily computed in logspace. $\square$

Corollary 9.5 *The set $\sim H_k$ is $\leq_{\mathrm{m}}^{\log}$-complete for Π_k^{p}.*

Oracle Machines and Relativized Complexity Classes

Originally, the polynomial-time hierarchy was defined in terms of *oracle Turing machines*. An *oracle TM* is like an ordinary TM, except it is equipped with an *oracle*, a means of answering membership questions about some set B in one step. It is useful for studying computation relative to free knowledge about the set B.

There are a few equivalent ways of defining oracle machines formally. In one definition, the machine is equipped with a special write-only *oracle query tape* on which finite strings can be written. The finite control also contains a special *oracle query state* as well as a special "yes" state and a special "no" state. When the machine enters the oracle query state, it is asking whether the string y currently written on the oracle tape is in B. The machine magically moves to the "yes" state if $y \in B$ and to the "no" state if $y \notin B$. There is no explicit representation of the set B. The oracle must give the same answer if queried with y again in the future. Note that the same oracle machine may be used with different oracles; the results of the oracle queries will be different.

Here is another equivalent definition. An oracle machine is a TM that in addition to its ordinary input and worktapes is equipped with a special one-way-infinite read-only input tape on which some infinite string is written. The extra tape is called the *oracle tape*, and the string written on it is called the *oracle*. The machine can move its oracle tape head one cell in either direction in each step and make decisions based on the symbols written on the oracle tape. Other than that, it behaves exactly like an ordinary Turing machine.

In the latter definition, we usually think of the oracle as a specification of a set $B \subseteq \mathbb{N}$. If the oracle is an infinite string over $\{0, 1\}$, then we can regard it as the characteristic function of B, where the nth bit of the oracle string is 1 iff $n \in B$.

Definition 9.6   *Let B be a set and let $\mathcal{C}$ be a complexity class.*

$$P^B \stackrel{\text{def}}{=} \{L(M) \mid M \text{ is a deterministic polynomial-time-bounded} \\ \text{oracle machine with oracle } B\},$$

$$NP^B \stackrel{\text{def}}{=} \{L(M) \mid M \text{ is a nondeterministic polynomial-time-} \\ \text{bounded oracle machine with oracle } B\},$$

$$P^{\mathcal{C}} \stackrel{\text{def}}{=} \bigcup_{B \in \mathcal{C}} P^B,$$

$$NP^{\mathcal{C}} \stackrel{\text{def}}{=} \bigcup_{B \in \mathcal{C}} NP^B.$$

Other relativized complexity classes such as $PSPACE^B$ and $PSPACE^{\mathcal{C}}$ can be defined similarly. These classes are called relativized complexity classes.

If B is $\leq_{\mathrm{m}}^{\mathrm{P}}$-complete for $\mathcal{C}$, then $P^{\mathcal{C}} = P^B$ and $NP^{\mathcal{C}} = NP^B$. This is because the reduction from some arbitrary $A \in \mathcal{C}$ to B can be performed before an oracle query is made. For example, $P^{NP} = P^{\text{SAT}}$ and $NP^{NP} = NP^{\text{SAT}}$.

Definition 9.7   *We write $A \leq^{\mathrm{p}}_{\mathrm{T}} B$ if $A \in P^B$. The relation $\leq^{\mathrm{p}}_{\mathrm{T}}$ is called* polynomial-time Turing reducibility *or* Cook reducibility.

Essentially, $A \leq^{\mathrm{p}}_{\mathrm{T}} B$ means that A can be computed in polynomial time, given free information about B. If $A \leq^{\mathrm{p}}_{\mathrm{m}} B$ then $A \leq^{\mathrm{p}}_{\mathrm{T}} B$, but the converse would imply $NP = \text{co-}NP$, because $\text{SAT} \leq^{\mathrm{p}}_{\mathrm{m}} \sim\text{SAT}$.

Definition of PH in Terms of Oracle Machines

Theorem 9.8   *Consider the hierarchy*

$$NP \;\subseteq\; NP^{NP} \;\subseteq\; NP^{NP^{NP}} \;\subseteq\; NP^{NP^{NP^{NP}}} \;\subseteq\; \cdots \;.$$

More formally, define

$$NP_1 \;\overset{\mathrm{def}}{=}\; NP,$$
$$NP_{k+1} \;\overset{\mathrm{def}}{=}\; NP^{NP_k}.$$

Then $NP_k = \Sigma^{\mathrm{p}}_k$ for all $k \geq 1$.

We prove this theorem next time.

Lecture 10

More on the Polynomial-Time Hierarchy

Recall from the last lecture the hierarchy

$$NP \subseteq NP^{NP} \subseteq NP^{NP^{NP}} \subseteq NP^{NP^{NP^{NP}}} \subseteq \cdots$$

defined in terms of oracle machines. Formally, we defined

$$NP_1 \stackrel{\text{def}}{=} NP$$
$$NP_{k+1} \stackrel{\text{def}}{=} NP^{NP_k}.$$

Recall also that Σ_k^{p} is the class of sets accepted by polynomial-time-bounded alternating TMs that make at most k alternations of $\vee$- and $\wedge$-configurations along any computation path, beginning with $\vee$. In this lecture we show that these two definitions characterize the same hierarchy of complexity classes.

Theorem 10.1 $NP_k = \Sigma_k^{\mathrm{p}}$ for all $k \geq 1$.

Proof. This is proved by induction on k. The basis $k = 1$ is immediate from the definitions: $NP_1 = \Sigma_1^{\mathrm{p}} = NP$.

For the induction step, we need only show that $NP^{\Sigma_k^{\mathrm{p}}} = \Sigma_{k+1}^{\mathrm{p}}$. We show the inclusion in both directions separately, the easier direction first.

($\supseteq$) For this inclusion, assume that we have a Σ_{k+1}-machine M running in time n^c, $A = L(M)$. We wish to show that $A \in NP^{\Sigma_k^{\mathrm{p}}}$. Let the configurations of M be encoded as strings over a finite alphabet Δ in some reasonable way. We can assume that all configurations reachable from the start configuration on any input x, $|x| = n$, are represented as strings in Δ^{n^c}.

Now consider the set

$$D \stackrel{\text{def}}{=} \{\alpha \mid \alpha \text{ is an } \wedge\text{-configuration of } M, |\alpha| = n^c, \text{ and } \alpha \text{ leads to}$$
$$\text{acceptance via a } \Pi_k \text{ computation in time at most } n^c\}.$$

Membership in D can be determined by a polynomial-time-bounded Π_k^{p} machine that just simulates M starting from the configuration α; therefore, $D \in \Pi_k^{\text{p}}$ and $\sim D \in \Sigma_k^{\text{p}}$. Moreover, M accepts x iff there exists a computation path leading from the start configuration through only $\vee$-configurations to some $\alpha \in D$.

Thus A can be accepted by a nondeterministic polynomial-time-bounded oracle machine with oracle $\sim D$ that on input x guesses a computation path from the start configuration of M through only $\vee$-configurations to some $\wedge$-configuration α, then consults the oracle to check whether $\alpha \in D$.

($\subseteq$) For this inclusion, assume that we have a nondeterministic n^c-time-bounded oracle machine M with oracle $B \in \Sigma_k^{\text{p}}$, and let $A = L(M)$. We wish to show that $A \in \Sigma_{k+1}^{\text{p}}$.

Build a Σ_{k+1} machine N that works as follows. On input x, N will begin by simulating M on input x, except that every time M wants to query the oracle on some string y, N nondeterministically guesses the answer that the oracle would return (that is, whether $y \in B$) and remembers y and the guessed answer. It continues the simulation until M arrives at an accept or reject state, which must happen by time n^c. If M wants to reject at that point, N just rejects. If M wants to accept, N has to verify that the guessed oracle answers were correct.

So far the computation tree of N looks like the computation tree of M, except that at each oracle query, N has a nondeterministic branch for the two possible responses of the oracle. Up to this point N has made only existential branches. At each leaf there is a process with a list of oracle queries and guessed responses that need to be verified. The lists may be different for different computation paths, because later queries may depend on the responses to earlier queries, but all lists are at most n^c in length.

Thus each process has a list $y_1, \ldots, y_m$ of oracle queries for which the guessed response from the oracle was positive and a list $z_1, \ldots, z_\ell$ for which the guessed response was negative. The total combined length of the y_i and z_j concatenated together is at most n^c, because the machine had to write them all down on its oracle query tape. It must now verify with a Σ_{k+1} computation in polynomial time that $y_i \in B$, $1 \leq i \leq m$, and $z_j \notin B$, $1 \leq j \leq \ell$. That Σ_{k+1} computation combined with the Σ_1 computation up to this point is still a Σ_{k+1} computation.

We have reduced the problem to showing that for $B \in \Sigma_k^p$, the set

$$\{y_1 \# \cdots \# y_m \# \# z_1 \# \cdots \# z_\ell \mid y_i \in B, \, 1 \le i \le m; \, z_j \notin B, \, 1 \le j \le \ell\}$$

is in Σ_{k+1}^p. Each guess $y_i \in B$ can be verified with a Σ_k computation, and each guess $z_j \notin B$ can be verified with a Π_k computation.

Our first thought might be to make an $(m + \ell)$-way $\wedge$-branch, each process taking a y_i or z_j and independently verifying $y_i \in B$ with a Σ_k computation and $z_j \notin B$ with a Π_k computation. Unfortunately, this does not work, because it results in a Π_{k+1} computation, which cannot in general be simulated by a Σ_{k+1} computation.

Instead, we do the first round of existential guessing for all the y_i sequentially. Let n^d be the time bound on a Σ_k-machine for B. Nondeterministically guess binary strings $w_1, \ldots, w_m$, each of length n^d. This takes time at most $mn^d \le n^{c+d}$. These strings will direct the first round of existential guesses in the Σ_k computations to verify $y_i \in B$. Now make an $(m + \ell)$-way universal branch, each process taking a positive query y_i or a negative query z_j. So far the computation is Σ_2. For each negative query z_j, we just verify $z_j \notin B$ using a Π_k computation for $\sim B$. This combines with the preceding Σ_2 computation to give a Σ_{k+1} computation. For each positive query y_i, we simulate the Σ_k computation for B to verify $y_i \in B$, but use the guessed w_i to direct the first level of existential branches, thus making the first level deterministic. The remaining computation is Π_{k-1}, so the total computation is Σ_{k+1}. $\square$

The trick we used at the end is essentially an instance of *skolemization*, a common technique used in logic:

$$\bigwedge_{i \in A} \bigvee_{w \in B} \varphi(i, w) \;=\; \bigvee_{f: A \to B} \bigwedge_{i \in A} \varphi(i, f(i)).$$

This is actually a generalized version of the distributive law of Boolean algebra. In our application, $A = \{1, 2, \ldots, m\}$, $B = \{0, 1\}^{n^d}$, and $\varphi(i, w)$ says that y_i is accepted by a Π_{k-1} machine that simulates the Σ_k computation of B, but uses w to direct the first level of nondeterministic choices.

Ordinarily an application of skolemization or the distributive law results in an exponential blowup in the size of the formula: if $|A| = m$ and $|B| = r$, then $|A \to B| = r^m$. However, here we are ok, because $|A| = m \le n^c$ and $|B| = 2^{n^d}$, so $|A \to B| = (2^{n^d})^m \le (2^{n^d})^{n^c} = 2^{n^{c+d}}$. The functions $f : A \to B$ in our application represent the possible choices of binary strings $w_1, \ldots, w_m$, and these can be guessed in time at most n^{c+d}.

Another characterization of the polynomial time hierarchy can be given in terms of quantified expressions. Let t be a numeric term. Define the bounded quantifiers $\exists^t$ and $\forall^t$ that limit the quantification to range over those strings whose length is bounded by the value of t. We can consider $\exists^t$ and $\forall^t$ to be abbreviations for

$$\exists^t y \; \varphi(y) \quad \stackrel{\text{def}}{\Longleftrightarrow} \quad \exists y \; |y| \leq t \wedge \varphi(y)$$

$$\forall^t y \; \varphi(y) \quad \stackrel{\text{def}}{\Longleftrightarrow} \quad \forall y \; |y| \leq t \rightarrow \varphi(y).$$

Theorem 10.2   *A set A is in Σ_k^{P} if and only if there is a deterministic polynomial-time computable $(k+1)$-ary predicate R and constant c such that*

$$A \;=\; \{x \mid \exists^{|x|^c} y_1 \; \forall^{|x|^c} y_2 \; \exists^{|x|^c} y_3 \;\cdots\; \mathsf{Q}^{|x|^c} y_k \; R(x, y_1, \ldots, y_k)\},$$

where $\mathsf{Q} = \exists$ if n is odd, $\forall$ if n is even.

Proof. Miscellaneous Exercise 32. $\square$

Lecture 11

Parallel Complexity

In this lecture we take a look at parallel computation from a complexity-theoretic point of view. Many machine models have been developed to study parallelism, perhaps the most prominent of which are parallel random access machines (PRAMs) and uniform families of circuits. Many of these models can simulate one another with relatively low overhead, so the parallel complexity classes they define are robust and natural.

Uniform Families of Circuits and NC

Uniform families of circuits were first defined and studied as a model of parallel complexity by Allan Borodin [22]. The most popular parallel complexity class defined in terms of uniform circuits is NC, for *Nick's class*, named after Nicholas Pippenger. The class first appeared in print under that name in a paper of Stephen Cook [32], who attributed the definition of the class to Pippenger, except for the uniformity condition.

NC lies between $NLOGSPACE$ and P. It is to parallel computation as P is to sequential computation—a robust and natural (albeit imperfect) approximation to the notion of *efficiently parallelizable*.

Many efficient NC algorithms have been developed for important problems. In particular, it has been shown that virtually all of linear algebra can be done in NC. Other problems have defied parallelization; for example, there is no known efficient parallel algorithm for the circuit value problem or for calculating integer greatest common divisors (gcd); however, polynomial gcd is in NC.

The $P = NP$ question asks whether a host of important combinatorial problems have efficient sequential solutions. The $NC = P$ question, in turn, asks whether a host of problems that are known to have efficient sequential solutions have efficient parallel solutions. Any problem that is $\leq_{\mathrm{m}}^{\log}$-complete for P, such as CVP, is in NC iff $P = NC$.

One popular definition of NC is in terms of PRAMs: a set is in NC if it has a $(\log n)^{O(1)}$-time (polylog-time) solution on a PRAM using polynomially many processors. This model is well suited to the analysis of parallel graph algorithms and other combinatorial problems. We concentrate on an alternative definition in terms of uniform families of circuits, which is more useful in algebraic problems. In this definition, a set is in NC if it has a logspace-uniform family of Boolean circuits of polylog depth and polynomial size. Formally:

Definition 11.1   (Cook [32])   *A family of Boolean circuits $C_0, C_1, C_2, \dots$ is a logspace-uniform family of Boolean circuits of polylog depth and polynomial size if:*

   (i)   *C_n has n input wires and is composed of $\vee$-, $\wedge$-, and $\neg$-gates;*

   (ii) *C_n is of depth at most $(\log n)^{O(1)}$ (polylog depth), where the depth is the length of the longest path from an input to an output;*

   (iii) *C_n is has no more than $n^{O(1)}$ gates (polynomial size); and*

   (iv) *the family is logspace-uniform, which means that there is a logspace transducer that produces the circuit C_n on input 0^n.*

A set $A \subseteq \{0,1\}^*$ is in NC if there exists such a family in which each C_n has one output wire, and for all $x \in \{0,1\}^*$, $x \in A$ iff $C_{|x|}(x) = 1$.

The uniformity condition (iv) is there mostly for technical reasons. It allows PRAMs and other models to simulate circuits. It is a very reasonable restriction, because in most applications the circuits can be constructed that easily. Note that without some kind of uniformity condition, there would exist trivial circuit families that decide undecidable problems; for example, take C_n to be a circuit with a single gate that outputs 1 if TM M_n halts on input n, 0 otherwise.

Boolean Matrix Multiplication

Here is an example of a family of logspace-uniform, polylog-depth circuits to compute the Boolean matrix product of two $n \times n$ matrices. The nth circuit has $2n^2$ input wires and n^2 output wires and computes the Boolean matrix product of two given $n \times n$ input matrices. (For the other input sizes, take some trivial circuit.) The size of the nth circuit will be $O(n^3)$ and the depth will be $\log n$.

The inputs supplied to the nth circuit are the entries of the two given matrices A, B. In the first step, the circuit computes $A_{ij} \wedge B_{jk}$ in parallel for all choices of i, j, k. This takes one step and n^3 $\wedge$-gates. Then for each i and k, the Boolean sums $\bigvee_j (A_{ij} \wedge B_{jk})$ are computed in parallel in a treelike fashion. This requires depth $\log n$ and $O(n)$ gates for each i and k, or $O(n^3)$ in all. The outputs are

$$(AB)_{ik} \;=\; \bigvee_j (A_{ij} \wedge B_{jk}). \tag{11.1}$$

The family is logspace-uniform, because the circuits are quite simple to describe and could be created by a logspace transducer. Note that the n^2 subcircuits that produce (11.1) are identical except for the indices i, k, which take only logspace to write down in binary.

Reflexive Transitive Closure

Here is another example. We describe a family of logspace-uniform, polylog-depth, and polynomial-size circuits to compute the reflexive transitive closure R^* of a given binary relation R on a set of size n. Recall that R^* consists of pairs (u, v) such that there exists an R-path of length zero or greater from u to v.

Suppose R is given by its $n \times n$ adjacency matrix. Note that

$$R^* \;=\; \bigvee_{i \geq 0} R^i \;=\; \bigvee_{i=0}^{n-1} R^i \;=\; (R \vee I)^{n-1}, \tag{11.2}$$

where $I = R^0$ is the $n \times n$ identity matrix. Note that $(R^i)_{uv} = 1$ if there is an R-path of length exactly i from u to v. We can limit the powers of R in the Boolean sum (11.2) to $n - 1$ (or any greater finite number) because if there exists a path at all from u to v, then there exists one of length at most $n - 1$ obtained by cutting out loops.

The circuit that computes R^* from R has n^2 input wires, on which are supplied the Boolean entries of the adjacency matrix of R. The circuit first computes $R \vee I$, then repeatedly squares the matrix $\lceil \log n \rceil$ times to obtain R^*:

$$(R \vee I)^2, \ (R \vee I)^4, \ (R \vee I)^8, \ (R \vee I)^{16}, \ \dots \ .$$

Each squaring step takes $\log n$ depth and polynomial size by the previous example, and there are $\log n$ squaring steps, so the total depth is $(\log n)^2$.

Again, the family is logspace-uniform, because the circuits are quite simple to describe and could be constructed by a logspace transducer.

Relation to Time-Space Classes

In order to describe the relationship of NC to conventional time and space complexity, we need to define a family of complexity classes for alternating machines that simultaneously keep track of time, space, and number of alternations.

Definition 11.2 *The class*

$$STA(S(n), T(n), A(n))$$

is the class of all sets accepted by ATMs that are simultaneously $S(n)$-space-bounded, $T(n)$-time-bounded, and make at most $A(n)$ alternations on inputs of length n. A $*$ in any position means "don't care." In other words, no bound is imposed. We also write Σ or Π in the third position if we need to specify that the alternations should start with an $\vee$ or $\wedge$, respectively.

For example,

$$
\begin{aligned}
LOGSPACE &= STA(\log n, *, 0), \\
NLOGSPACE &= STA(\log n, *, \Sigma 1), \\
P &= STA(\log n, *, *) = STA(*, n^{O(1)}, 0), \\
NP &= STA(*, n^{O(1)}, \Sigma 1), \\
\Sigma_k^{\mathrm{p}} &= STA(*, n^{O(1)}, \Sigma k), \\
\Pi_k^{\mathrm{p}} &= STA(*, n^{O(1)}, \Pi k), \\
PSPACE &= STA(*, n^{O(1)}, *) = STA(n^{O(1)}, *, 0),
\end{aligned}
$$

and so on.

The following theorem of Ruzzo relates NC to more conventional complexity classes.

Theorem 11.3 (Ruzzo [106]) $\qquad NC = STA(\log n, *, (\log n)^{O(1)}).$

One can see from this theorem that $NLOGSPACE \subseteq NC \subseteq P$, because

$$
\begin{aligned}
NLOGSPACE &= STA(\log n, *, \Sigma 1), \\
NC &= STA(\log n, *, (\log n)^{O(1)}), \\
P &= STA(\log n, *, *).
\end{aligned}
$$

We prove this theorem next time.

Lecture 12

Relation of NC to Time-Space Classes

Recall that NC is the class of sets computable by polylog-depth, polynomial-size, logspace-uniform families of Boolean circuits $C_0, C_1, \ldots$. The following theorem relates this class to more conventional time-space classes and places NC between $NLOGSPACE$ and P.

Theorem 12.1   (Ruzzo [106]) $NC = STA(\log n, *, (\log n)^{O(1)})$.

Proof. We show the inclusion in both directions.

($\subseteq$) We said before that *logspace-uniform* means that there is a logspace transducer M that on input 0^n produces the nth circuit C_n. Let us be a little more careful about what we mean by this.

Because C_n has only polynomially many gates, there is a naming scheme in which each gate can be named with a string of length $O(\log n)$. We reserve the names $1, 2, \ldots, n$ in binary for the names of the n input ports. On input 0^n, the logspace transducer M must produce

(i)   an enumeration of the names of all gates in the circuit C_n;

(ii)   for each gate c, a tag indicating whether c is an $\wedge$-gate, an $\vee$-gate, a $\neg$-gate, or an input port;

(iii)   a list of all wires (c, d) in the circuit, where c and d are legal gate names enumerated in (i); and

(iv)   for one of the gates c, a tag indicating that c is the output gate.

We assume for simplicity that $\vee$- and $\wedge$-gates have exactly two inputs, $\neg$-gates exactly one, and input ports none. Circuits with unbounded fan-in can be simulated with at most an $O(\log n)$ factor increase in depth and at most double the size.

Because we can construct the dual circuit in logspace (the construction is similar to the proof of Lemma 7.3), we can assume without loss of generality that C_n has no negate gates other than those applied immediately to inputs. That is, if d is a negate gate and (c, d) is the unique wire coming into d, then c is an input port.

Now we design an alternating logspace machine N to simulate this family of circuits. On input x of length n, the machine N will use M to produce C_n on the fly and evaluate $C_n(x)$. First N runs M to find the name of the output gate and its type, which it writes on its worktape. Now suppose some process of N has the name and type of a gate d written on its worktape.

- If d is an $\wedge$- or an $\vee$-gate, it starts M from scratch, enumerating wires to find the two wires (c, d), (c', d) coming into d. When it has found these wires, it makes an existential or universal branch according as d is an $\vee$- or $\wedge$-gate, respectively. Each of the two subprocesses takes one of c and c' and repeats.

- If d is an input port $1 \le d \le n$, N accepts or rejects according as the dth bit of the input x is 1 or 0, respectively.

- If d is a $\neg$-gate, N runs M to find the input port c such that (c, d) is the unique wire coming into d. It accepts or rejects according as the cth bit of the input x is 0 or 1, respectively.

The machine N needs no more than logarithmic space to do any of these tasks, because all it has to remember at any time is the name of the current gate c it is visiting. Moreover, the number of alternations that N makes is bounded by the depth of the circuit it is simulating, which is $(\log n)^{O(1)}$.

($\supseteq$) For this inclusion, assume that we are given an alternating logspace machine N making at most $(\log n)^c$ alternations on inputs of length n. We wish to construct a logspace-uniform family of circuits of polylog depth and polynomial size simulating N.

As argued in Theorem 7.4, we can assume without loss of generality that N runs in polynomial time. Encoding configurations in some reasonable way as strings over a finite alphabet, each configuration reachable from the start configuration on an input of length n can be represented as a string of length $\log n$, and there are only n^c such configurations for some constant c.

For inputs $x \in \{0,1\}^n$, we can represent the next-configuration relation of N on input x as an $n^c \times n^c$ Boolean matrix R_x whose rows and columns are indexed by configurations; thus $R_x(\alpha, \beta) = 1$ if configuration β is an immediate successor of configuration α on input x according to the transition rules of N.

The circuit C_n will first compute the entries of R_x. Because configurations are encoded as strings of length $O(\log n)$, the pairs (α, β) can also serve as the names of the gates for purposes of generating the circuit uniformly in logspace. A logspace machine can easily compare α and β and determine whether there is a possible transition from α to β and build a trivial circuit to output the value of $R_x(\alpha, \beta)$. This will depend on the input symbol being scanned in configuration α, so the circuit will access the ith input port of C_n, where i is the position of N's input head encoded in the configuration α. The depth of the circuit constructed so far is 1.

Once C_n has computed R_x, it constructs matrices for the relations

$$S_x \overset{\text{def}}{=} \{(\alpha, \beta) \mid R_x(\alpha, \beta) = 1 \text{ and } \operatorname{type}(\alpha) = \operatorname{type}(\beta)\},$$

$$T_x \overset{\text{def}}{=} \{(\alpha, \beta) \mid R_x(\alpha, \beta) = 1 \text{ and } \operatorname{type}(\alpha) \neq \operatorname{type}(\beta)\},$$

$$S_x^* \overset{\text{def}}{=} \text{reflexive transitive closure of } S_x,$$

and the product $S_x^* T_x$. A pair (α, β) is in S_x^* if there is a computation path in N of length zero or greater on input x from α to β such that all configurations on the path have the same type. A pair (α, β) is in $S_x^* T_x$ if there is a computation path from α to β such that all configurations on the path except β have the same type as α, and β has a different type.

The matrices S_x and T_x are just the componentwise Boolean meet of R_x with the relation $\{(\alpha, \beta) \mid \operatorname{type}(\alpha) = \operatorname{type}(\beta)\}$ and its complement, respectively. The reflexive transitive closure S_x^* and the product $S_x^* T_x$ can be computed using constructions given in Lecture 11.

Now think of the computation tree of N on input x, $|x| = n$, as divided into $(\log n)^c$ levels. Within each level, either all configurations are existential or all are universal.

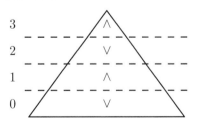

Number the levels $0, 1, 2, \ldots$ from bottom to top. An existential configuration α at level $i + 1$ is an accepting configuration iff there exists a configuration β at level i such that $S_x^* T_x(\alpha, \beta) = 1$ and β is an accepting configuration. A universal configuration α at level $j + 1$ is an accepting configuration iff for all existential configurations β at level j such that $S_x^* T_x(\alpha, \beta) = 1$, β is an accepting configuration.

Now the circuit computes a series of Boolean vectors $b_0, b_1, \ldots$, each of length n^c and indexed by configurations, such that $b_i(\alpha) = 1$ iff α is an accepting configuration at the ith level.

For existential levels $i + 1$, the circuit computes the matrix–vector product

$$b_{i+1} \quad := \quad S_x^* T_x b_i.$$

For universal levels $i + 1$, the circuit computes

$$b_{i+1} \quad := \quad \neg(S_x^* T_x(\neg b_i)),$$

where $\neg$ applied to a vector is interpreted componentwise.

We can take the initial vector b_0 to be the zero vector. Recalling our convention that an accept configuration is an $\wedge$-configuration with no successors and a reject configuration is an $\vee$-configuration with no successors, this does the right thing. For example, an $\wedge$-configuration α at level 1 is an accepting configuration if there does not exist a computation path leading to an $\vee$-configuration, in which case $b_1(\alpha) = 1$.

One can show by induction that a configuration α of the appropriate type is an accepting configuration at the ith level iff $b_i(\alpha) = 1$. The output value is $b_{(\log n)^c}(\mathbf{start})$, where $\mathbf{start}$ is the start configuration.

There are at most $(\log n)^c$ levels, and each level requires $\log n$ depth and polynomial size for the matrix–matrix and matrix–vector products and other Boolean operations. Moreover, the circuit is uniform at each level and can be generated by a logspace transducer. $\square$

Lecture 13

Probabilistic Complexity

There are many instances of problems with efficient randomized or probabilistic algorithms for which no good deterministic algorithms are known. In the next few lectures we take a complexity-theoretic look at probabilistic computation. We define a simple model of randomized computation, the *probabilistic Turing machine*, define some basic probabilistic complexity classes, and outline the relationship of these classes to conventional time and space classes. Our main result, which we prove next time, is that the class *BPP* of sets accepted by polynomial-time probabilistic algorithms with error probability bounded below $\frac{1}{2}$ is contained in $\Sigma_2^p \cap \Pi_2^p$ [112].

Many probabilistic algorithms have only a one-sided error; that is, if the input string is in the set, then the algorithm accepts with high probability; but if the string is not in the set, then the algorithm rejects always. The corresponding probabilistic complexity class is known as *RP* and is called *random polynomial time*.

Discrete Probability

Before we begin, let us recall some basic concepts from discrete probability theory.

Law of Sum   The *law of sum* says that if $\mathcal{A}$ is a collection of pairwise disjoint events, that is, if $A \cap B = \varnothing$ for all $A, B \in \mathcal{A}$, $A \neq B$, then the probability that at least one of the events in $\mathcal{A}$ occurs is the sum of the probabilities:

$$\Pr(\bigcup \mathcal{A}) \;=\; \sum_{A \in \mathcal{A}} \Pr(A).$$

Expectation   The *expected value* $\mathcal{E}X$ of a discrete random variable X is the weighted sum of its possible values, each weighted by the probability that X takes on that value:

$$\mathcal{E}X \;=\; \sum_{n} n \cdot \Pr(X = n).$$

For example, consider the toss of a coin. Let

$$X \;=\; \begin{cases} 1, & \text{if the coin turns up heads} \\ 0, & \text{otherwise.} \end{cases} \tag{13.1}$$

Then $\mathcal{E}X = \frac{1}{2}$ if the coin is unbiased. This is the expected number of heads in one flip. Any function $f(X)$ of a discrete random variable X is a random variable with expectation

$$\mathcal{E}f(X) \;=\; \sum_{n} n \cdot \Pr(f(X) = n) \;=\; \sum_{m} f(m) \cdot \Pr(X = m).$$

It follows immediately from the definition that the expectation function $\mathcal{E}$ is linear. For example, if X_i are the random variables (13.1) associated with n coin flips, then

$$\mathcal{E}(X_1 + X_2 + \cdots + X_n) \;=\; \mathcal{E}X_1 + \mathcal{E}X_2 + \cdots + \mathcal{E}X_n,$$

and this gives the expected number of heads in n flips. The X_i need not be independent; in fact, they could all be the same flip.

Conditional Probability and Conditional Expectation   The *conditional probability* $\Pr(A \mid B)$ is the probability that event A occurs given that event B occurs. Formally,

$$\Pr(A \mid B) \;=\; \frac{\Pr(A \cap B)}{\Pr(B)}.$$

The conditional probability is undefined if $\Pr(B) = 0$.

The *conditional expectation* $\mathcal{E}(X \mid B)$ is the expected value of the random variable X given that event B occurs. Formally,

$$\mathcal{E}(X \mid B) \;=\; \sum_{n} n \cdot \Pr(X = n \mid B).$$

If the event B is that another random variable Y takes on a particular value m, then we get a real-valued function $\mathcal{E}(X \mid Y = m)$ of m. Composing this function with the random variable Y itself, we get a new random variable, denoted $\mathcal{E}(X \mid Y)$, which is a function of the random variable Y. The random variable $\mathcal{E}(X \mid Y)$ takes on value n with probability

$$\sum_{\mathcal{E}(X \mid Y=m)=n} \Pr(Y = m),$$

where the sum is over all m such that $\mathcal{E}(X \mid Y = m) = n$. The expected value of $\mathcal{E}(X \mid Y)$ is just $\mathcal{E}X$:

$$
\begin{aligned}
\mathcal{E}(\mathcal{E}(X \mid Y)) &= \sum_m \mathcal{E}(X \mid Y = m) \cdot \Pr(Y = m) \\
&= \sum_m \sum_n n \cdot \Pr(X = n \mid Y = m) \cdot \Pr(Y = m) \\
&= \sum_n n \cdot \sum_m \Pr(X = n \wedge Y = m) \qquad (13.2) \\
&= \sum_n n \cdot \Pr(X = n) \\
&= \mathcal{E}X
\end{aligned}
$$

(see [39, p. 223]).

Independence and Pairwise Independence   A set of events $\mathcal{A}$ are *independent* if for any subset $\mathcal{B} \subseteq \mathcal{A}$,

$$\Pr\left(\bigcap \mathcal{B}\right) = \prod_{A \in \mathcal{B}} \Pr(A).$$

They are *pairwise independent* if for every $A, B \in \mathcal{A}$, $A \neq B$,

$$\Pr(A \cap B) = \Pr(A) \cdot \Pr(B).$$

For example, the probability that two successive flips of a fair coin both come up heads is $\frac{1}{4}$.

Pairwise independent events need not be independent. Consider the following three events:

- The first flip gives heads.

- The second flip gives heads.

- Of the two flips, one is heads and one is tails.

The probability of each pair is $\frac{1}{4}$, but the three cannot happen simultaneously.

If A and B are independent, then $\Pr(A \mid B) = \Pr(A)$.

Inclusion–Exclusion Principle   It follows from the law of sum that for any events A and B, disjoint or not,

$$\Pr(A \cup B) \;=\; \Pr(A) + \Pr(B) - \Pr(A \cap B).$$

More generally, for any collection $\mathcal{A}$ of events,

$$\Pr(\bigcup \mathcal{A})$$
$$= \sum_{A \in \mathcal{A}} \Pr(A) - \sum_{\substack{\mathcal{B} \subseteq \mathcal{A} \\ |\mathcal{B}|=2}} \Pr(\bigcap \mathcal{B}) + \sum_{\substack{\mathcal{B} \subseteq \mathcal{A} \\ |\mathcal{B}|=3}} \Pr(\bigcap \mathcal{B}) - \cdots \pm \Pr(\bigcap \mathcal{A}).$$

This equation is often used to estimate the probability of a join of several events. The first term alone gives an upper bound and the first two terms give a lower bound:

$$\Pr(\bigcup \mathcal{A}) \;\leq\; \sum_{A \in \mathcal{A}} \Pr(A)$$

$$\Pr(\bigcup \mathcal{A}) \;\geq\; \sum_{A \in \mathcal{A}} \Pr(A) - \sum_{\substack{A,B \in \mathcal{A} \\ A \neq B}} \Pr(A \cap B).$$

Probabilistic Turing Machines

Intuitively, we can think of a probabilistic Turing machine as an ordinary deterministic TM, except that at certain points in the computation it can flip a fair coin and make a binary decision based on the outcome. The probability of acceptance is the probability that its computation path, directed by the outcomes of the coin tosses, leads to an accept state.

Formally, we define a *probabilistic Turing machine* to be an ordinary deterministic TM with an extra semi-infinite read-only tape containing a binary string called the *random bits*. The machine runs as an ordinary deterministic TM, consulting its random bits in a read-only fashion. We write $M(x, y)$ for the outcome, either accept or reject, of the computation of M on input x with random bits y. We say that M is $T(n)$ time bounded (respectively, $S(n)$ space bounded) if for every input x of length n and every random bit string, it runs for at most $T(n)$ steps (respectively, uses at most $S(n)$ worktape cells).

In this model, the probability of an event is measured with respect to the uniform distribution on the space of all sequences of random bits. This is the measure that would result if a fair coin were flipped infinitely many times with the ith random bit determined by the outcome of the ith coin flip.

In practice, we consider only time-bounded computations, in which case the machine can look at only finitely many random bits. This makes the calculation of the probabilities of events easier. For example, if M is $T(n)$ time bounded, then the probability that M accepts its input string x is

$$\Pr_y(M(x,y) \text{ accepts}) \quad = \quad \frac{|\{y \in \{0,1\}^k \mid M(x,y) \text{ accepts}\}|}{2^k},$$

where k is any number exceeding $T(|x|)$. The notation $\Pr_y(E)$ refers to the probability of event E with a bit string y chosen uniformly at random among all strings of length k.

Randomness can be regarded as a computational resource, much like time and space. One can measure the number of random bits consulted in a computation. We show some examples of this in Lectures 18 to 20.

The following are two basic complexity classes defined for probabilistic Turing machines.

Definition 13.1   *A set A is in RP if there is a probabilistic Turing machine M with polynomial time bound n^c such that*

- *if $x \in A$, then $\Pr_y(M(x,y)$ accepts$) \geq \frac{3}{4}$; and*

- *if $x \notin A$, then $\Pr_y(M(x,y)$ accepts$) = 0$.*

The definition of BPP is the same, except we replace the second condition with:

- *if $x \notin A$, then $\Pr_y(M(x,y)$ accepts$) \leq \frac{1}{4}$.*

Equivalently, a set A is in BPP if there is a probabilistic Turing machine M with time bound n^c such that for all inputs x,

$$\Pr_y(M(x,y) \text{ errs in deciding whether } x \in A) \quad \leq \quad \tfrac{1}{4}.$$

We have used $\frac{1}{4}$ and $\frac{3}{4}$ in the definition of *RP* and *BPP*, but actually any $\frac{1}{2} - \varepsilon$ and $\frac{1}{2} + \varepsilon$ will do. It matters only that the probabilities be bounded away from $\frac{1}{2}$ by a positive constant ε independent of the input size.

Also, as previously observed, the length of the random bit string is not important; any set of strings of sufficient length will do, as long as the machine has access to as many random bits as it needs.

It is easy to see that $P \subseteq RP \subseteq NP$, $RP \subseteq BPP$, and *BPP* is closed under complement. We show next time that $BPP \subseteq \Sigma_2^p \cap \Pi_2^p$.

Other classes such as *RPSPACE* and *RNC* can be defined similarly.

Probabilistic Tests with Polynomials

Here is an example of a probabilistic test for which no equally efficient deterministic test is known: determining whether a given multivariate polynomial $p(x_1, \dots, x_n)$ of low degree with integer coefficients is identically 0.

We assume that p is given in the form of a straight-line program with operations $+$, $\cdot$, and scalar operations, or equivalently, in the form of an arithmetic circuit. We could check deterministically if p is identically 0 by multiplying it out to represent it as a sum of terms and checking whether all the terms cancel, but this would take exponential time in general.

Alternatively, we can evaluate p on some $a_1, \ldots, a_n$ chosen at random from some sufficiently large set of integers. If p is identically 0, then $p(a_1, \ldots, a_n) = 0$. If not, then $p(a_1, \ldots, a_n) \neq 0$ with high probability. The reason for this is that the zero set of a nonzero polynomial is sparse.

This works even over finite fields, provided the field is large enough and the degree of the polynomial is not too large. Here we can use the following result, commonly known as the *Schwartz–Zippel lemma*, but also discovered independently by DeMillo and Lipton [36, 110, 127] (see [75] for a proof):

Theorem 13.2   **(Schwartz–Zippel Lemma)**   *Let $\mathbb{F}$ be a field and let $S \subseteq \mathbb{F}$ be an arbitrary subset of $\mathbb{F}$. Let $p(\overline{x})$ be a nonzero polynomial of n variables $\overline{x} = x_1, \ldots, x_n$ and total degree[1] d with coefficients in $\mathbb{F}$. Then the equation $p(\overline{x}) = 0$ has at most $d \cdot |S|^{n-1}$ solutions in S^n.*

Corollary 13.3   *Let $p(x_1, \ldots, x_n)$ be a nonzero polynomial of total degree d with coefficients in a field $\mathbb{F}$, and let $S \subseteq \mathbb{F}$. If p is evaluated on an element $(s_1, \ldots, s_n)$ chosen uniformly at random from S^n, then*

$$\Pr(p(s_1, \ldots, s_n) = 0) \;\leq\; \frac{d}{|S|}.$$

This lemma is useful in quite a number of combinatorial applications. Here are two examples.

Example 13.4   A *perfect matching* in a bipartite graph is a subset M of the edges such that

(i) no two edges in M share a common vertex, and

(ii) every vertex is the endpoint of some edge in M.

It is known how to test for the existence of a perfect matching in a bipartite graph G and find one if it exists in polynomial time [62]. It is unknown whether this problem is in NC. However, the following approach, based on an observation of Lovász [80], gives a random NC algorithm.

Assign to each edge (i, j) of G an indeterminate x_{ij} and consider the $n \times n$ bipartite adjacency matrix X with these indeterminates instead of 1. For example,

---

[1] Maximum degree of any term.

The determinant $\det X$ is a polynomial of degree n in the indeterminates x_{ij} with one term for each perfect matching, and none of these terms cancel. For example, the graph above has two perfect matchings

corresponding to the two terms of the determinant

$$\det X = x_{12}x_{23}x_{31} - x_{11}x_{23}x_{32}.$$

Thus G has a perfect matching iff $\det X$ does not vanish identically. This is difficult to test deterministically, because $\det X$ may be quite large.

However, the determinant of an integer matrix can be calculated in NC using Csanky's algorithm [34] (see [75] for a proof), and we can use this to test in RNC whether $\det X$ is identically 0. We can simply assign randomly chosen elements of a large enough finite field (say $\mathbb{Z}_p$, where p is some prime greater than $2n$) to the x_{ij}, then ask whether the determinant evaluated at those random elements is 0. This will happen with probability 1 if $\det X$ is indeed identically 0, and with probability at most $\frac{n}{2n} = \frac{1}{2}$ if not, by Corollary 13.3.

Given the ability to test for the existence of a perfect matching, we can then find one by deleting edges and their endpoints one by one and testing for the existence of a perfect matching without that edge.   □

Example 13.5   Here is an efficient probabilistic test for deciding whether two unordered[2] directed trees of height h and size n are isomorphic. Associate with each vertex v a polynomial f_v in the variables $x_0, x_1, \ldots, x_h$ inductively, as follows. For each leaf v, set $f_v = x_0$. For each internal node v of height k with children $v_1, \ldots, v_m$, set

$$f_v = (x_k - f_{v_1})(x_k - f_{v_2}) \cdots (x_k - f_{v_m}).$$

The degree of f_v is equal to the number of leaves in the subtree rooted at v. Using the fact that polynomial factorization is unique, it can be shown that two trees are isomorphic iff the polynomials associated with the roots of

---

[2]A directed tree is *ordered* if the left-to-right order of each node's children is given.

the trees are equal. This gives an efficient probabilistic test for isomorphism of unordered trees: test whether the difference of these two polynomials is identically zero by evaluating it on a random input. □

Another example of a problem with an efficient probabilistic solution is primality testing: given a positive integer, is it prime? For many years, this problem was known to be in P only under the assumption of the *extended Riemann hypothesis*, an unproved conjecture of analytic number theory [86], but was known to be in RP via the *Miller–Rabin test* [86, 100] (see [75]). This test is quite efficient and always answers "prime" if the given number is prime and "composite" with high probability if the given number is composite. An improved probabilistic primality test was given recently by Agrawal and Biswas [2]. The problem was also known to be in $NP \cap$ co-NP (Theorem C.4).

Quite recently, Agarwal, Kayal, and Saxena have shown that primality testing is in P unconditionally [3]. However, their algorithm runs in time $O(n^{12})$ and is currently not yet competitive with the best probabilistic methods in practice.

Lecture 14

$BPP \subseteq \Sigma_2^{\mathrm{p}} \cap \Pi_2^{\mathrm{p}}$

In this lecture we prove that $BPP \subseteq \Sigma_2^{\mathrm{p}} \cap \Pi_2^{\mathrm{p}}$. It suffices to show that $BPP \subseteq \Sigma_2^{\mathrm{p}}$, because BPP is closed under complement. This result is due to Sipser [112].

Amplification

By repeating trials in an RP or BPP computation, we can cause the probability of error to diminish exponentially.

Lemma 14.1 (Amplification Lemma)   *If $A \in RP$, then for any polynomial n^d there is a probabilistic polynomial-time-bounded Turing machine M such that for inputs x of length n,*

(i)   if $x \in A$, then $\mathrm{Pr}_y(M(x,y)\ accepts) \geq 1 - 2^{-n^d}$; and

(ii)   if $x \notin A$, then $\mathrm{Pr}_y(M(x,y)\ accepts) = 0$.

If $A \in BPP$, then for any polynomial n^d there is a probabilistic polynomial-time-bounded Turing machine M such that for inputs x of length n,

$$\mathrm{Pr}_y(M(x,y)\ errs\ in\ deciding\ whether\ x \in A) \leq 2^{-n^d}.$$

Proof. For *RP*, let M be a probabilistic polynomial-time-bounded TM such that for all x,

- if $x \in A$ then $\Pr_y(M(x, y) \text{ accepts}) \geq \frac{3}{4}$; and

- if $x \notin A$ then $\Pr_y(M(x, y) \text{ accepts}) = 0$.

Build another probabilistic TM N that on input x just runs M on x n^d times, using a new block of random bits for each trial, and accepts if *any one* of the trials accepts. If M is time-bounded by n^c, hence uses at most n^c random bits, then N will be n^{c+d}-time-bounded and use at most n^{c+d} random bits, which is still polynomial. If $x \notin A$, then M always rejects, so N rejects; and if $x \in A$, then the probability that N errs is

$$
\begin{aligned}
\Pr_{y_1, \dots, y_{n^d}}(N(x, y_1, \dots, y_{n^d}) \text{ rejects}) &= \prod_{i=1}^{n^d} \Pr_{y_i}(M(x, y_i) \text{ rejects}) \\
&= \Pr_y(M(x, y) \text{ rejects})^{n^d} \\
&\leq 4^{-n^d}.
\end{aligned}
$$

For *BPP*, the construction of N is exactly the same, except that to decide whether to accept or reject, N does n^{d+1} trials and picks the majority outcome. The probability of error is the probability that at most half of the n^{d+1} outcomes are correct; this is bounded by

$$
\begin{aligned}
\sum_{k=0}^{n^{d+1}/2} \binom{n^{d+1}}{k} \left(\frac{3}{4}\right)^k \left(\frac{1}{4}\right)^{n^{d+1}-k} &\leq 4^{-n^{d+1}} 3^{n^{d+1}/2} \sum_{k=0}^{n^{d+1}/2} \binom{n^{d+1}}{k} \\
&= 4^{-n^{d+1}} 3^{n^{d+1}/2} 2^{n^{d+1}-1} \\
&\leq \left(\frac{3}{4}\right)^{n^{d+1}/2} \\
&\leq 2^{-n^{d+1}/6} \\
&\leq 2^{-n^d}
\end{aligned}
$$

for sufficiently large n. $\square$

$BPP \subseteq \Sigma_2^p \cap \Pi_2^p$

Let $A \in BPP$. By the amplification lemma (Theorem 14.1), there is a $c > 1$ and a deterministic polynomial-time-bounded TM M running in time n^c such that for all inputs x,

- if $x \in A$, then $\Pr_y(M(x, y) \text{ accepts}) \geq 1 - 2^{-n}$, and

- if $x \notin A$, then $\Pr_y(M(x,y) \text{ accepts}) \leq 2^{-n}$,

where $\Pr_y(E)$ denotes the probability of the event E taken over all strings y chosen uniformly at random from the set $\{0,1\}^{n^c}$.

Fix an input string x of length n, and let $m = n^c$. Define

$$A_x \overset{\text{def}}{=} \{y \in \{0,1\}^m \mid M(x,y) \text{ accepts}\}$$
$$R_x \overset{\text{def}}{=} \{y \in \{0,1\}^m \mid M(x,y) \text{ rejects}\} = \{0,1\}^m - A_x.$$

Then for $x \in A$,

$$|A_x| \geq 2^m - 2^{m-n} \quad \text{and} \quad |R_x| \leq 2^{m-n},$$

and for $x \notin A$,

$$|R_x| \geq 2^m - 2^{m-n} \quad \text{and} \quad |A_x| \leq 2^{m-n}.$$

Claim 14.2   *The string x is in A if and only if there exist m strings $z_1, \ldots, z_m$, each of length m, such that*

$$\{y \oplus z_j \mid 1 \leq j \leq m, \ y \in A_x\} = \{0,1\}^m,$$

where $\oplus$ denotes exclusive-or or bitwise mod 2 sum.

The idea here is that if $x \in A$, then the set A_x is so big that by mapping it around with some small set of permutations of the form $y \mapsto y \oplus z$, we hit every string; and if $x \notin A$, then A_x is so small that no small collection of such permutations does this.

If the claim is true, then this is all we need to show $A \in \Sigma_2^p$: to determine whether $x \in A$,

(i)   guess $z_1, \ldots, z_m$ using existential branching;

(ii)   generate all w of length m using universal branching; and

(iii)   check that $w \in \{y \oplus z_j \mid 1 \leq j \leq m, \ y \in A_x\}$, or equivalently that $\{w \oplus z_j \mid 1 \leq j \leq m\}$ intersects A_x, by running $M(x, w \oplus z_j)$ for all $1 \leq j \leq m$.

Part (iii) of the computation can be done deterministically, so this is a Σ_2^p computation.

Proof of Claim 14.2. We can prove this lemma just by counting. First, assume that $x \notin A$. Then $|A_x| \leq 2^{m-n}$. For any choice of $z_1, \ldots, z_m$,

$$\{y \oplus z_j \mid 1 \leq j \leq m, \ y \in A_x\} = \bigcup_{j=1}^{m} \{y \oplus z_j \mid y \in A_x\}.$$

An upper bound on the size of this set is

$$\sum_{j=1}^{m} |\{y \oplus z_j \mid y \in A_x\}| \;=\; \sum_{j=1}^{m} |A_x| \;\leq\; m 2^{m-n} \;<\; 2^m$$

for sufficiently large n. Thus

$$\{y \oplus z_j \mid 1 \leq j \leq m, \; y \in A_x\} \;\neq\; \{0,1\}^m.$$

Now suppose $x \in A$. Then $|R_x| \leq 2^{m-n}$. Let us call $z_1, \ldots, z_m$ *bad* if for some w,

$$\{w \oplus z_j \mid 1 \leq j \leq m\} \;\subseteq\; R_x,$$

good otherwise. We wish to show that there exists a good $z_1, \ldots, z_m$. But each bad $z_1, \ldots, z_m$ is determined by a subset of R_x of size m, of which there are at most $(2^{m-n})^m$, and a string $w \in \{0,1\}^m$, of which there are 2^m. Thus an upper bound on the number of bad $z_1, \ldots, z_m$ is

$$(2^{m-n})^m 2^m \;=\; 2^{m(m-n+1)} \;<\; 2^{m^2},$$

and the right-hand side is the total number of choices of $z_1, \ldots, z_m$, so some $z_1, \ldots, z_m$ must be good. $\qquad\Box$

Supplementary Lecture B

Chinese Remaindering

The following is a very useful theorem with many applications in computer science. It says that a large number can be faithfully represented as a sequence of remainders modulo a list of small relatively prime moduli.

Let $\mathbb{Z}_n$ denote the ring of integers modulo n. Recall that two positive integers are *relatively prime* if they have no common factor except 1.

Theorem B.1   (Chinese Remainder Theorem)   *Let $n_1, \ldots, n_k$ be pairwise relatively prime positive integers, and let $n = \prod_{i=1}^{k} n_i$. The ring $\mathbb{Z}_n$ and the direct product of rings $\mathbb{Z}_{n_1} \times \cdots \times \mathbb{Z}_{n_k}$ are isomorphic under the function*

$$\sigma : \mathbb{Z}_n \;\to\; \mathbb{Z}_{n_1} \times \cdots \times \mathbb{Z}_{n_k}$$

$$\sigma(x) \;\overset{\text{def}}{=}\; (x \bmod n_1, \ldots, x \bmod n_k).$$

This just says that the numbers modulo n and the k-tuples of numbers modulo n_i, $1 \le i \le k$, are in one-to-one correspondence, and that arithmetic is preserved under the map f. For example, the following table compares $\mathbb{Z}_{15}$ to $\mathbb{Z}_3 \times \mathbb{Z}_5$.

x	0	1	2	3	4	5	6	7	8	9	10	11	12	13	14
$x \bmod 3$	0	1	2	0	1	2	0	1	2	0	1	2	0	1	2
$x \bmod 5$	0	1	2	3	4	0	1	2	3	4	0	1	2	3	4

Note that each pair in $\mathbb{Z}_3 \times \mathbb{Z}_5$ occurs exactly once. This is because 3 and 5 are relatively prime. Arithmetic is preserved as well: for example, 4 and 7 correspond to the pairs $(1, 4)$ and $(1, 2)$, respectively; multiplying these pairwise gives the pair $(1, 3)$ (modulo 3 and 5, respectively), which occurs under 13; and $4 \times 7 = 28 \equiv 13 \pmod{15}$.

Also, σ and σ^{-1} are computable in polynomial time. To compute $\sigma(x)$, we just reduce x modulo $n_1, \dots, n_k$. To compute $\sigma^{-1}(x_1, \dots, x_k)$, we first compute, for each $1 \leq i \leq k$, positive integers s and t such that $s(n/n_i) - tn_i = 1$ and take $u_i \stackrel{\text{def}}{=} sn/n_i$. Because n_i and n/n_i are relatively prime, the numbers s and t exist and are available as a byproduct of the Euclidean algorithm (Miscellaneous Exercise 42). For each $1 \leq i, j \leq k$, $u_i \equiv 1 \pmod{n_i}$ and $u_i \equiv 0 \pmod{n_j}$, $i \neq j$. Then

$$\sigma^{-1}(x_1, \dots, x_k) \quad = \quad x_1 u_1 + \cdots + x_k u_k \pmod{n}.$$

For further details see [4, pp. 289ff].

A Stronger Version

The Chinese remainder theorem actually holds in a much more general form. We need this stronger form for Berlekamp's factoring algorithm in Lecture D.

Recall from algebra that an *ideal* of a commutative ring R is a set I such that

- $0 \in I$,

- if $x, y \in I$ then $x + y \in I$, and

- if $x \in I$ and $y \in R$, then $xy \in I$.

Ideals are the *kernels* of ring homomorphisms (the set of elements mapped to 0). It is easy to check that if $h : R \to R'$ is a ring homomorphism, then $\{x \in R \mid h(x) = 0\}$ is an ideal; conversely, if I is an ideal, then there is a homomorphism $[\ ]_I : R \to R/I$ whose kernel is I. The ring R/I is the *quotient ring* consisting of equivalence classes $[x]_I \stackrel{\text{def}}{=} \{y \in R \mid x \equiv_I y\}$, where $x \equiv_I y$ iff $x - y \in I$.

Let us call two ideals *relatively prime* if the smallest ideal containing them both is all of R.

For example, in $\mathbb{Z}$, the set of multiples of $n \geq 1$ forms an ideal, and it is the kernel of the ring homomorphism $x \mapsto x \bmod n$. The quotient ring is $\mathbb{Z}_n$. The ideals generated by m and n in this way are relatively prime as ideals iff m and n are relatively prime as integers in the usual sense.

Theorem B.2   *Let $I_1, I_2, \ldots, I_k$ be pairwise relatively prime ideals of commutative ring R, and let $I = \bigcap_{i=1}^{k} I_i$. There is an isomorphism*

$$R/I \;\cong\; R/I_1 \times \cdots \times R/I_k$$

given by the map

$$\sigma : R/I \;\rightarrow\; R/I_1 \times \cdots \times R/I_k$$
$$\sigma([x]_I) \;\overset{\text{def}}{=}\; ([x]_{I_1}, \ldots, [x]_{I_k}).$$

Proof. We must first argue that the map σ is well defined on $\equiv_I$-equivalence classes; that is, the action of σ on $[x]_I$ does not depend on the choice of the representative x. But this is true, because if $x \equiv_I y$, then $x \equiv_{I_i} y$ for $1 \le i \le k$ by definition of I.

It is a routine matter to check that σ is a ring homomorphism. Also, σ is one-to-one, because if $x \equiv_{I_i} y$, $1 \le i \le k$, then $x \equiv_I y$. The only difficulty is showing that σ is onto.

To show that σ is onto, by analogy with the proof of Theorem B.1, it suffices to show that for each $1 \le i \le k$ there exists $x \in R$ such that $x \equiv_{I_i} 1$ and $x \equiv_{I_j} 0$ for $j \ne i$. Equivalently, we must show that there exists $x \in R$ such that $x - 1 \in I_i$ and $x \in \bigcap_{j \ne i} I_j$. Without loss of generality, assume $i = 1$.

By the assumption of relative primality, for each $j \ge 2$ there exist $x_j \in I_1$ and $y_j \in I_j$ such that $1 = x_j + y_j$ (note that the smallest ideal containing both I and J is the set of sums $\{x + y \mid x \in I \text{ and } y \in J\}$, and this ideal is all of R iff it contains 1). Now we proceed by induction. Suppose we have constructed u_m and v_m such that

- $u_m \in I_1$,

- $v_m \in \bigcap_{j=2}^{m} I_j$, and

- $1 = u_m + v_m$.

Let

$$u_{m+1} \;\overset{\text{def}}{=}\; x_{m+1} + y_{m+1} u_m$$
$$v_{m+1} \;\overset{\text{def}}{=}\; y_{m+1} v_m.$$

Then $u_{m+1} \in I_1$, $v_{m+1} \in \bigcap_{j=1}^{m+1} I_j$, and

$$\begin{aligned}
u_{m+1} + v_{m+1} &= x_{m+1} + y_{m+1} u_m + y_{m+1} v_m \\
&= x_{m+1} + y_{m+1}(u_m + v_m) \\
&= 1.
\end{aligned}$$

When done, we have $v_k \in \bigcap_{j=2}^{k} I_j$ and $v_k - 1 = u_k \in I_1$, thus v_k is the desired element.   □

In our application in Lecture D, R is a ring of polynomials over a finite field.

Supplementary Lecture C

Complexity of Primality Testing

Testing whether a given binary integer is prime was for a long time one of the few known examples of a problem in $NP \cap$ co-NP that was not known to be in P. For many years, this problem was known to be in P only under the assumption of the *extended Riemann hypothesis*, an unproved conjecture of analytic number theory [86]. It was known to be in co-RP via the *Miller–Rabin test* [86, 100] (see [75]) and also known to be in RP by results of Adleman and Huang [1]. In 2002, Agarwal, Kayal, and Saxena showed that primality testing is in P unconditionally [3]. An improved probabilistic primality test was given recently by Agrawal and Biswas [2].

For our applications in Lecture 16, it suffices to show that the problem is in $NP \cap$ co-NP. This was first shown by Pratt in 1975 [97]. We give Pratt's proof in this lecture (Theorem C.4).

Let PRIMES be the set of binary representations of primes. It is easy to see that ~PRIMES, the set of *composite numbers*, is in NP: guess two nontrivial factors of the given number and multiply them. Thus PRIMES is in co-NP. It is much harder to show that PRIMES is also in NP.

Pratt's NP algorithm for primality is based on *Fermat's theorem*. If $1 \le a \le n - 1$ and there exists an $m \ge 1$ such that $a^m \equiv 1 \pmod{n}$, then we define the *order* of a modulo n to be the least such m. For example, the first few powers of 3 modulo 10 are 1, 3, 9, 7, 1, therefore the order of 3 modulo 10 is 4. If n is prime, then all numbers in the range $1, \ldots, n - 1$

have an order modulo n, but not so if n is composite. For example, all powers of 5 are 5 modulo 10.

In general, a positive integer a has an order modulo n iff a is relatively prime to n (that is, if $\gcd(a, n) = 1$) iff a is invertible modulo n; that is, if there exists b, $1 \leq b \leq n - 1$, such that $ab \equiv 1 \pmod{n}$ (Miscellaneous Exercise 43).

Lemma C.1   (Fermat's Theorem)   *A number $p \geq 2$ is prime if and only if some element of $\{1, \ldots, p - 1\}$ has order $p - 1$ modulo p.*

To prove Fermat's theorem, we need a few basic facts of number theory and finite fields.

Define

$$\mathbb{Z}_n^* \stackrel{\text{def}}{=} \{a \mid 1 \leq a \leq n - 1, \ \gcd(a, n) = 1\}.$$

For example, $\mathbb{Z}_{10}^* = \{1, 3, 7, 9\}$. Because $\mathbb{Z}_n^*$ consists of the invertible elements of $\mathbb{Z}_n$, it forms a group under multiplication modulo n. This is known as the *group of units* modulo n. The size of $\mathbb{Z}_n^*$ is denoted $\varphi(n)$. The function φ is known as the *Euler φ function*.

We can calculate $\varphi(n)$ for any n if we know the prime factorization of n. For a prime power p^k, $\varphi(p^k) = p^{k-1}(p - 1)$, because a number is relatively prime to p^k iff it is not a multiple of p, and there are $p^{k-1} - 1$ multiples of p in $\{1, \ldots, p^k - 1\}$. By the Chinese remainder theorem (Theorem B.1), if m and n are relatively prime, then the map $x \mapsto (x \bmod m, x \bmod n)$ is a ring isomorphism $\mathbb{Z}_{mn} \to \mathbb{Z}_m \times \mathbb{Z}_n$, therefore a group isomorphism $\mathbb{Z}_{mn}^* \to \mathbb{Z}_m^* \times \mathbb{Z}_n^*$, thus $\varphi(mn) = \varphi(m)\varphi(n)$. Combining these facts, it follows that if $n = p_1^{k_1} \cdots p_m^{k_m}$ is the prime factorization of n, then

$$\varphi(n) \ = \ \prod_{i=1}^{m} p_i^{k_i - 1}(p_i - 1). \tag{C.1}$$

We must show that for prime p, the group $\mathbb{Z}_p^*$ has an element of order $p - 1$. Because $\varphi(p) = p - 1$ is the size of the whole group $\mathbb{Z}_p^*$, that is the same as saying that $\mathbb{Z}_p^*$ is cyclic; that is, $\{1, 2, \ldots, p - 1\} = \{1, a, a^2, \ldots, a^{p-2}\}$ (modulo p) for some $1 \leq a \leq p - 1$. Such an element a is called a *cyclic generator* of $\mathbb{Z}_p^*$.

It is actually true that the multiplicative group of any finite field is cyclic. This is no harder to show for arbitrary finite fields than for the fields $\mathbb{Z}_p$, so that is what we do.

Recall that the *characteristic* of a finite field $\mathbb{F}$ is the smallest number p such that $\underbrace{1 + \cdots + 1}_{p} = 0$. This must be a prime, because otherwise $\mathbb{F}$ would contain zero divisors. The *prime subfield* of $\mathbb{F}$ is the smallest subfield of $\mathbb{F}$ and is isomorphic to $\mathbb{Z}_p$, where p is the characteristic. The number of elements in $\mathbb{F}$ must be p^k for some k, because $\mathbb{F}$ is a vector space over its prime subfield of some dimension k, therefore is isomorphic (as a vector space, not as a field) to $\mathbb{Z}_p^k$.

In fact, up to isomorphism there is only one finite field of cardinality q for each prime power q, and it is called GF_q (for the *Galois field on q elements*). It is not $\mathbb{Z}_q$ if q is not prime, because $\mathbb{Z}_q$ has zero divisors.

Write $m \mid n$ if m divides n.

Lemma C.2 $n = \sum_{m \mid n} \varphi(m)$.

Proof. By induction on the prime factorization of n. If n is a prime power p^k, then

$$\sum_{m \mid n} \varphi(m) \quad = \quad \sum_{i=0}^{k} \varphi(p^i) \quad = \quad 1 + \sum_{i=1}^{k}(p^i - p^{i-1}) \quad = \quad p^k.$$

If $n = n_1 n_2$ where n_1 and n_2 are relatively prime, then

$$\sum_{m \mid n} \varphi(m) \quad = \sum_{m \mid n_1 n_2} \varphi(m) \quad = \sum_{\substack{m_1 \mid n_1 \\ m_2 \mid n_2}} \varphi(m_1 m_2) \quad = \sum_{\substack{m_1 \mid n_1 \\ m_2 \mid n_2}} \varphi(m_1)\varphi(m_2)$$

$$= \quad (\sum_{m_1 \mid n_1} \varphi(m_1))(\sum_{m_2 \mid n_2} \varphi(m_2)) \quad = \quad n_1 n_2.$$

$\square$

The nonzero elements of GF_q are all invertible and thus form a group GF_q^* of size $q-1$. Moreover, all elements of GF_q^* have order dividing $q-1$, thus all are roots of the polynomial $x^{q-1} - 1$. Because $x^{q-1} - 1$ can have at most $q-1$ roots, that is all of them. Thus

$$\prod_{a \in \mathrm{GF}_q^*} (x - a) \quad = \quad x^{q-1} - 1. \tag{C.2}$$

Let $\Phi_k(x)$ be the polynomial whose roots are the elements of GF_q^* of order k:

$$\Phi_k(x) \quad \overset{\text{def}}{=} \quad \prod_{\substack{a \in \mathrm{GF}_q^* \\ \text{order } a = k}} (x - a).$$

By (C.2),

$$x^{q-1} - 1 \quad = \quad \prod_{k \mid q-1} \Phi_k(x).$$

Lemma C.3   *For $k \mid q-1$, the degree of $\Phi_k(x)$ is $\varphi(k)$.*

Proof. By induction. For $k = 1$, $\Phi_1(x) = x - 1$ and $\deg \Phi_1(x) = 1 = \varphi(1)$. For $k > 1$,

$$
\begin{aligned}
k &= \deg(x^k - 1) \\
&= \deg \prod_{m \mid k} \Phi_m(x) \\
&= \sum_{m \mid k} \deg \Phi_m(x) \\
&= \deg \Phi_k(x) + \sum_{\substack{m \mid k \\ m < k}} \deg \Phi_m(x) \\
&= \deg \Phi_k(x) + \sum_{\substack{m \mid k \\ m < k}} \varphi(m) \qquad \text{induction hypothesis} \\
&= \deg \Phi_k(x) + \sum_{m \mid k} \varphi(m) - \varphi(k) \\
&= \deg \Phi_k(x) + k - \varphi(k) \qquad \text{Lemma C.2,}
\end{aligned}
$$

therefore $\deg \Phi_k(x) = \varphi(k)$. $\qquad\qquad\qquad\qquad\qquad\qquad\qquad$ □

Proof of Lemma C.1. For any prime power q, the roots of $\Phi_{q-1}(x)$ are the cyclic generators of $\mathbb{GF}_q^*$. By Lemma C.3, there are exactly $\varphi(q - 1) = \deg \Phi_{q-1}(x)$ of them. This number is nonzero for any prime power $q \geq 2$, therefore $\mathbb{GF}_q^*$ has a cyclic generator. In the special case q is prime, this gives a number with order $q - 1$ modulo q.

Conversely, unless n is prime, the number $\varphi(n)$ as defined in (C.1) is strictly less than $n - 1$. Moreover, any element $a \in \{1, \ldots, n - 1\}$ that has an order modulo n, being a member of the group $\mathbb{Z}_n^*$, must have order dividing $\varphi(n)$. $\qquad\qquad\qquad\qquad\qquad\qquad\qquad\qquad\qquad$ □

Primality Testing

Now we show how to use Fermat's theorem to get an *NP* algorithm for primality.

Theorem C.4 (Pratt [97])   PRIMES $\in$ *NP* $\cap$ *co-NP*.

Proof. We have already observed that PRIMES is in co-*NP*. To show that it is in *NP*, we perform the following nondeterministic computation on input $p \geq 2$.

1. If $p = 2$, accept. If $p > 2$ and p is even, reject.

2. Guess the prime factorization of $p - 1$, say $p - 1 = p_1^{k_1} \cdots p_m^{k_m}$. Verify by multiplication that this equation holds.

3. Guess $a \in \{2, \ldots, p - 1\}$ and verify by modular arithmetic that $a^{p-1} \equiv 1 \pmod{p}$.

4. Verify for each $1 \le i \le m$ that $a^{(p-1)/p_i} \not\equiv 1 \pmod{p}$.

5. Recursively verify that $p_1, \ldots, p_m$ are prime.

Steps 3 and 4 together imply that the order of a modulo p is $p - 1$. Once we have verified 3, we know that the order of a exists and must divide $p - 1$. This is because if $a^k \equiv 1 \pmod{p}$ and $a^m \equiv 1 \pmod{p}$, then $a^{\gcd(k,m)} \equiv 1 \pmod{p}$, because by the Euclidean algorithm, there exist s, t such that $\gcd(k, m) = sk - tm$. Thus if the order of a were strictly less than $p - 1$, then it would divide $(p - 1)/q$ for some prime q dividing $p - 1$, in which case we would have $a^{(p-1)/q} \equiv 1 \pmod{p}$. The check 4 verifies that this does not happen.

Step 1 is constant time. Step 2 takes time polynomial in $\log p$. Steps 3 and 4 can be performed in time polynomial in $\log p$ by repeated squaring and reducing modulo p. Thus the time taken is polynomial in $\log p$ plus the time taken by the recursive calls in step 5.

To analyze the total time taken by this recursive algorithm, rather than solving a recurrence relation, it is easier just to look at the whole tree of recursive calls. Each node in the tree is labeled with the parameter of the recursive call represented by that node. If a node is labeled q and the guessed prime factorization of $q - 1$ is $q_1^{k_1} \cdots q_m^{k_m}$, then that node has m children labeled $q_1, \ldots, q_m$. The leaves of the tree are labeled 2. Because 2 is a prime factor of every $q - 1$, every node has at least two children unless its label is of the form $2^k + 1$, in which case it has one child and that child is a leaf.

One can show inductively that the product of the labels of all the leaves of a subtree with root label q is at most q. Because the leaves are all labeled 2, this says that the number of leaves of that subtree is at most $\log_2 q$. Moreover, because the tree is at least binary branching except at the lowest level, one can show inductively that the total number of nodes in any subtree is at most $3\ell - 1$, where ℓ is the number of leaves in that subtree. Thus, if p is the label of the root of the entire tree, the total number of nodes is at most $3 \log_2 p - 1$.

Thus the total amount of time taken by the algorithm is the time to perform steps 1–4 times the number of nodes in the tree, which is polynomial in $\log p$. $\qquad\square$

Supplementary Lecture D

Berlekamp's Algorithm

Here is an efficient probabilistic algorithm due to Berlekamp [13] for factoring a given univariate polynomial over a finite field of large characteristic. No deterministic polynomial time algorithm is known. However, one can give an efficient deterministic irreducibility test, and the factorization problem can be reduced deterministically to the special case in which the given polynomial is guaranteed to split into linear factors over the prime subfield.

This is an example of a *Las Vegas* algorithm: the expected running time is polynomial with a small probability that it may run for longer, but the answer is guaranteed to be correct. There are also *Monte Carlo* probabilistic algorithms, in which the answer is always produced quickly and is very probably correct, but there is no absolute guarantee of correctness.

For this result, we need a few more basic facts about finite fields beyond those presented in Supplementary Lecture C. Recall that there exists a finite field GF_q of q elements iff q is a prime power $q = p^k$. The field $\mathrm{GF}_p \cong \mathbb{Z}_p$ is the smallest subfield of GF_q and is called the *prime subfield* of GF_q, and p is called the *characteristic* of GF_q. The field GF_q is the unique field of q elements, not just up to isomorphism, but absolutely unique among subfields of a given algebraic closure $\overline{\mathrm{GF}_p}$ of GF_p. This is because the elements of GF_q are exactly the roots of the polynomial $x^q - x$. As we observed in Supplementary Lecture C, the nonzero elements all have order dividing $q - 1$, the size of GF_q^*, thus they are all roots of $x^{q-1} - 1$;

and throwing in 0 gives $x^q - x$. Thus

$$\mathbb{GF}_q = \{a \in \overline{\mathbb{GF}}_p \mid a^q = a\}.$$

The field $\mathbb{GF}_q$ is a subfield of $\mathbb{GF}_r$ iff r is a power of q. In particular, $\mathbb{GF}_q$ and $\mathbb{GF}_r$ have the same characteristic. For finite extensions of $\mathbb{GF}_q$, this says that $\mathbb{GF}_{q^m} \subseteq \mathbb{GF}_{q^n}$ iff m divides n. Thus the lattice of finite extensions of $\mathbb{GF}_q$ in $\overline{\mathbb{GF}}_p$ is isomorphic to the lattice of nonnegative integers under divisibility.

The Galois group of the extension $\mathbb{GF}_{q^m} : \mathbb{GF}_q$ (the group of automorphisms of $\mathbb{GF}_{q^m}$ fixing $\mathbb{GF}_q$ pointwise) is cyclic and generated by the automorphism $a \mapsto a^q$. Because $\mathbb{GF}_{q^m}$ is unique, the extension is normal (all automorphisms fix $\mathbb{GF}_{q^m}$ setwise).

Representation

For computational purposes, the field $\mathbb{GF}_q$ for $q = p^k$ is usually represented as a quotient $\mathbb{Z}_p[y]/h$ of the ring of univariate polynomials with coefficients in $\mathbb{Z}_p$ modulo an irreducible polynomial $h \in \mathbb{Z}_p[y]$ of degree k. Thus if f is a polynomial of degree m with coefficients in $\mathbb{GF}_q$, the coefficients of f would be represented as polynomials in $\mathbb{Z}_p[y]$ modulo h. Then f is represented as an element of $\mathbb{Z}_p[x,y]/h$. This representation allows us to perform the usual ring operations using polynomial arithmetic modulo h.

The Factorization Problem

Let f be a polynomial of degree m, not necessarily irreducible, with coefficients in a finite field $\mathbb{GF}_q$ of prime characteristic p. As described above, f is represented as a polynomial in $\mathbb{Z}_p[x,y]/h$, where h is an irreducible polynomial in $\mathbb{Z}_p[y]$.

We wish to find the irreducible factors of f. We can assume that f is squarefree (no multiple roots) by taking the gcd of f with its formal derivative with respect to x, except when the formal derivative vanishes identically. But in that case, all terms of f have degree a multiple of p, so f has a nontrivial factorization

$$f(x) = \sum_{i=0}^{m/p} a_i x^{ip} = \left(\sum_{i=0}^{m/p} a_i^{q/p} x^i \right)^p.$$

Suppose f factors into irreducible factors $f_1, \ldots, f_n$ over $\mathbb{GF}_q$. Each $\mathbb{GF}_q[x]/f_i$ is a finite extension of $\mathbb{GF}_q$ isomorphic to $\mathbb{GF}_{q^{m_i}}$, where m_i is the degree of f_i. Because the f_i are pairwise relatively prime, by the

Chinese remainder theorem there is an isomorphism

$$\mathrm{GF}_q[x]/f \;\cong\; \mathrm{GF}_q[x]/f_1 \times \cdots \times \mathrm{GF}_q[x]/f_n$$
$$\cong\; \mathrm{GF}_{q^{m_1}} \times \cdots \times \mathrm{GF}_{q^{m_n}} \qquad\qquad (\mathrm{D}.1)$$

given by the map

$$\sigma : \mathrm{GF}_q[x]/f \;\rightarrow\; \mathrm{GF}_q[x]/f_1 \times \cdots \times \mathrm{GF}_q[x]/f_n$$
$$\sigma(g) \;\overset{\mathrm{def}}{=}\; (g \bmod f_1, \ldots, g \bmod f_n).$$

The elements $g \in \mathrm{GF}_q[x]/f$ are represented as polynomials in $\mathbb{Z}_p[x,y]$ modulo h and f.

Consider the subalgebra GF_p^n of (D.1), where p is the characteristic. This is an algebra of dimension n over GF_p and contains exactly p^n elements. Its preimage under σ is the subalgebra

$$A \;\overset{\mathrm{def}}{=}\; \{g \mid g \bmod f_i \in \mathrm{GF}_p,\; 1 \le i \le n\} \;=\; \{g \mid g^p \equiv g \;(\bmod\; f)\}$$

of $\mathrm{GF}_q[x]/f$. This follows from the fact that the automorphism $a \mapsto a^p$ fixes every element of GF_p and moves every element not in GF_p, and multiplication in (D.1) is componentwise, so this map applied to (D.1) fixes exactly the elements of GF_p^n. Because the coefficients of g as a polynomial in $\mathrm{GF}_p[x,y]$ are fixed by $a \mapsto a^p$, this is equivalent to the condition $g(x,y)^p = g(x^p, y^p)$. By computing all the powers $x^{ip} y^{jp}$ modulo f and h for $i \le m$ and $j \le k$ by repeated squaring and reducing modulo f and h, then equating the coefficients of $g(x,y)$ and $g(x^p, y^p)$, we obtain a system of linear equations in the indeterminate coefficients of $g(x,y)$ over $\mathbb{Z}_p$, which we know how to solve in polynomial time by Gaussian elimination. This allows us to determine n and compute a basis for the subspace A over $\mathbb{Z}_p$. In particular, f is already irreducible iff $n = 1$.

If $n > 1$, for any polynomial $g \in A$, we have

$$\sigma(g) \;=\; (a_1, \ldots, a_n)$$

with $a_i \in \mathrm{GF}_p$, $1 \le i \le n$. We do not know the a_i, but we know that they exist. If we can get our hands on a $g \in A$ such that at least one $a_i = 0$ and at least one $a_j \ne 0$, then we get a nontrivial factorization, because $a_i \equiv g \bmod f_i = 0$ iff f_i divides g, so f and g would have a nontrivial common factor, which can be found by computing the gcd.

If the characteristic is small enough, we can solve the problem deterministically: just pick an element $g \in A$ of degree at least 1, so that it is not in $\mathbb{Z}_p$, then consider the elements $g - a$ for $a \in \mathbb{Z}_p$. One of these must work. Because $g \notin \mathbb{Z}_p$, there must be some $a_i \ne a_j$ in $\sigma(g)$, thus $\sigma(g - a_i)$ is 0 in position i and nonzero in position j. This gives a nontrivial factorization.

If the characteristic is large, here is how to get such a g with high probability. Pick an $\ell \in A$ at random and take

$$g \;=\; \ell^{(q-1)/2} - 1 \bmod f.$$

We pick $\ell \in A$ at random by picking a random linear combination of the basis for A. We compute the appropriate power of ℓ modulo f by repeated squaring and reducing modulo f.

We now show that the chances are about even or better that g and f have a nontrivial common factor. Because the characteristic is odd, exactly half the nonzero elements of $\mathbb{GF}_q$ are squares; that is, elements of the form b^2 for some $b \in \mathbb{GF}_q$. The squares are exactly the roots of $x^{(q-1)/2} - 1$. If ℓ is chosen at random, and if $\sigma(\ell) = (a_1, \dots, a_n)$, then for each i we have about an even chance that a_i is a square, and these events are independent. Thus we have about an even chance or better that at least one a_i is a square and at least one is not. Computing $\ell^{(q-1)/2} - 1$, we get 0 in location i iff a_i was a square.

To reduce the problem deterministically to the case in which f splits into linear factors over $\mathbb{GF}_p$, pick $g \in A$ of degree at least 1. Thus if $\sigma(g) = (a_1, \dots, a_n)$, not all the a_i are equal. The elements a_i are exactly the elements $a \in \mathbb{GF}_p$ such that $g - a$ and f have a root in common, because

$$\sigma(g - a) \;=\; (a_1 - a, \dots, a_n - a),$$

which has a 0 in position i iff $a = a_i$ iff f_i divides $g - a$. Thus the a_i are all the roots of the resultant

$$r(z) \;=\; \mathsf{Res}_x(g(x) - z, f(x))$$

lying in $\mathbb{GF}_p$ (see [64, 75] for an introduction to resultants). We can pick out the roots of r lying in $\mathbb{GF}_p$ by taking the gcd of $r(z)$ and $z^{q-1} - 1$. This is done by computing $r(z)$ first, which is of degree m, then computing z^{q-1} modulo r by repeated squaring and reducing modulo r to get a low-degree polynomial s, then taking the gcd of $s - 1$ and r.

After removing repeated roots as above, the resulting polynomial is of degree at most m and splits into linear factors over $\mathbb{GF}_p$. Moreover, if a is any root of this polynomial, then the gcd of f and $g - a$ will be a nontrivial factor of f. We have thus reduced the problem of factoring a polynomial of degree m to the problem of factoring a polynomial of degree at most m, all of whose roots lie in the prime field.

Lecture 15

Interactive Proofs

In the next few lectures we take a look at a model of computation involving interactive protocols between two agents. One of the agents wants to convey some information to the other, and the other wants to be convinced with a high degree of certainty that the information is correct. Such protocols arise in cryptography and message authentication.

Polynomial-time interactive protocols give rise to a complexity class *IP*. Whereas we can think of a set in *NP* as a set of theorems admitting *short proofs*, we can think of a set in *IP* as a set of theorems having *efficient interactive proofs*.

Interactive Proof Systems

The machine model consists of two independent Turing machines P (the *prover*) and V (the *verifier*). The two machines share a common read-only input tape and a read/write communication tape, but otherwise operate independently. Each machine has its own private worktape. We assume in addition that

- V has access to a private string of random bits;

- V runs in polynomial time; and

- P is not bounded in time or space, but must halt on all inputs and may only write strings of polynomial length on the communication tape. In fact, we could even have defined P to be some kind of oracle or black box computing some noncomputable function—the exact nature of P does not really matter.

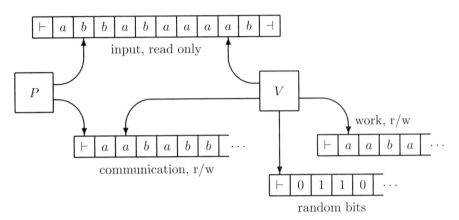

The two machines alternate. When it is V's turn, it runs for polynomial time, accessing its random bits whenever it needs to make a probabilistic decision. At some point it writes a message to P on the communication tape and enters a special state that causes control to transfer to P. The machine P then runs for as long as it likes, reading the message from V and eventually writing a message back to V of polynomial length on the communication tape and transferring control back to V. Control passes back and forth between the two agents for some polynomial number of rounds determined by V, after which V decides whether to accept or reject.

We can think of P as trying to convince V that the input string satisfies some property, and of V as trying to verify that that information is correct. When that property is indeed true of the input string, it should be possible for P to convince V of that fact with high probability. By accepting, V indicates that it is convinced that the property is true.

Now imagine substituting an evil intruder P' for P. Let us say that V runs its part of the protocol as before, but P' may behave differently from P. In particular, P' might try to convince V that the property of interest holds of the input string when in fact it does not. The verifier V should be able to detect this type of dishonest behavior with high probability and reject.

The formal definition is as follows.

Definition 15.1 *The protocol (P, V) is an* interactive proof system *for a set of strings A if for all x,*

- *if $x \in A$, then*

$$\mathrm{Pr}_y((P,V) \; accepts \; x) \;\; \geq \;\; \tfrac{3}{4},$$

- *and if $x \notin A$, then for any P',*

$$\mathrm{Pr}_y((P',V) \; accepts \; x) \;\; \leq \;\; \tfrac{1}{4},$$

where the probability is with respect to the random bits y chosen uniformly from the set of all strings of some length exceeding the polynomial time bound of the verifier.

A set A is in IP if it has an interactive proof system.

The constants $\tfrac{3}{4}$ and $\tfrac{1}{4}$ in Definition 15.1 are inconsequential. By amplification, we could specify $1 - \varepsilon$ and ε for any $\varepsilon > 0$ (Miscellaneous Exercise 44).

Examples of Interactive Proofs

The best way to get a feel for this model is to look at some examples.

Example 15.2   Boolean satisfiability (and in fact, any set $A \in NP$) is in IP, because there is a one-round protocol that does not use any random bits. The prover would like to convince the verifier that a given Boolean formula written on the input tape is satisfiable. The prover just sends the verifier a satisfying assignment if one exists. Because the prover is not restricted by any time bound, it can search exhaustively for a satisfying assignment and send the first one it finds to V. If no satisfying assignment exists, it just sends something arbitrary. The verifier then evaluates the input formula on the truth assignment communicated by P. If the formula is satisfied, then V accepts; if not, it rejects.

Now if the given formula really is satisfiable, then (P,V) accepts with probability 1. The verifier is completely convinced, because it has a proof in the form of a satisfying assignment. On the other hand, if the formula is not satisfiable, then no prover P', regardless of how clever or malicious it is, can convince V otherwise. In this case, (P',V) accepts with probability 0.   □

Example 15.3   Here is an example due to Goldreich, Micali, and Wigderson [48] that is a bit more complicated: graph *non*isomorphism is in IP. This is interesting, because graph isomorphism is in NP (given a pair of graphs, guess an isomorphism and verify that it is an isomorphism in polynomial time), but it is not known to be in co-NP.

Here is an IP protocol for graph nonisomorphism. The input is an encoding of two graphs G, H on n vertices. The following procedure is executed k times.

(i)   The verifier V chooses a random permutation of $\{1, 2, \ldots, n\}$. This requires roughly $n \log n$ random bits. It applies the permutation to the vertices of the graphs G and H to get G' and H', respectively. Then V flips a coin. If it comes up heads, it sends G' to P, and if it comes up tails, it sends H' to P.

(ii)   The prover P checks whether G on its input tape and the graph sent to it by V are isomorphic, say by exhautively searching for an isomorphism, and communicates its finding—isomorphic or nonisomorphic—honestly to V.

(iii) V takes the following action based on its coin flip and the prover's response.

If the flip is	and P responds	then do this
heads	isomorphic	continue
heads	nonisomorphic	reject immediately
tails	isomorphic	reject immediately
tails	nonisomorphic	continue

If the protocol makes it all the way through k rounds, then V accepts, convinced with a high degree of certainty that G and H are not isomorphic.

Now let us argue that V has good reason to be convinced. Suppose that G and H really are not isomorphic. The prover P, because it plays honestly, will always answer "isomorphic" when passed a permutation of G and "nonisomorphic" when passed a permutation of H, so the second and third rows of the table will never occur. The protocol will make it through all k rounds successfully and V will accept with probability 1.

On the other hand, suppose G and H are isomorphic. In each round, V will send the prover a random permutation of G or H, depending on the result of its coin flip, but the prover cannot tell the difference. It cannot see V's random bits or worktape; it must make its decisions purely on the basis of the message from V. A dishonest prover P', trying to fool V into erroneously accepting, must respond "nonisomorphic" whenever V's flip was tails and "isomorphic" whenever V's flip was heads, but there is no way it can tell which of these two events occurred. The chances of accidentally choosing the correct alternative k times in a row are 2^{-k}.   □

In the next two lectures we prove a remarkable theorem: $IP = PSPACE$.

Interactive proof systems and the class IP were defined by Goldwasser, Micali, and Rackoff [49]. A related model, called *Arthur–Merlin games*, was defined by Babai [8].

Lecture 16

$PSPACE \subseteq IP$

In this lecture we show that any set in $PSPACE$ has an IP protocol. This result is due to Shamir [111], based on work of Lund, Fortnow, Karloff, and Nisan [81].

It suffices to show that the QBF problem has an IP protocol. That is, given any quantified Boolean formula B, if B is true then the prover P can convince the verifier V of that fact with high probability, and if B is false then *no* prover P' can convince V that B is true with more than negligible probability.

The proof consists of several steps.

1. We first show how to transform the given formula into a special *simple form*.

2. We then transform a simple Boolean formula B into an arithmetic expression A by replacing the Boolean operators with arithmetic operators. The arithmetic expression A represents a nonzero value iff the Boolean formula B is true. This step is called *arithmetization*.

3. We reduce the problem to determining whether an arithmetic expression A is zero modulo a sufficiently large prime p. This step uses the Chinese remainder theorem and the fact that primality is in NP, as shown previously in Lectures B and C.

4. Finally, we describe a protocol for the prover to convince the verifier with high probability that an arithmetic expression A vanishes modulo p. The protocol consists of several rounds and is inductive on the structure of A.

Step 1

Definition 16.1   *A quantified Boolean formula B is* simple *if negations are applied only to variables, and for every subformula of the form $\exists x\, C(x)$ or $\forall x\, C(x)$ of B, any free (unquantified) occurrence of x in $C(x)$ occurs in the scope of at most one universal quantifier $\forall y$ in $C(x)$. In other words, there is at most one $\forall y$ between any occurrence of a variable and its point of quantification.*

For example, the Boolean formula

$$\forall x\, \forall y\, \exists z\, ((x \vee y) \wedge \forall w\, (\neg y \vee z \vee w)) \tag{16.1}$$

is simple, whereas the formula

$$\forall x\, \forall y\, \exists z\, ((x \vee y) \wedge \forall w\, (\neg x \vee z \vee w)) \tag{16.2}$$

is not, because the second occurrence of x occurs in the scope of both the $\forall w$ and the $\forall y$, both of which occur in the scope of the $\forall x$. In the formula

$$\forall x\, \forall y\, \forall z\, F(x, y, z),$$

if x occurs free in $F(x, y, z)$, then the formula is not simple.

Lemma 16.2   *Every quantified Boolean formula can be put into simple form by a logspace transducer with at most a quadratic blowup in size.*

Proof. First move all negations inward using the two De Morgan laws $\neg(x \vee y) = \neg x \wedge \neg y$ and $\neg(x \wedge y) = \neg x \vee \neg y$, the law of double negation $\neg\neg x = x$, and the quantifier rules $\neg\forall x\varphi = \exists x\neg\varphi$ and $\neg\exists x\varphi = \forall x\neg\varphi$. This allows us to assume without loss of generality that all negations are applied only to variables.

Now for every subformula of the form

$$\forall x\, C(x, y_1, \ldots, y_n), \tag{16.3}$$

where $y_1, \ldots, y_n$ are the free variables of (16.3) (that is, they are quantified outside the subformula (16.3)), introduce new variables $y_1', \ldots, y_n'$ and replace (16.3) with

$$\forall x\, (\exists y_1' \ldots \exists y_n' \bigwedge_{i=1}^{n} (y_i \leftrightarrow y_i') \wedge C(x, y_1', \ldots, y_n')) \tag{16.4}$$

in the original formula. Do this from the outside in, larger subformulas first. For example, applying this transformation to (16.2) would yield

$$\forall x \; \forall y \; \exists x'((x \leftrightarrow x') \wedge \exists z \; ((x' \vee y)$$
$$\wedge \; \forall w \; \exists x'' \; \exists z' \; (x'' \leftrightarrow x') \wedge (z' \leftrightarrow z) \wedge (\neg x'' \vee z' \vee w))).$$

In (16.4), the variables y_i now occur exactly once each, just inside the $\forall x$. Occurrences of y_i in C are replaced by y_i', which are quantified *inside* the $\forall x$.   □

The reason for this transformation becomes apparent when we do the arithmetization. It leads to polynomials of low degree.

Step 2   This is the arithmetization step. Suppose we have a simple quantified Boolean formula B. Change this to an arithmetic formula A with variables ranging over the integers $\mathbb{Z}$ as follows.

- Replace each negative literal $\neg x$ with $1 - x$.

- Keep each positive literal x as it is.

- Replace each $\wedge$ by $\cdot$ (multiplication).

- Replace each $\vee$ by $+$.

- Replace each $\exists x$ by $\sum_{x \in \{0,1\}}$. The expression $\sum_{x \in \{0,1\}} C(x)$ is just a succinct way of writing $C(0) + C(1)$ without having to duplicate the subexpression C.

- Replace each $\forall x$ by $\prod_{x \in \{0,1\}}$. The expression $\prod_{x \in \{0,1\}} C(x)$ is just a succinct way of writing $C(0) \cdot C(1)$.

Let A be the resulting arithmetic expression. For example, if B is the simple formula (16.1), then A would be

$$\prod_{x \in \{0,1\}} \prod_{y \in \{0,1\}} \sum_{z \in \{0,1\}} ((x+y) \cdot \prod_{w \in \{0,1\}} (1 - y + z + w)). \qquad (16.5)$$

Each such expression denotes a number obtained by evaluating the expression. For example, (16.5) evaluates to

$$\prod_{x \in \{0,1\}} \prod_{y \in \{0,1\}} \sum_{z \in \{0,1\}} ((x+y) \cdot \prod_{w \in \{0,1\}} (1-y+z+w))$$

$$= \prod_{x \in \{0,1\}} \prod_{y \in \{0,1\}} \sum_{z \in \{0,1\}} (x+y)(1-y+z)(2-y+z)$$

$$= \prod_{x \in \{0,1\}} \prod_{y \in \{0,1\}} (x+y)(1-y)(2-y) + (x+y)(2-y)(3-y)$$

$$= \prod_{x \in \{0,1\}} 16x(x+1)$$

$$= 0.$$

Because the arithmetic formula evaluates to 0, the original quantified Boolean formula (16.1) was false.

Step 3   Because of the operators $\sum_x$ and $\prod_x$, these arithmetic expressions can be too costly to evaluate directly. The value of the expression can be as big as 2^{2^n}, where n is the size of the original formula; for example,

$$\prod_{x_1} \prod_{x_2} \cdots \prod_{x_n} 2 \quad = \quad 2^{2^n}.$$

However, we can also prove by induction on the depth of the expression that 2^{2^n} is an upper bound on the value. Thus, using the Chinese remainder theorem (Theorem B.1), we can show that the expression is nonzero iff it is nonzero modulo some prime p, where p can be written in binary using polynomially many bits.

The task for the prover is therefore reduced to providing a small (n^c-bit) prime p and a nonzero value $a \in \mathbb{Z}_p^* = \mathbb{Z}_p - \{0\}$ and convincing the verifier that

(i) p is prime, and

(ii)   the arithmetic expression $A - a$ vanishes modulo p.

Proving that p is prime is easily done in one round using the fact that primes are in NP (Theorem C.4).

Step 4   We must show how to convince the verifier that an arithmetic expression A vanishes modulo p, where p is a prime of polynomially many bits. Let $\mathbb{Z}_p$ denote the field of integers modulo p. We can also assume without loss of generality that $p \gg n$.

For an expression of the form $\sum_x B(x)$ or $\prod_x B(x)$, the subexpression $B(x)$ reduces modulo p to a polynomial in x with coefficients in $\mathbb{Z}_p$. For example, removing the outermost $\prod_x$ from (16.5) yields

$$\prod_{y \in \{0,1\}} \sum_{z \in \{0,1\}} ((x+y) \cdot \prod_{w \in \{0,1\}} (1 - y + z + w)) = 16x(x+1),$$

and if $p = 7$, this would be $2x(x+1)$. In general:

Lemma 16.3 *If $\sum_x D(x)$ or $\prod_x D(x)$ is an arithmetic formula derived from a simple Boolean formula, then $D(x)$ is equivalent modulo p to a polynomial $d(x)$ of linear degree with coefficients in $\mathbb{Z}_p$.*

Proof. Every occurrence of x in $D(x)$ is in the scope of at most one $\prod_y$. Replace each subexpression $\prod_y E(x, y)$ with $E(x, 0) \cdot E(x, 1)$. Because all these subexpressions are disjoint, this at most doubles the size of the expression. Now a simple inductive argument shows that the resulting polynomial is of degree at most linear in the size of the expression. $\square$

Lemma 16.4 *Let $d(x) \in \mathbb{Z}_p[x]$ be a polynomial of degree at most n. The probability that d vanishes on a randomly chosen element of $\mathbb{Z}_p$ is at most n/p.*

Proof. The polynomial $d(x)$ has at most n roots in $\mathbb{Z}_p$. $\square$

Now say the prover wants to persuade the verifier that some expression A vanishes modulo p. Write the expression as

$$B(\prod_{z_1} D_1(z_1), \sum_{z_2} D_2(z_2), \dots, \prod_{z_m} D_m(z_m)),$$

where the $\sum_{z_i} D_i(z_i)$ and $\prod_{z_i} D_i(z_i)$ are the maximal subexpressions of B of this form; in other words, $B(y_1, \dots, y_m)$ has no occurrence of $\sum_z$ or $\prod_z$. As we have argued, each of the subexpressions $D_i(z_i)$ is equivalent to a polynomial $d_i(z_i) \in \mathbb{Z}_p[z_i]$ of low degree. The prover sends the verifier

$$d_1(z), \dots, d_m(z) \ \in \ \mathbb{Z}_p[z],$$

asserting that

$$D_i(z) \ \equiv \ d_i(z) \ (\mathrm{mod}\ p), \ 1 \le i \le m. \tag{16.6}$$

The verifier checks that

$$B(\prod_{z_1} d_1(z_1), \sum_{z_2} d_2(z_2), \dots, \prod_{z_m} d_m(z_m)) \ \equiv \ 0 \ (\mathrm{mod}\ p)$$

by direct evaluation. Then the prover needs to prove (16.6) to the verifier. The verifier picks random elements $a_i \in \mathbb{Z}_p$ and asks the prover to verify

$$D_i(a_i) \ \equiv \ d_i(a_i) \ (\mathrm{mod}\ p), \ 1 \le i \le m,$$

or in other words

$$D_i(a_i) - d_i(a_i) \quad \equiv \quad 0 \pmod{p}, \ 1 \le i \le m.$$

We are back to the beginning of Step 4 with a strictly simpler expression.

Lecture 17

$IP \subseteq PSPACE$

In this lecture we show that $IP \subseteq PSPACE$. The interesting thing about this result is that no time or space bounds are assumed about the prover.

Recall that IP protocols are defined in terms of two communicating agents P and V, the *prover* and the *verifier*, respectively. Each has read-only access to the input and read/write access to its own private worktape. They communicate by means of communication tapes. In addition, V has access to a source of random bits.

Computation proceeds in rounds. The verifier runs for at most polynomial time, composing a message m_1 which it sends to P, then transfers control to P. The prover may then run for as long as it wants, after which it sends a message ℓ_1 of polynomial length back to V, then transfers control back to V, and the process is repeated, generating the messages m_2, ℓ_2, and so on. After polynomially many (say N) rounds, V indicates in its last message its acceptance ($m_N = 1$) or rejection ($m_N = 0$).

The set L is said to be *accepted* by this protocol if on input x, $|x| = n$,

(i) if $x \in L$, then the probability that V accepts is at least $3/4$, assuming all random bit strings are equally likely; and

(ii) if $x \notin L$, then the probability that V accepts is at most $1/4$; moreover, this is still true even if P is replaced by an arbitrarily malicious impostor.

By amplification (Miscellaneous Exercise 44), we can replace the probabilities $3/4$ and $1/4$ in (i) and (ii) with $1 - \varepsilon$ and ε, respectively.

Intuitively, if $x \in L$, then P can convince V of this fact with high probability; whereas if $x \notin L$, then *no one* can convince V otherwise with more than negligible probability.

Suppose (i) and (ii) are true. Let $N = n^c$ be a time bound on the entire protocol (excluding P's private computation). Thus all messages are bounded in length by N, there are at most N rounds, and V uses at most N random bits on inputs of length n.

First let us assume that the prover P runs in $PSPACE$. In this case it is easy to decide membership in L in $PSPACE$. On input x, $|x| = n$, we just cycle through all possible strings of random bits of length N sequentially, simulating the entire protocol on each and counting the number of times V accepts. We accept the input x if this number is at least 2^{N-1}, which guarantees probability at least $1/2$ of acceptance. By (i) and (ii), this occurs iff $x \in L$. The entire computation can be done in $PSPACE$.

Now let us drop the assumption that P runs in $PSPACE$. In fact, we do not place any requirements at all on P's behavior except that it be deterministic and its messages polynomially bounded. Formally, P is simply a function that takes the history of messages $m_1, \ldots, m_k$ previously received from V and the input string x and produces a new message $\ell_k = P(x, m_1, \ldots, m_k)$ to send to V. The function need not even be computable!

Now because there exists a protocol P satisfying (i) and (ii), on any input the prover might as well choose its messages so as to maximize the probability of V's acceptance. Then (i) and (ii) will still be true if the prover plays this optimizing strategy. Moreover, as we show below, such a strategy (call it P_{opt}) can be computed in $PSPACE$.

We can assume without loss of generality that all of the prover's messages are just one bit in length; if necessary, the protocol can easily be modified so that the verifier asks for the prover's responses one bit at a time. Thus we can think of P as an oracle that, whenever queried on the string $x\#m_1\# \cdots \#m_k$ consisting of the input x and the history $m_1, \ldots, m_k$ of previous messages from V, returns a single bit: 1 if $x\#m_1\# \cdots \#m_k$ is in the oracle set and 0 if not. We also assume for technical reasons that V's random tape head moves only to the right; that is, V reads each random bit at most once. If V needs to remember a random bit, it just saves it on the worktape.

Under these assumptions, the protocol is described by a computation tree T. The vertices of T are labeled by configurations describing V's current state and the contents and head positions of V's input, work, and message tapes (but not the contents or head position of V's random tape). A query to the random tape is modeled by a binary branch in T, determined by the value of the new random bit just read. We call such a branch a *random branch*. The tree T also has another kind of binary branch called an *oracle branch* corresponding to calls to the oracle, which model P's one-bit responses to messages from V. The tree is of depth at most N and all of its vertex labels can be described by strings of length N over a finite alphabet Δ.

For each fixed strategy P, that is, for each fixed oracle, the oracle branches are completely determined by P. This allows us to prune one subtree from every oracle branch to obtain a tree T_P with only random branches. The probability of V's acceptance under strategy P is the sum of the probabilities of all paths in T_P leading to acceptance ($m_N = 1$). The probability of a path is the product of the edge probabilities along that path, where the probability of an edge out of a random branch is $1/2$ and out of any other vertex is 1.

Now on input x, P_{opt} must calculate the optimal response ℓ_i to each oracle query from V that maximizes the probability of V's acceptance. The prover P_{opt} has complete knowledge of V's program, but it does not know the random bits, except for whatever information is contained in the sequence of messages from V up to now. The verifier's messages m_i are random variables ranging over strings of length at most N and depend on the previous random branches in the tree, the input string, and the prover's responses to previous messages.

Denote by $\text{Pr}_{\text{opt}}(E)$ the probability that event E occurs, assuming that the prover always behaves optimally; that is, that the prover's messages $\ell_1, \ldots, \ell_N$ are always chosen whenever possible so as to maximize the probability of acceptance. More generally, denote by $\text{Pr}_{\text{opt}}(E \mid F)$ the conditional probability that event E occurs given that event F occurs, assuming that the prover always behaves optimally (subject to any constraints imposed by F). Formally,

$$\text{Pr}_{\text{opt}}(E \mid F) \quad \overset{\text{def}}{=} \quad \begin{cases} \dfrac{\text{Pr}_{\text{opt}}(E \wedge F)}{\text{Pr}_{\text{opt}}(F)}, & \text{if } \text{Pr}_{\text{opt}}(F) \neq 0 \\ \\ \text{undefined}, & \text{otherwise.} \end{cases}$$

We use the fact that if F_i are disjoint events and $F = \bigvee_i F_i$, then

$$\text{Pr}_{\text{opt}}(E \mid F) \quad = \quad \sum_i \text{Pr}_{\text{opt}}(E \mid F_i) \cdot \text{Pr}_{\text{opt}}(F_i \mid F) \tag{17.1}$$

(Miscellaneous Exercise 73).

The acceptance condition is the event $m_N = 1$. For any $y_1, \ldots, y_i \in \{0, 1\}^N$ and $z_1, \ldots, z_i \in \{0, 1\}$, let R_i and S_i denote the events

$$R_i \overset{\text{def}}{=} \bigwedge_{j=1}^{i} m_j = y_j \qquad\qquad S_i \overset{\text{def}}{=} \bigwedge_{j=1}^{i} \ell_j = z_j.$$

The events R_i and S_i are thus functions of $y_1, \ldots, y_i$ and $z_1, \ldots, z_i$, respectively, although we do not make this dependence explicit. Then

$$\Pr_{\mathrm{opt}}(m_N = 1 \mid R_{i-1} \wedge S_{i-1})$$

$$= \sum_{y_i \in \{0,1\}^N} \Pr_{\mathrm{opt}}(m_N = 1 \mid R_i \wedge S_{i-1}) \cdot \Pr_{\mathrm{opt}}(R_i \mid R_{i-1} \wedge S_{i-1})$$

$$= \sum_{y_i \in \{0,1\}^N} \max_{z_i \in \{0,1\}} \Pr_{\mathrm{opt}}(m_N = 1 \mid R_i \wedge S_i) \cdot \Pr_{\mathrm{opt}}(R_i \mid R_{i-1} \wedge S_{i-1}).$$

$$(17.2)$$

The first equation in (17.2) is from (17.1) and the second reflects the prover's desire to maximize the probability of acceptance. The quantities

$$\Pr_{\mathrm{opt}}(R_i \mid R_{i-1} \wedge S_{i-1}) \quad = \quad \Pr_{\mathrm{opt}}(m_i = y_i \mid R_{i-1} \wedge S_{i-1})$$

can be calculated directly in *PSPACE* by simulating V's computation on all random bit strings of length N, supplying $z_1, \ldots, z_{i-1}$ as responses to the oracle queries, discarding all computations for which V does not generate messages $y_1, \ldots, y_{i-1}$ in that order, and calculating the fraction of those remaining computations for which V generates the message $m_i = y_i$. Using this *PSPACE* computation as a subroutine, the values of

$$\Pr_{\mathrm{opt}}(m_N = 1 \mid R_i \wedge S_i) \tag{17.3}$$

can be calculated by depth-first search on the computation tree using (17.2). Of course, the probability of acceptance is

$$\Pr_{\mathrm{opt}}(m_N = 1) \quad = \quad \Pr_{\mathrm{opt}}(m_N = 1 \mid R_0 \wedge S_0).$$

The entire computation can be done in *PSPACE*. The prover can also calculate its optimal move at any point in the protocol in *PSPACE* by calculating (17.3) for $z_i \in \{0, 1\}$ and choosing ℓ_i to be the value that gives the maximum.

This result is attributed to Paul Feldman in a paper of Goldwasser and Sipser [50] and also follows from results in their paper.

Lecture 18

Probabilistically Checkable Proofs

In the next few lectures we take a look at some complexity-theoretic results about the relationship between interactive protocols and approximation algorithms. The model we use is known as *probabilistically checkable proofs* (*PCP*). This model of computation is essentially the same as the interactive proof model introduced in Lecture 15 with the minor modification that we do not allow errors in the positive case. Thus we amend the definition slightly to say that a set L has *probabilistically checkable proofs* if there is an interactive protocol (P, V) such that V runs in polynomial time and

(i)   if $x \in L$, then (P, V) accepts with probability 1; and

(ii)   if $x \notin L$, then for any P', (P', V) accepts with probability at most $\frac{1}{2}$.

The $\frac{1}{2}$ in (ii) is inconsequential. Using amplification (Miscellaneous Exercise 44), we could require any $\varepsilon > 0$ without loss of generality.

As in Lecture 17, we can assume without loss of generality that V includes the input x and all previous messages in its queries to P, and that P's messages consist of a single bit. Thus we can view P as an oracle; that is, the prover's strategy is nonadaptive. In the literature on PCP, one normally regards the oracle as encoding a binary string constituting a proof that $x \in L$ and the queries as the extraction of bits of this string.

We also place bounds on the number of random bits V uses and the number of oracle queries it makes. We say that an interactive protocol is $(r(n), q(n))$-*bounded* if on inputs of length n, the verifier uses at most $r(n)$ random bits and makes at most $q(n)$ oracle queries along any computation path. We denote by $PCP(r(n), q(n))$ the family of sets having $(O(r(n)), O(q(n)))$-bounded protocols.

The main result in this area is:

Theorem 18.1   (Arora et al. [6]) $NP = PCP(\log n, 1)$.

Theorem 18.1 is quite powerful and is the culmination of a long string of related research [6, 7, 9, 11, 19, 38, 57, 58, 69, 81, 105]. It says that all sets in NP have interactive protocols that use at most $O(\log n)$ random bits and query at most a constant number of bits of the proof independent of the size of the input.

We do not prove Theorem 18.1 in this course—it would take more time that we have!—however, in Lectures 19 and 20, we prove a weaker version of it that contains many of the main ideas, namely that $NP \subseteq PCP(n^3, 1)$. First, however, we indicate some consequences of this theorem in the realm of approximation algorithms.

PCP and Hardness of Approximation

Probabilistically checkable proofs are strongly related to approximation algorithms for combinatorial problems. We have seen various decision problems with yes/no answers such as 3SAT and MAZE. Often these problems have an associated optimization problem that can be a maximization problem or a minimization problem. For example, the problem MAX-3SAT asks for a truth assignment to a given Boolean formula in 3CNF that maximizes the number of satisfied clauses. Another typical example is the problem MAX-CLIQUE, which asks for a maximum clique (complete subgraph) of a given undirected graph.

An α-*approximation algorithm* for an optimization problem is an algorithm that produces an answer that is within a fixed constant ratio α of the optimal. For a maximization problem such as MAX-3SAT or MAX-CLIQUE, this means producing an answer of size at least α times the size of the maximum solution for some fixed $0 < \alpha \leq 1$, independent of the size of the input. The constant α is called the *approximation ratio*.

A *polynomial-time approximation scheme* (PTAS) for an optimization problem is a family of polynomial-time algorithms that allow the answer to be approximated to within any prechosen approximation ratio at the cost of a higher (but still polynomial) running time. Thus the running time for any algorithm in the scheme is $O(n^{f(\alpha)})$, where the exponent $f(\alpha)$ may depend on the desired approximation ratio.

The following lemma shows that there is a polynomial-time 7/8-approximation algorithm for MAX-3SAT.

Lemma 18.2 (Johnson [66]) *There is a polynomial-time algorithm that, given a Boolean formula B in 3CNF with n variables and m clauses such that each clause contains three distinct variables, finds a truth assignment that satisfies $7m/8$ clauses.*

The assumption that each clause contain three distinct variables is necessary. For example, in the formula

$$(x \vee y \vee y) \wedge (x \vee \overline{y} \vee \overline{y}) \wedge (\overline{x} \vee y \vee y) \wedge (\overline{x} \vee \overline{y} \vee \overline{y}),$$

it is impossible to satisfy more than $3/4$ of the clauses.

Proof. Choose a random truth assignment $r_1, \ldots, r_n$ to the variables $x_1, \ldots, x_n$ by flipping a fair coin independently for each variable. Let S_i and S be the random variables

$$S_i \overset{\text{def}}{=} \begin{cases} 1, & \text{if } r_1, \ldots, r_n \text{ satisfies clause } i \\ 0, & \text{otherwise} \end{cases}$$

$$S \overset{\text{def}}{=} S_1 + \cdots + S_m.$$

Then S is the number of clauses satisfied by the random assignment. The expected value of S_i is $7/8$, the probability that clause i is satisfied. By linearity of expectation, the expected value of S is $7m/8$. Because this is the expected number of satisfied clauses, there must be an assignment that satisfies at least this many clauses.

This argument shows that there exists an assignment that satisfies at least $7/8$ of the clauses, but does not tell us how to find one. However, a greedy algorithm does it. We assign truth values to $x_1, \ldots, x_n$ in that order. Suppose we have already determined a partial assignment $a_1, \ldots, a_{k-1}$ to $x_1, \ldots, x_{k-1}$. Calculate the expected number of clauses satisfied by assigning $x_k = 0$ and the remaining truth values $x_{k+1}, \ldots, x_n$ randomly. Do the same for $x_k = 1$. Let a_k be the value for x_k that gives the maximum. If E_k is the event $\bigwedge_{i=1}^{k} r_i = a_i$, then the conditional expectation $\mathcal{E}(S \mid E_k)$ is the expected number of clauses satisfied by assigning $a_1, \ldots, a_k$ to $x_1, \ldots, x_k$ and the remaining variables randomly. One can show that by choice of a_k, the sequence of $\mathcal{E}(S \mid E_k)$ is nondecreasing with k, therefore the final number of satisfied clauses is

$$\mathcal{E}(S \mid E_n) \geq \mathcal{E}(S \mid E_0) = \mathcal{E}(S) = 7m/8.$$

Further implementation details and a proof of correctness are left as exercises (Miscellaneous Exercise 45). □

Is it possible to do better? Barring $P = NP$, the next best situation would be if MAX-3SAT had a polynomial-time approximation scheme, which would allow it to be approximated in polynomial time to any desired approximation ratio. For certain other combinatorial optimization problems, this is provably impossible unless $P = NP$. In fact, MAX-CLIQUE has no polynomial-time approximation algorithm to any nontrivial approximation ratio unless $P = NP$:

Theorem 18.3   (Feige et al. [38])   *There is a polynomial-time α-approximation algorithm for* MAX-CLIQUE *for some $0 < \alpha \leq 1$ if and only if $P = NP$.*

   Proof. If $P = NP$, then MAX-CLIQUE can be solved exactly in polynomial time (Miscellaneous Exercise 46(b)). Conversely, suppose there were an α-approximation algorithm for MAX-CLIQUE, $0 < \alpha \leq 1$. Let L be an arbitrary set in NP. By the *PCP* theorem (Theorem 18.1), L has a *PCP* protocol that uses $c \log n$ random bits and k oracle queries on inputs of length n, where c, k are constants. By amplification (Miscellaneous Exercise 44), we can assume that the acceptance probability for $x \notin L$ is strictly less than α.

   For input string x, each random bit string $y \in \{0,1\}^{n^c}$ and sequence of oracle responses $a \in \{0,1\}^k$ determine a unique sequence of oracle queries $z_1, \dots, z_k$. The first oracle query z_1 is determined by y alone. After the verifier sees the response a_1, that and y uniquely determine the next query z_2, and so on. Build an undirected graph $G = (V, E)$ with

$$V \overset{\text{def}}{=} \{(y, a) \in \{0,1\}^{n^c} \times \{0,1\}^k \mid \text{the computation path determined by } (y, a) \text{ accepts } x\}$$

$$E \overset{\text{def}}{=} \{((y, a), (y', a')) \mid (y, a) \text{ and } (y', a') \text{ are consistent}\},$$

where (y, a) and (y', a') are *consistent* if $a_i = a'_j$ whenever $z_i = z'_j$. Note that (y, a) and (y, b) are not consistent if $a \neq b$. One can show that if $x \in L$, then the maximum clique of G is of size n^c, whereas if $x \notin L$, the maximum clique of G is of size strictly less than αn^c. Thus the α-approximation algorithm could be used to decide membership in L: if $x \in L$, then the approximation algorithm will give a clique of size at least αn^c, whereas if $x \notin L$, it will give a clique of size strictly less than αn^c. Further details are left as an exercise (Miscellaneous Exercise 47).   □

   It is unknown whether MAX-3SAT has a polynomial-time approximation scheme. However, it turns out that this question is equivalent to $P = NP$.

Theorem 18.4   (Arora et al. [6])   *There is a polynomial-time approximation scheme for* MAX-3SAT *if and only if $P = NP$.*

   Proof. If $P = NP$, then MAX-3SAT can be solved exactly in polynomial time (Miscellaneous Exercise 46(a)). For the converse, we use the *PCP* theorem (Theorem 18.1). Suppose $L \in NP$. Because $NP \subseteq PCP(\log n, 1)$, there is a PCP protocol (P, V) for L using at most $c \log n$ random bits and k oracle queries on inputs x of length n for constants c, k. Normally, a reduction from L to 3SAT (such as the one given in Theorem 6.2) would produce, for a given x, a Boolean formula φ_x in 3CNF such that φ_x has a satisfying assignment iff $x \in L$. However, because we have a PCP protocol for L, we can get a reduction satisfying the stronger property:

(i)   If $x \in L$, then φ_x is satisfiable.

(ii) If $x \notin L$, then no truth assignment satisfies more than a $1 - \varepsilon$ fraction of the clauses of φ_x, where $\varepsilon = (k-2)^{-1} 2^{-(k+1)}$.

Thus if we could approximate the maximum number of clauses to within an arbitrary constant ratio, then we could distinguish between (i) and (ii).

As argued in Lecture 17, the PCP protocol for L determines a computation tree T for V containing random branches and oracle branches. The random branches are determined by the random bits of V and the oracle branches are determined by the responses to the oracle queries. There are only n^c possible random bit strings with $c \log n$ random bits. For each such random bit string y, let T_y be the subtree of T obtained by determinizing all the random branches according to y. Thus T_y has only oracle branches.

Along each path in T_y, at most k oracle queries are made, thus there are at most 2^k possible paths. However, there may be as many as $2^k - 1$ distinct queries to the oracle in T_y, because later queries can depend on the responses to earlier queries. Let $z_1, \ldots, z_m$ be a list of all the oracle queries in T_y. Associating a Boolean variable with each z_i, we can write down a formula in CNF describing the oracle responses that lead to acceptance. For example, suppose T_y looks like

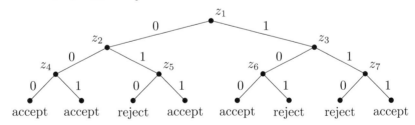

First write down a DNF formula that describes the query responses that lead to rejection. For the example pictured above, this would give

$$(\overline{z}_1 \wedge z_2 \wedge \overline{z}_5) \vee (z_1 \wedge \overline{z}_3 \wedge z_6) \vee (z_1 \wedge z_3 \wedge \overline{z}_7).$$

Negating and applying the De Morgan laws yields a formula in CNF describing the query responses that lead to acceptance:

$$(z_1 \vee \overline{z}_2 \vee z_5) \wedge (\overline{z}_1 \vee z_3 \vee \overline{z}_6) \wedge (\overline{z}_1 \vee \overline{z}_3 \vee z_7).$$

In this example, we already have a 3CNF formula, but for larger k we can reduce to 3CNF using $k - 3$ applications of the resolution rule of propositional logic:

$$(w_1 \vee \cdots \vee w_k) \quad \mapsto$$
$$(w_1 \vee w_2 \vee x_1) \wedge (\overline{x}_1 \vee w_3 \vee x_2) \wedge \cdots \wedge (\overline{x}_{k-3} \vee w_{k-1} \vee w_k),$$

where $x_1, \ldots, x_{k-3}$ are new variables. Denote the resulting CNF formula by C_y. Then C_y consists of at most $(k-2)2^k$ clauses.

Let φ_x be the conjunction of all the constraints C_y for all n^c random bit strings y. Then φ_x is a 3CNF formula with at most $n^c(k-2)2^k$ clauses. If $x \in L$, then (P, V) accepts x with probability 1, so for all random bit strings y with oracle queries determined by P, the tree T_y accepts, and the corresponding truth assignment satisfies C_y. In this case φ_x is satisfied. On the other hand, if $x \notin L$, then no (P', V) accepts x with probability more than $1/2$, so with oracle queries determined by P', at least half of the trees T_y reject, and the corresponding constraints C_y are not satisfied. For this to occur, at least one clause of each of these constraints is not satisfied, which is at least a $(k-2)^{-1}2^{-(k+1)} = \varepsilon$ fraction of all clauses in φ_x.

Now if MAX-3SAT had a polynomial-time α-approximation algorithm with approximation ratio $\alpha > 1 - \varepsilon$, then we could decide membership in L deterministically in polynomial time: if $x \in L$, our approximation algorithm would give an assignment satisfying at least an $\alpha > 1 - \varepsilon$ fraction of the clauses, whereas if $x \notin L$, the largest fraction of clauses that can be simultaneously satisfied is at most $1 - \varepsilon$. $\qquad \square$

Lecture 19

$NP \subseteq PCP(n^3, 1)$

In this lecture and the next, we show that $NP \subseteq PCP(n^3, 1)$. This is still a far cry from the stronger result $NP \subseteq PCP(\log n, 1)$, but many of the main ideas used in that proof are already present in this weaker version.

The proof of $NP \subseteq PCP(n^3, 1)$ breaks down into several steps:

1. Arithmetization of Boolean formulas over $\mathbb{Z}_2$;

2. Probabilistically testing whether a vector is the zero vector;

3. Probabilistic linearity testing;

4. Random self-correction to avoid a sparse set of errors.

Arithmetization

Let B be a Boolean formula in 3CNF with m clauses over Boolean variables $x = x_1, \ldots, x_n$. We encode B as an m-vector of polynomials $p_i \in \mathbb{Z}_2[x]$, $1 \leq i \leq m$, each p_i of degree 3. Each clause of B is represented as a product of linear polynomials x_i or $1 - x_i$, the former if $\neg x_i$ appears in the clause and the latter if x_i appears in the clause. For example, the clause $x_1 \vee \neg x_2 \vee x_3$ would become the polynomial $(1 - x_1)x_2(1 - x_3)$. Then for any truth assignment $a = a_1, \ldots, a_n$, a satisfies the clause (regarding a as a vector of Boolean values) iff the corresponding polynomial evaluated at

a is 0 (regarding a as a vector of elements of $\mathbb{Z}_2$). This argument uses the fact that $\mathbb{Z}_2$ is a field, therefore has no zero divisors; in our example above, $(1 - a_1)a_2(1 - a_3) = 0$ iff $1 - a_1 = 0$, $a_2 = 0$, or $1 - a_3 = 0$.

We thus have a vector $p(x) = p_1(x), \ldots, p_m(x)$ of m degree-3 polynomials over $x = x_1, \ldots, x_n$ with coefficients in $\mathbb{Z}_2$. The original formula is satisfiable iff there is a vector of values $a = a_1, \ldots, a_n \in \mathbb{Z}_2^n$ such that $p_i(a) = 0$, $1 \leq i \leq m$. How can the verifier be convinced of this fact after only a constant number of queries to the prover? If the verifier could ask the prover for all n bits of a, then it could evaluate the polynomials $p_i(a)$ and check that they are all 0. However, this would take more than a constant number of queries.

Instead, the verifier can ask the prover for the value of $r \bullet p(a)$, where $r = r_1, \ldots, r_m$ is a vector of m elements of $\mathbb{Z}_2$ chosen uniformly at random and $\bullet$ is the inner product

$$r \bullet p(a) \quad \stackrel{\text{def}}{=} \quad \sum_{i=1}^{m} r_i \cdot p_i(a). \tag{19.1}$$

If $p(a) = 0$, then surely $r \bullet p(a) = 0$, and if $p(a) \neq 0$, then $r \bullet p(a) = 0$ with probability $1/2$. To see this, note that we are testing whether r lies in the kernel of the linear map $r \mapsto r \bullet p(a)$, which is a subspace of the m-dimensional vector space $\mathbb{Z}_2^m$ over $\mathbb{Z}_2$. Recall from linear algebra that the dimension of the domain of a linear map is equal to the sum of the dimensions of its kernel and its image. Thus the kernel is of dimension m if $p(a) = 0$ and $m - 1$ if $p(a) \neq 0$, and a vector space of dimension n over $\mathbb{Z}_2$ is isomorphic to $\mathbb{Z}_2^n$, thus contains 2^n elements.

The good news is that each test (19.1) requires only one query. The bad news is that we have no guarantee that the answers returned by these tests in any way correspond to the linear map $r \mapsto r \bullet p(a)$. This brings us to steps 3 and 4 of the proof, which we develop below. But for now, suppose we are convinced that the answers returned by the prover on queries r are indeed $r \bullet p(a)$ for some a. If we perform the test (19.1) k times and the answer is always 0, then we can accept with high confidence, because the probability of that happening by accident when $p(a) \neq 0$ is 2^{-k}. If the answer is ever 1, then we reject immediately. If indeed $p(a) = 0$, then we accept with probability 1; and if $p(a) \neq 0$, then we accept erroneously with probability 2^{-k}.

Linearity Testing

We would eventually like to show how the verifier can convince itself that the prover is returning values corresponding to some linear map $r \mapsto r \bullet p(a)$. First, we show how the verifier can convince itself that the values returned correspond to some linear map $r \mapsto r \bullet b$. Recall that a function

$f : V \to \mathbb{F}$ from a finite-dimensional vector space V over a field of scalars $\mathbb{F}$ is *linear* if for all $a, b \in \mathbb{F}$ and $u, v \in V$, $f(au + bv) = af(u) + bf(v)$. For $V = \mathbb{F}^m$, the linear maps $f : \mathbb{F}^m \to \mathbb{F}$ are exactly the maps of the form $r \mapsto r \bullet b$ for some $b \in \mathbb{F}^m$. This is because over the standard basis in $\mathbb{F}^m$, any such linear map is represented by multiplication by a $1 \times m$ matrix, which is the same as the inner product with an m-vector. This is a good start, although we still need to verify that $b = p(a)$ for some a.

If the scalar field $\mathbb{F}$ is $\mathbb{Z}_p$, linearity is equivalent to *additivity*: $f(u+v) = f(u) + f(v)$ for all $u, v \in V$.[1] Here is a proposal for a linearity test in this case.

1. Pick $u, v \in V$ uniformly at random. Query the prover for $f(u)$, $f(v)$, and $f(u + v)$.

2. If $f(u) + f(v) \neq f(u+v)$, halt immediately and reject. If $f(u) + f(v) = f(u + v)$, keep going.

3. Repeat steps 1 and 2 k times. If we have not rejected after k trials, accept.

Random Self-Correction

Note that the protocol above requires only a constant number ($3k$) of queries. Unfortunately, this is not enough to verify with certainty, or even with high probability, that the map f is linear. The function f could agree with a linear function on all but one input, and if we did not query f on that input, we would never find out that it is not linear, and the chances of missing that one input are pretty high.

However, the protocol does convince the verifier that there is a linear function g that agrees with f on a very large fraction of the inputs. We argue this below (Lemma 19.1). This is good enough for our purposes. For suppose there is such a g. When we are asking the prover for the value of $f(u)$, what we really want is the value of $g(u)$. So instead of asking the prover for the value of $f(u)$ directly, we can spend two queries and ask the prover for the values of $f(u + v)$ and $f(v)$ for some randomly chosen v. Chances are pretty good that we will get $g(u + v)$ and $g(v)$, because f and g agree on a very large fraction of the inputs, and $g(u + v) - g(v) = g(u)$ because g is linear. That way we can avoid errors in f with high probability. This trick is known as *random self-correction*.

---

[1] This is false for finite fields not of the form $\mathbb{Z}_p$. For $n \geq 2$, the map $x \mapsto x^p$ is a linear map of the n-dimensional vector space $\mathbb{GF}_{p^n}$ over $\mathbb{GF}_p \cong \mathbb{Z}_p$, therefore additive; but it is not a linear map of $\mathbb{GF}_{p^n}$ as a one-dimensional vector space over itself, because for $x \in \mathbb{GF}_{p^2} - \mathbb{GF}_p$, $x^p \neq x \cdot 1^p$.

Lemma 19.1 *Let $\mathbb{F} = \mathbb{Z}_p$ for some prime p, and let V be an n-dimensional vector space over $\mathbb{F}$. If $\varepsilon < 1/6$ and $f : V \to \mathbb{F}$ satisfies*

$$\Pr_{u,v}(f(u+v) = f(u) + f(v)) > 1 - \varepsilon, \tag{19.2}$$

then there exists a linear map $g : V \to \mathbb{F}$ that agrees with f on at least a $1 - 3\varepsilon$ fraction of the inputs.

Proof. It follows from the law of sum that for any events E and F,

$$\Pr(E \vee F) \leq \Pr(E) + \Pr(F) \tag{19.3}$$

$$\Pr(E \wedge F) \geq \Pr(E) + \Pr(F) - 1. \tag{19.4}$$

$$E \to F \Rightarrow \Pr(E) \leq \Pr(F). \tag{19.5}$$

Assume that f satisfies (19.2). We define $g(u)$ to be the majority value of $f(u+w) - f(w)$ over all choices of $w \in V$. We must first show that the majority exists. Fix $u \in V$. For $v, w \in V$, define

$$v \equiv_u w \stackrel{\text{def}}{\Longleftrightarrow} f(u+v) - f(v) = f(u+w) - f(w).$$

Let $\delta_1, \ldots, \delta_k$ be the sizes of the equivalence classes of $\equiv_u$ as a fraction of the size of V, listed in increasing order. Thus the size of the ith equivalence class is $\delta_i |V|$ and $\sum_{i=1}^k \delta_i = 1$. The probability that two randomly chosen elements of V are $\equiv_u$-equivalent is

$$\Pr_{v,w}(v \equiv_u w) = \frac{\sum_i (\delta_i |V|)^2}{|V|^2}. \tag{19.6}$$

The numerator of the right-hand side of (19.6) is the number of ways of choosing two equivalent elements of V, and the denominator is the total number of ways of choosing two elements of V. Simplifying (19.6), we get

$$\frac{\sum_i (\delta_i |V|)^2}{|V|^2} = \sum_i \delta_i^2 \leq \left(\sum_i \delta_i\right)\delta_k = \delta_k. \tag{19.7}$$

We also have

$$
\begin{aligned}
\Pr_{v,w}(v \equiv_u w) &= \Pr_{v,w}(f(u+v) - f(v) = f(u+w) - f(w)) \\
&= \Pr_{v,w}(f(u+v) + f(w) = f(u+w) + f(v)) \\
&\geq \Pr_{v,w}(f(u+v) + f(w) = f(u+v+w) \\
&\qquad \wedge f(u+w) + f(v) = f(u+v+w)) \\
&\geq \Pr_{v,w}(f(u+v) + f(w) = f(u+v+w)) \\
&\qquad + \Pr_{v,w}(f(u+w) + f(v) = f(u+v+w)) - 1 \\
&\geq 1 - 2\varepsilon.
\end{aligned}
$$

The three inequalities follow from (19.5), (19.3), and (19.2), respectively. Combining this with (19.6) and (19.7), we have that $1 - 2\varepsilon \le \delta_k$, and because $\varepsilon < 1/4$, we have that $\delta_k > 1/2$. Thus the largest $\equiv_u$-class contains a majority of the elements of V.

Let Δ_u be the majority equivalence class of $\equiv_u$. We have shown

$$\Pr_w(w \in \Delta_u) \quad = \quad \delta_k \quad \ge \quad 1 - 2\varepsilon. \tag{19.8}$$

Now we show that under the assumption $\varepsilon < 1/6$, the function g is linear; that is, for all $u, v \in V$, $g(u + v) = g(u) + g(v)$. Note that if there exists $w \in V$ such that $w \in \Delta_{u+v} \cap \Delta_v$ and $v + w \in \Delta_u$, then

$$
\begin{aligned}
g(u + v) &= f(u + v + w) - f(w) & &\text{because } w \in \Delta_{u+v} \\
g(u) &= f(u + v + w) - f(v + w) & &\text{because } v + w \in \Delta_u \\
g(v) &= f(v + w) - f(w) & &\text{because } w \in \Delta_v,
\end{aligned}
$$

and the desired equation $g(u + v) = g(u) + g(v)$ would follow. It therefore suffices to show the existence of such a w. But by (19.4) and (19.8), the probability that a randomly chosen w satisfies the desired properties is

$$
\begin{aligned}
&\Pr_w(w \in \Delta_{u+v} \wedge w \in \Delta_v \wedge v + w \in \Delta_u) \\
&\ge \quad \Pr_w(w \in \Delta_{u+v}) + \Pr_w(w \in \Delta_v) + \Pr_w(v + w \in \Delta_u) - 2 \\
&\ge \quad 1 - 6\varepsilon \;>\; 0.
\end{aligned}
$$

The probability is nonzero, therefore such a w must exist.

Finally, we show that f and g agree on a large fraction of their inputs. By (19.5), (19.4), and (19.2),

$$
\begin{aligned}
\Pr_u(f(u) = g(u)) &= \Pr_{u,v}(f(u) = g(u)) \\
&\ge \Pr_{u,v}(f(u) = g(u) \wedge v \in \Delta_u) \\
&= \Pr_{u,v}(f(u) = f(u + v) - f(v) \wedge v \in \Delta_u) \\
&\ge \Pr_{u,v}(f(u) = f(u + v) - f(v)) + \Pr_{u,v}(v \in \Delta_u) - 1 \\
&\ge 1 - 3\varepsilon.
\end{aligned}
$$

$\square$

By Lemma 19.1, if there is no linear function g that agrees with f on at least a $1 - 3\varepsilon$ fraction of the inputs, then the probability that the linearity test $f(u + v) = f(u) + f(v)$ succeeds for k rounds and we continue erroneously is at most $(1 - \varepsilon)^k$, which can be made as small as we like by picking k large. Thus if we pass the linearity test, we can be confident that such a g exists.

We complete the proof next time.

Lecture 20

More on PCP

In this lecture we complete the proof that $NP \subseteq PCP(n^3, 1)$.

Let $\mathbb{F} = \mathbb{Z}_2$ and let $p(x) = p_1(x), \dots, p_m(x)$ be the vector of degree-3 polynomials over variables $x = x_1, \dots, x_n$ obtained in the arithmetization step (Lecture 19). We wish to be convinced that there exists an assignment $a = a_1, \dots, a_n$ to the variables x such that $p_i(a) = 0$, $1 \le i \le m$.

Let $y = y_1, \dots, y_m$ be a new set of variables, and form the polynomial $y \bullet p(x)$. This is a polynomial of degree 3 in the x's and linear in the y's. Collecting terms of like degree in the x's, rewrite this as

$$y \bullet p(x) \quad = \quad A(y) + \sum_i B_i(y)x_i + \sum_{i,j} C_{ij}(y)x_i x_j + \sum_{i,j,k} D_{ijk}(y)x_i x_j x_k.$$

(20.1)

We can simplify this expression further. Let $B(y)$, $C(y)$, and $D(y)$ be the following n-, n^2-, and n^3-vectors, respectively.

$$B(y) \quad \overset{\text{def}}{=} \quad B_1(y), \dots, B_n(y)$$

$$C(y) \quad \overset{\text{def}}{=} \quad C_{11}(y), C_{12}(y), \dots, C_{1n}(y), C_{21}(y), \dots, C_{nn}(y)$$

$$D(y) \quad \overset{\text{def}}{=} \quad D_{111}(y), D_{112}(y), \dots, D_{11n}(y), D_{121}(y), \dots, D_{nnn}(y).$$

For vectors $u = u_1, \dots, u_k$ and $v = v_1, \dots, v_\ell$, define the *Kronecker (tensor) product* of u and v to be the $k\ell$-vector

$$u \otimes v \quad \overset{\text{def}}{=} \quad u_1 v_1, u_1 v_2, \dots, u_1 v_\ell, u_2 v_1, \dots, u_k v_\ell.$$

Then (20.1) can be rewritten

$$y \bullet p(x) \quad = \quad A(y) \; + \; B(y) \bullet x \; + \; C(y) \bullet (x \otimes x) \; + \; D(y) \bullet (x \otimes x \otimes x).$$

Evaluating at $r \in \mathbb{F}^m$ and $a \in \mathbb{F}^n$, we get

$$r \bullet p(a) \quad = \quad A(r) \; + \; B(r) \bullet a \; + \; C(r) \bullet (a \otimes a) \; + \; D(r) \bullet (a \otimes a \otimes a).$$
$$(20.2)$$

We wish to be convinced that there exists $a \in \mathbb{F}^n$ such that (20.2) vanishes for all $r \in \mathbb{F}^m$.

Instead of asking the prover to provide values of $r \bullet p(a)$ for random r, we instead ask for the values of three different functions $f : \mathbb{F}^n \to \mathbb{F}$, $g : \mathbb{F}^{n^2} \to \mathbb{F}$, and $h : \mathbb{F}^{n^3} \to \mathbb{F}$ purporting to be the linear functions

$$r \; \mapsto \; r \bullet a \qquad s \; \mapsto \; s \bullet (a \otimes a) \qquad t \; \mapsto \; t \bullet (a \otimes a \otimes a), \qquad (20.3)$$

respectively, for some $a \in \mathbb{F}^n$. Our queries to the prover consist of a specification of which of f, g, or h we wish to query and an input to that function.

Suppose we can convince ourselves that f, g, and h are very close to linear functions of the form (20.3). Using random self-correction as described in Lecture 19, we can assume that for any input, we get the values of the functions (20.3) with very high probability. Then we can convince ourselves that $p(a) = 0$ using the following protocol.

1. Pick a random $r \in \mathbb{F}^m$ and compute $A(r)$, $B(r)$, $C(r)$, and $D(r)$.

2. Ask the prover for the values of $B(r) \bullet a$, $C(r) \bullet (a \otimes a)$, and $D(r) \bullet (a \otimes a \otimes a)$ by querying f, g, and h, respectively, using random self-correction. For example, to find out the value of $C(r) \bullet (a \otimes a)$, we would ask the prover for the values of $g(C(r) + v)$ and $g(v)$ for a randomly chosen v and subtract.

3. Compute the sum of these values and $A(r)$. This should be $r \bullet p(a)$ as in (20.2). If the result is nonzero, reject immediately, otherwise keep going.

4. Repeat steps 1–3 for k rounds. If we make it all the way through without rejecting, accept.

This protocol uses $3k$ queries and $O(n^3)$ random bits. Note that even though r is uniformly distributed, $B(r)$, $C(r)$, and $D(r)$ may not be; but the random self-correction ensures that we obtain accurate values of the linear functions (20.3) in step 2.

It remains to show how to convince ourselves that f, g, and h are close to linear functions of the form (20.3). By the linearity test of Lecture 19, we can convince ourselves that f, g and h are close to linear functions of the form

$$r \;\mapsto\; r \bullet a \qquad s \mapsto s \bullet b \qquad t \mapsto t \bullet c, \tag{20.4}$$

respectively, for some $a \in \mathbb{F}^n$, $b \in \mathbb{F}^{n^2}$, and $c \in \mathbb{F}^{n^3}$, but we would like to know that $b = a \otimes a$ and $c = a \otimes a \otimes a$.

To verify that $b = a \otimes a$, perform the following test k times: choose $u, v \in \mathbb{F}^n$ at random and check that

$$(u \bullet a)(v \bullet a) \;=\; (u \otimes v) \bullet b. \tag{20.5}$$

Here we are using the law

$$(u \bullet c)(v \bullet d) \;=\; (u \otimes v) \bullet (c \otimes d) \tag{20.6}$$

relating inner products and Kronecker products; both sides are equal to $\sum_{i,j} u_i c_i v_j d_j$. The values of $u \bullet a$, $v \bullet a$, and $(u \otimes v) \bullet b$ are obtained by querying f and g using random self-correction. If this test succeeds k times, accept; if it ever fails, reject.

To see why this works, note that because $s \mapsto s \bullet b$ is linear, the function $(\mathbb{F}^n)^2 \to \mathbb{F}$ defined by

$$(u, v) \;\mapsto\; (u \otimes v) \bullet b$$

is bilinear (linear in both variables). Over the standard basis in $\mathbb{F}^n$, any bilinear function $(\mathbb{F}^n)^2 \to \mathbb{F}$ can be written

$$(u, v) \;\mapsto\; u^T B v,$$

where B is an $n \times n$ matrix. Here u, v are considered column vectors and u^T denotes the transpose of u, which is a row vector. The matrix B is just b rearranged:

$$B_{ij} \;=\; e_i^T B e_j \;=\; (e_i \otimes e_j) \bullet b,$$

where e_i is the basis vector consisting of 1 in position i and 0 elsewhere. To say that $b = a \otimes a$ is the same as saying $B = aa^T$.

Thus we can rewrite (20.5) as

$$(u^T a)(a^T v) \;=\; u^T B v.$$

Each step of our protocol tests for randomly chosen u, v whether this equation holds, or equivalently, whether

$$u^T(aa^T - B)v \;=\; 0.$$

Surely if $B = aa^T$, the test succeeds. It remains only to show that if $B \neq aa^T$, then the test fails with positive probability independent of n.

But for any nonzero $n \times n$ matrix C,

$$
\begin{aligned}
&\mathrm{Pr}_{u,v}(u^T C v = 0) \\
&= \mathrm{Pr}_{u,v}(Cv = 0 \vee (Cv \neq 0 \wedge u^T C v = 0)) \\
&= \mathrm{Pr}_{u,v}(Cv = 0) + \mathrm{Pr}_{u,v}(Cv \neq 0 \wedge u^T C v = 0) \\
&= \mathrm{Pr}_{u,v}(Cv = 0) \\
&\qquad + \mathrm{Pr}_{u,v}(u^T C v = 0 \mid Cv \neq 0) \cdot \mathrm{Pr}_{u,v}(Cv \neq 0).
\end{aligned}
\tag{20.7}
$$

As argued in Lecture 19, the conditional probability in (20.7) is

$$
\mathrm{Pr}_{u,v}(u^T C v = 0 \mid Cv \neq 0) = \tfrac{1}{2},
$$

because this is the event that u lies in the kernel of the linear map $u \mapsto u \bullet Cv$, a subspace of dimension $n - 1$. Also, if $C \neq 0$, then the rank of C (number of linearly independent columns) is at least 1, and the probability that $Cv = 0$ is the probability that v lies in the kernel of C, which is a subspace of dimension $n - \mathrm{rank}\, C$. Thus

$$
\begin{aligned}
\mathrm{Pr}(Cv = 0) &= \mathrm{Pr}(v \in \ker C) = |\mathbb{F}|^{\dim \ker C} / |\mathbb{F}|^n \\
&= |\mathbb{F}|^{n - \mathrm{rank}\, C} / |\mathbb{F}|^n = |\mathbb{F}|^{-\mathrm{rank}\, C} \leq \tfrac{1}{2},
\end{aligned}
$$

because $|\mathbb{F}| \geq 2$ and $\mathrm{rank}\, C \geq 1$. Combining these observations with (20.7), we have

$$
\begin{aligned}
&\mathrm{Pr}_{u,v}(Cv = 0) + \mathrm{Pr}_{u,v}(u^T C v = 0 \mid Cv \neq 0) \cdot \mathrm{Pr}_{u,v}(Cv \neq 0) \\
&= \mathrm{Pr}_{u,v}(Cv = 0) + \tfrac{1}{2}(1 - \mathrm{Pr}_{u,v}(Cv = 0)) \\
&= \tfrac{1}{2}\mathrm{Pr}_{u,v}(Cv = 0) + \tfrac{1}{2} \\
&\leq \tfrac{3}{4}.
\end{aligned}
$$

The protocol to verify that $c = a \otimes a \otimes a$ is similar (Miscellaneous Exercise 48).

Supplementary Lecture E

A Crash Course in Logic

In this lecture we present the basics of first-order logic as background for the lectures on the complexity of logical theories (Lectures 21–25).

What Is Logic?

Logic typically has three parts: *syntax*, *semantics*, and *deductive apparatus*. *Syntax* is concerned with the correct formation of expressions—whether the symbols are in the right places. *Semantics* concerns itself with meaning—how to interpret syntactically correct expressions as meaningful statements about something. Finally, the *deductive apparatus* gives rules for deriving theorems mechanically. We do not concern ourselves much with deductive apparatus here.

Relational Structures

First-order logic is good for expressing and reasoning about basic mathematical properties of algebraic and combinatorial structures. Examples of such structures are: groups, rings, fields, vector spaces, graphs, trees, ordered sets, and the natural numbers.

Such a structure $\mathcal{A}$ typically consists of a set A, called the *domain* or *carrier* of $\mathcal{A}$, along with some distinguished n-ary functions $f^{\mathcal{A}} : A^n \to A$

for various n, constants $c^A \in A$ (which can be viewed as 0-ary functions), and n-ary relations $R^A \subseteq A^n$ for various n. The number of inputs n is called the *arity* of the function or relation. Functions or relations of arity 0, 1, 2, 3, and n are called *nullary, unary, binary, ternary*, and *n-ary*, respectively.

The list of distinguished functions and relations of $\mathcal{A}$ along with their arities is called the *signature* of $\mathcal{A}$. It is usually represented by an alphabet Σ of function and relation symbols, one for each distinguished function or relation of $\mathcal{A}$, each with a fixed associated arity.

Example E.1 The structure $\mathbb{N}$ of number theory consists of the set $\omega = \{0, 1, 2, \dots\}$, the natural numbers, along with the binary operations of addition and multiplication, constant additive and multiplicative identity elements, and the binary equality relation. The signature of number theory is $(+, \cdot, 0, 1, =)$, where $+$ and $\cdot$ are binary function symbols, 0 and 1 are constant symbols, and $=$ is a binary relation symbol.

A *group* is any structure consisting of a set with a binary multiplication operation, a unary inverse operation, a constant identity element, and a binary equality relation, satisfying certain properties. The signature of group theory is $(\cdot, {}^{-1}, 1, =)$, where $\cdot$ is a binary function symbol, ${}^{-1}$ is a unary function symbol, 1 is a constant symbol, and $=$ is a binary relation symbol.

A *partial order* is any set with a binary inequality relation and a binary equality relation satisfying certain properties. The signature of the theory of partial orders is $(\leq, =)$, where $\leq$ and $=$ are binary relation symbols. $\square$

When discussing structures in general, we usually assume a fixed but arbitrary signature Σ. We usually use $f, g, \dots$ to denote function symbols of arity at least one, $c, d, \dots$ to denote constant symbols, and $R, S, \dots$ to denote relation symbols. The functions and relations they represent in the structure $\mathcal{A}$ are denoted f^A, c^A, R^A, and so on.

At the risk of confusion, when working in a specific structure, we often use the same symbol for both the symbol of Σ and the semantic object it denotes; for example, in number theory, we might use $+$ to denote both the symbol of the signature of number theory and the addition operation on the natural numbers.

Syntax

The syntax of first-order logic can be separated into two parts, the first application-specific and the second application-independent. The application-specific part specifies the correct formation of *terms* from the symbols of Σ. The application-independent part specifies the correct formation of *formulas* from propositional connectives $\vee, \wedge, \neg, \rightarrow, \leftrightarrow, 0$ (falsity), and 1 (truth), variables $x, y, z, \ldots$, quantifiers $\forall$ and $\exists$, and parentheses. These symbols are part of every first-order language.

Terms

Fix a signature Σ, and let X be a set of *variables*. A *term* is a well-formed expression built from the function symbols of Σ and variables X, regarding elements of X as symbols of arity 0. Here *well formed* means that the arities of all the symbols are respected. For example, if f is a binary function symbol, g is a unary function symbol, c, d are constant symbols, and x, y are variables, then

$$c \quad x \quad f(g(x), f(c, g(y))) \quad g(f(g(x), c), f(d, g(y)))$$

are typical terms.

Depending on custom, terms involving binary function symbols are sometimes written in infix notation, as in $(x + 1) \cdot y$, and those involving unary function symbols are sometimes written in postfix notation, as in x^{-1}.

Valuations and the Meaning of Terms

A *valuation* over a structure $\mathcal{A}$ with domain A is a map from variables to values:

$$u : X \quad \rightarrow \quad A.$$

These maps are often called *environments* in programming language semantics. Any valuation extends uniquely by induction to a map

$$u : \{\text{terms}\} \quad \rightarrow \quad A$$

as follows: for any terms $t_1, \ldots, t_n$ and n-ary function symbol f,

$$u(f(t_1, \ldots, t_n)) \quad \overset{\text{def}}{=} \quad f^{\mathcal{A}}(u(t_1), \ldots, u(t_n)).$$

This definition also includes the case $n = 0$: for constants c, $u(c) = c^{\mathcal{A}}$. A term with no variables is called a *ground term*. Note that for ground terms t, the value $u(t)$ is independent of u. For this reason we often write $t^{\mathcal{A}}$ instead of $u(t)$ for ground terms t.

If u is a valuation, x is a variable, and $a \in A$, we denote by $u[x/a]$ the valuation that agrees with u except on variable x, on which it takes the value a. In other words,

$$u[x/a](y) \quad \overset{\text{def}}{=} \quad \begin{cases} u(y), & \text{if } y \neq x, \\ a, & \text{otherwise.} \end{cases}$$

The operator $[x/a]$ is called a *rebinding operator*.

Formulas and Sentences

An *atomic formula* is either a Boolean constant 0 or 1 or an expression of the form $R(t_1, \ldots, t_n)$, where R is an n-ary relation symbol of the signature and $t_1, \ldots, t_n$ are terms. Depending on the application, atomic formulas involving binary relation symbols are sometimes written in infix notation, as in $g(x) = y$.

Formulas are defined inductively:

- Every atomic formula is a formula;

- If φ and ψ are formulas and x is a variable, then the following are formulas: $\varphi \wedge \psi$, $\varphi \vee \psi$, $\varphi \to \psi$, $\varphi \leftrightarrow \psi$, $\neg\varphi$, $\exists x\ \varphi$, and $\forall x\ \varphi$.

We use parentheses in ambiguous situations when it is not clear how to parse the formula. Quantifiers may appear more than once with the same variable in the same formula.

For example,

$$\exists x\ ((\forall z\ y \le z) \to x \le y) \tag{E.1}$$

is a typical formula of the first-order language of ordered structures.

Scope, Free and Bound Occurrences of Variables

Suppose the formula φ has an occurrence of a subformula of the form $\mathbf{Q}x\ \psi$, where $\mathbf{Q}$ is a quantifier, either $\exists$ or $\forall$. The *scope* of the $\mathbf{Q}x$ in that occurrence of $\mathbf{Q}x\ \psi$ is that occurrence of ψ. (We have to say "occurrence" because quantifiers and subformulas can have more than one occurrence in a given formula.)

Consider an occurrence of a variable x in a formula φ (as a term, not as part of a quantifier expression $\mathbf{Q}x$). Such an occurrence of x is called *bound* if it is in the scope of a quantifier $\mathbf{Q}x$, *free* if not. A bound occurrence of x is *bound to* the occurrence of $\mathbf{Q}x$ with the smallest scope in which that occurrence of x occurs.

For example, in (E.1), the scope of the $\exists x$ is $((\forall z\ y \le z) \to x \le y)$, and the scope of the $\forall z$ is $y \le z$. The single occurrence of x is bound to the $\exists x$, the single occurrence of z is bound to the $\forall z$, and the two occurrences of y are free. In

$$\exists x\ ((\forall y\ y \le z) \to x \le y), \tag{E.2}$$

on the other hand, the single occurrence of x is bound to the $\exists x$, the first occurrence of y is bound to the $\forall y$, and the single occurrence of z and the second occurrence of y are free.

A *sentence* is a formula with no free variables.

It is customary to write $\varphi(x_1, \ldots, x_n)$ to indicate that all free variables of φ are among $x_1, \ldots, x_n$.

Interpretation of Formulas and Sentences

Given a structure $\mathcal{A}$ and a valuation of variables u over $\mathcal{A}$, every formula has a truth value defined inductively as follows. We write

$$\mathcal{A}, u \ \vDash \ \varphi$$

and say "φ is true in $\mathcal{A}$ under valuation u" if the truth value associated with the formula φ is 1 (true) under the inductive definition we are about to give.

For atomic formulas, $\mathcal{A}, u \vDash 1$ always, $\mathcal{A}, u \vDash 0$ never, and

$$\mathcal{A}, u \vDash R(t_1, \ldots, t_n) \ \overset{\text{def}}{\Longleftrightarrow} \ R^{\mathcal{A}}(u(t_1), \ldots, u(t_n)).$$

For compound formulas,

$$\mathcal{A}, u \vDash \varphi \wedge \psi \ \overset{\text{def}}{\Longleftrightarrow} \ \mathcal{A}, u \vDash \varphi \text{ and } \mathcal{A}, u \vDash \psi$$

$$\mathcal{A}, u \vDash \varphi \vee \psi \ \overset{\text{def}}{\Longleftrightarrow} \ \mathcal{A}, u \vDash \varphi \text{ or } \mathcal{A}, u \vDash \psi$$

$$\mathcal{A}, u \vDash \neg\varphi \ \overset{\text{def}}{\Longleftrightarrow} \ \text{it is not the case that } \mathcal{A}, u \vDash \varphi$$

$$\mathcal{A}, u \vDash \exists x \ \varphi \ \overset{\text{def}}{\Longleftrightarrow} \ \text{there exists } a \in \mathcal{A} \text{ such that } \mathcal{A}, u[x/a] \vDash \varphi$$

$$\mathcal{A}, u \vDash \forall x \ \varphi \ \overset{\text{def}}{\Longleftrightarrow} \ \text{for all } a \in \mathcal{A}, \mathcal{A}, u[x/a] \vDash \varphi.$$

Whether $\mathcal{A}, u \vDash \varphi$ depends only on the values that u assigns to the free variables of φ. In other words, if u and v agree on all variables with a free occurrence in φ, then $\mathcal{A}, u \vDash \varphi$ iff $\mathcal{A}, v \vDash \varphi$. This can be shown by induction on the structure of φ. In particular, for sentences (formulas with no free variables), whether $\mathcal{A}, u \vDash \varphi$ does not depend on u at all. In this case we omit the u and write $\mathcal{A} \vDash \varphi$ and say "φ is true in $\mathcal{A}$" if the sentence φ is true in $\mathcal{A}$ under any valuation (hence all valuations).

If Φ is a set of sentences, we write $\mathcal{A} \vDash \Phi$ if $\mathcal{A} \vDash \varphi$ for all $\varphi \in \Phi$.

If the free variables of φ are all among $x_1, \ldots, x_n$, that is, if $\varphi = \varphi(x_1, \ldots, x_n)$, and if $a_1, \ldots, a_n \in \mathcal{A}$, it is common to abuse notation by writing

$$\mathcal{A} \ \vDash \ \varphi(a_1, \ldots, a_n) \tag{E.3}$$

for

$$\mathcal{A}, u \ \vDash \ \varphi(x_1, \ldots, x_n), \tag{E.4}$$

where u is some valuation such that $u(x_i) = a_i$, $1 \leq i \leq n$. This is an abuse of notation because it is mixing syntactic objects (φ) with semantic objects ($a_1, \ldots, a_n$). Some authors deal with this by including a constant for each element of the domain of $\mathcal{A}$ and substituting the constant a_i for x_i in the definition of truth. Please just remember that anytime you see (E.3), although strictly speaking it is a type error, it really should be taken as an abbreviation for (E.4).

Prenex Form

There are semantics-preserving rules for transforming first-order formulas to a semantically equivalent special form called *prenex form*. In prenex form, all quantifiers occur first, followed by a quantifier free part. The rules are

$$\varphi \wedge \forall x \; \psi(x) \quad \Leftrightarrow \quad \forall x \; (\varphi \wedge \psi(x))$$
$$\varphi \vee \forall x \; \psi(x) \quad \Leftrightarrow \quad \forall x \; (\varphi \vee \psi(x))$$
$$\varphi \wedge \exists x \; \psi(x) \quad \Leftrightarrow \quad \exists x \; (\varphi \wedge \psi(x))$$
$$\varphi \vee \exists x \; \psi(x) \quad \Leftrightarrow \quad \exists x \; (\varphi \vee \psi(x))$$
$$\neg \forall x \; \psi(x) \quad \Leftrightarrow \quad \exists x \; \neg\psi(x)$$
$$\neg \exists x \; \psi(x) \quad \Leftrightarrow \quad \forall x \; \neg\psi(x),$$

provided x does not occur free in φ. If x does occur free in φ, one can change the bound variable by applying the rule

$$\forall x \; \psi(x) \quad \Leftrightarrow \quad \forall y \; \psi(y),$$

where y is a new variable. To transform a formula to prenex form, the rules would be applied from left to right.

First-Order Theories

The *first-order theory of a structure* $\mathcal{A}$, denoted $\mathrm{Th}(\mathcal{A})$, is the set of sentences in the first-order language of $\mathcal{A}$ that are true in $\mathcal{A}$:

$$\mathrm{Th}(\mathcal{A}) \stackrel{\text{def}}{=} \{\varphi \mid \mathcal{A} \vDash \varphi\}.$$

For example, *first-order number theory* is the set of first-order sentences true in $\mathbb{N}$.

If $\mathcal{C}$ is a class of structures all of the same signature, the *first-order theory of* $\mathcal{C}$, denoted $\mathrm{Th}(\mathcal{C})$, is the set of sentences in appropriate first-order language that are true in all structures in $\mathcal{C}$:

$$\mathrm{Th}(\mathcal{C}) \stackrel{\text{def}}{=} \bigcap_{\mathcal{A} \in \mathcal{C}} \mathrm{Th}(\mathcal{A}).$$

For example, first-order group theory is the set of sentences in the language of groups that are true in all groups.

Axiomatization

If Φ is a set of sentences over some signature Σ, the class of *models of* Φ is the class of structures of signature Σ that satisfy all the sentences of Φ.

This class is denoted $\text{Mod}(\Phi)$:

$$\text{Mod}(\Phi) \stackrel{\text{def}}{=} \{\mathcal{A} \mid \mathcal{A} \vDash \Phi\}.$$

A sentence φ is called a *logical consequence* of a set of sentences Φ if it is true in all models of Φ. In other words, the set of logical consequences of Φ is the set $\text{Th}(\text{Mod}(\Phi))$.

We often specify a class of structures by giving a set of *axioms*, which are just first-order sentences. The class being specified is defined to be the class of models of those sentences.

Example E.2   A *group* is a structure of signature $(\cdot, 1, ^{-1}, =)$ satisfying the first-order group axioms

$$\forall x \; \forall y \; \forall z \; x(yz) = (xy)z$$
$$\forall x \; x1 = x$$
$$\forall x \; 1x = x$$
$$\forall x \; xx^{-1} = 1$$
$$\forall x \; x^{-1}x = 1$$

and the axioms of equality

$$\forall x \; x = x \qquad\qquad \forall x \; \forall y \; \forall z \; x = y \rightarrow xz = yz$$
$$\forall x \; \forall y \; x = y \rightarrow y = x \qquad\qquad \forall x \; \forall y \; \forall z \; x = y \rightarrow zx = zy$$
$$\forall x \; \forall y \; \forall z \; (x = y \wedge y = z) \rightarrow x = z \qquad \forall x \; \forall y \; x = y \rightarrow x^{-1} = y^{-1}.$$

$\square$

The Decision Problem

The *decision problem* for a first-order theory is to determine whether a given sentence is an element of the theory. For a theory of a structure such as $\mathbb{N}$, this is just the problem of deciding whether a given sentence in the language of number theory is true in $\mathbb{N}$. For the theory of a class of structures $\mathcal{C}$, the decision problem is to determine whether a given sentence is true in all structures in the class; that is, whether it is in $\text{Th}(\mathcal{C})$. For a set of first-order axioms Φ, the decision problem is to determine whether a given sentence is a logical consequence of Φ.

Lecture 21

Complexity of Decidable Theories

Here we outline a general treatment of the complexity of first-order theories in terms of *Ehrenfeucht–Fraïssé games*. A good general introduction to this technique is given in Ferrante and Rackoff's monograph [41].

Ehrenfeucht–Fraïssé Games

A *game* is specified by a tuple (BOARDS, MOVE), where BOARDS is a set of boards and MOVE is a binary relation giving the legal moves. Often the decision problem for a logical theory can be reduced to a finite game, where the board positions represent arrangements or equivalence classes of k-tuples of elements obtained by eliminating quantifiers one by one, and the next move relation represents the different new arrangements that can be obtained by picking the next element. We show that in many cases, finding an efficient decision procedure amounts to finding a finite game of a particular size and shape.

Dense Linear Order Without Endpoints

The motivating example is the theory of dense linear order without endpoints. The signature for this theory is $(\leq, =)$. We write $s < t$ as an abbreviation for $s \leq t \wedge s \neq t$ and $s \geq t$ for $t \leq s$.

Recall that a binary relation $\leq$ on a set A is a *partial order* if it is reflexive, antisymmetric, and transitive. These properties are expressed by the first-order formulas

$$\forall x \; x \leq x$$
$$\forall x \; \forall y \; (x \leq y \wedge y \leq x) \leftrightarrow x = y$$
$$\forall x \; \forall y \; \forall z \; (x \leq y \wedge y \leq z) \rightarrow x \leq z,$$

respectively. The usual axioms of equality

$$\forall x \; x = x$$
$$\forall x \; \forall y \; x = y \rightarrow y = x$$
$$\forall x \; \forall y \; \forall z \; (x = y \wedge y = z) \rightarrow x = z$$

follow from these.

In addition, $\leq$ is a *linear* or *total order* if

$$\forall x \; \forall y \; x \leq y \vee y \leq x.$$

Thus the elements of the set A are lined up in a straight line. A linear order $\leq$ is *dense* if there is a point strictly between any pair of distinct points:

$$\forall x \; \forall z \; (x < z \quad \rightarrow \quad \exists y \; x < y \wedge y < z).$$

The order has no endpoints if

$$\forall x \; (\exists y \; x < y \wedge \exists z \; z < x).$$

The rational numbers $\mathbb{Q}$ and real numbers $\mathbb{R}$ with their natural orders are examples of dense linear orders without endpoints. The integers $\mathbb{Z}$ with their natural order are not, as they are not dense.

Back and Forth

The first thing to notice about dense linear orders without endpoints is that the countable ones all look the same. There is a theorem of logic called the *Löwenheim–Skolem theorem* that states that any first-order theory over a countable language that has a model of some infinite cardinality has a countable model, so we can without loss of generality restrict our attention to countable models. But for dense linear orders without endpoints, there is only one countable model up to isomorphism, and it looks like the rationals $\mathbb{Q}$. There are no finite models.

Theorem 21.1   *Any two countable dense linear orders without endpoints are isomorphic. More generally, if $\mathcal{D} = (D, \leq)$ and $\mathcal{E} = (E, \leq)$ are countable dense linear orders without endpoints, and if $a_0, \ldots, a_{k-1}$ and $b_0, \ldots, b_{k-1}$ are k-tuples of elements of $\mathcal{D}$ and $\mathcal{E}$, respectively, such that the map $f : a_i \mapsto b_i$ is a local isomorphism (that is, a one-to-one function such that $a_i \leq a_j$ iff $b_i \leq b_j$), then f extends to an isomorphism $f : \mathcal{D} \rightarrow \mathcal{E}$.*

Proof. This is a standard *back-and-forth argument*. Pick the first element $\mathcal{D}$ in some enumeration of $\mathcal{D}$ that is not among the a_i, $0 \leq i \leq k-1$, and call it a_k. It lies in some relation to all the a_i; that is, for each a_i, either $a_k > a_i$ or $a_k < a_i$. Because $\mathcal{E}$ is dense and has no endpoints, there must exist an element b_k that lies in the same relationship to the b_i, $0 \leq i \leq k-1$. Pick the first such b_k in some enumeration of $\mathcal{E}$ and set $f(a_k) = b_k$. We have extended the domain of f by one and preserved the fact that it is a local isomorphism.

Now do the same thing from the other side: pick the first unmatched element b_{k+1} of $\mathcal{E}$ and find an element a_{k+1} of $\mathcal{D}$ to match with it (that is, make $f(a_{k+1}) = b_{k+1}$) preserving local isomorphism. Go back and forth like this forever. Every element of $\mathcal{D}$ and $\mathcal{E}$ is eventually matched with something on the other side, so we get an isomorphism.   □

A Decision Procedure

Let $\mathcal{D}$ be a dense linear order without endpoints. (By the result of the previous section, we might as well take $\mathcal{D}$ to be $\mathbb{Q}$.) To decide the sentence

$$\exists x \, \forall y \, \exists z \; x < y \wedge (x \geq z \vee y = z),$$

for example, it suffices to pick an arbitrary point $a \in \mathcal{D}$ and then check whether

$$\forall y \, \exists z \; a < y \wedge (a \geq z \vee y = z). \tag{21.1}$$

It does not matter which a is chosen, because all such a look the same: by Theorem 21.1, for every $a, a' \in \mathcal{D}$, there is an automorphism of $\mathcal{D}$ mapping a to a'. Now to check (21.1), we wish to check whether for all $b \in \mathcal{D}$,

$$\exists z \; a < b \wedge (a \geq z \vee b = z). \tag{21.2}$$

However, there are *essentially* only three choices for b: either less than a, equal to a, or greater than a. If we pick any b_0, b_1, and b_2 representing these three respective possibilities respectively, then any other pair (a', b') of elements of $\mathcal{D}$ looks like one of (a, b_0), (a, b_1), or (a, b_2). Therefore it suffices to check (21.2) for $b \in \{b_0, b_1, b_2\}$. Finally, to eliminate the last quantifier $\exists z$ in (21.2), we pick a $c \in \mathcal{D}$ in one of finitely many ways. For (a, b_0), because $b_0 < a$, there are essentially five ways to pick c: either $c < b_0$, $c = b_0$, $b_0 < c < a$, $c = a$, or $c > a$. For (a, b_1), there are essentially three ways to pick c, and for (a, b_2) there are five.

Once we have chosen three elements $a, b, c \in \mathcal{D}$ and know their relative order in $\mathcal{D}$, we have enough information to determine the truth or falsity of the quantifier-free part

$$a < b \wedge (a \geq c \vee b = c). \tag{21.3}$$

We did not really need to pick out actual elements a, b, c of $\mathcal{D}$. We could do the whole procedure *symbolically*. Define a *linear arrangement* of the variables $x_1, \ldots, x_k$ to be a list of the variables $x_1, \ldots, x_k$ in some order, with either $<$ or $=$ in between each adjacent pair. The linear arrangements will be the board positions of our game.

A linear arrangement determines for each pair $i, j \in \{1, \ldots, k\}$ whether $x_i < x_j$, $x_i = x_j$, or $x_i > x_j$ by transitivity. It therefore determines the truth of all atomic formulas $x_i \leq x_j$ and $x_i = x_j$. These in turn determine the truth of any quantifier-free formula with variables among $x_1, \ldots, x_k$.

For a linear arrangement α of $x_1, \ldots, x_k$ and a quantifier-free formula φ with all free variables among $x_1, \ldots, x_k$, we write

$$\alpha \models \varphi \tag{21.4}$$

if φ is true under arrangement α.

We now extend the relation $\models$ of (21.4) to formulas φ with quantifiers. For any linear arrangement α of $x_1, \ldots, x_k$, let $\text{MOVE}(\alpha)$ be the set of arrangements β of $x_1, \ldots, x_k, x_{k+1}$ obtained from α by inserting x_{k+1} in all possible ways. The set $\text{MOVE}(\alpha)$ could be as large as $2k + 1$ or as small as 3, depending on how many equal signs ($=$) appear in α.

For a formula of the form

$$\exists x_{k+1} \; Q_{k+2} x_{k+2} \; \cdots \; Q_n x_n \; \psi(x_1, \ldots, x_k, x_{k+1}, \ldots, x_n),$$

where ψ is quantifier-free and contains only variables among $x_1, \ldots, x_n$, and for α a linear arrangement of $x_1, \ldots, x_k$, define

$$\alpha \models \exists x_{k+1} \; Q_{k+2} x_{k+2} \; \cdots \; Q_n x_n \; \psi(x_1, \ldots, x_k, x_{k+1}, \ldots, x_n)$$

if there exists a $\beta \in \text{MOVE}(\alpha)$ such that

$$\beta \models Q_{k+2} x_{k+2} \; \cdots \; Q_n x_n \; \psi(x_1, \ldots, x_k, x_{k+1}, \ldots, x_n).$$

Similarly, define

$$\alpha \models \forall x_{k+1} \; Q_{k+2} x_{k+2} \; \cdots \; Q_n x_n \; \psi(x_1, \ldots, x_k, x_{k+1}, \ldots, x_n)$$

if for all $\beta \in \text{MOVE}(\alpha)$,

$$\beta \models Q_{k+2} x_{k+2} \; \cdots \; Q_n x_n \; \psi(x_1, \ldots, x_k, x_{k+1}, \ldots, x_n).$$

The basis of this inductive definition is given by (21.4).

Let α_0 be the null arrangement, and let φ be a prenex formula of the form

$$Q_1 x_1 \; \cdots \; Q_n x_n \; \psi(x_1, \ldots, x_n),$$

where ψ is quantifier-free. It is not difficult to show by induction that $\alpha_0 \models \varphi$ iff φ is true in all dense linear orders without endpoints. This gives an alternating polynomial-time algorithm for the theory of dense linear order without endpoints, as follows. The initial process will attempt to check whether

$$\alpha_0 \models Q_1 x_1 \cdots Q_n x_n \ \psi(x_1, \ldots, x_n).$$

Subsequently, a process attempting to check whether

$$\alpha \models \exists x_{k+1} \ Q_{k+2} x_{k+2} \cdots Q_n x_n \ \psi(x_1, \ldots, x_n)$$

will produce all $\beta \in \text{MOVE}(\alpha)$ using existential branching, and for each such β, check whether

$$\beta \models Q_{k+2} x_{k+2} \cdots Q_n x_n \ \psi(x_1, \ldots, x_n).$$

To check whether

$$\alpha \models \forall x_{k+1} \ Q_{k+2} x_{k+2} \cdots Q_n x_n \ \psi(x_1, \ldots, x_n),$$

universal branching is used. Once all quantifiers are eliminated, the truth of the quantifier-free part is easily determined from the arrangement of $x_1, \ldots, x_n$ in polynomial time.

We have given an alternating polynomial-time algorithm for deciding the first-order theory of dense linear orders without endpoints. The theory is also *PSPACE*-hard, as is any nontrivial first-order theory (Miscellaneous Exercise 49). We have shown

Theorem 21.2   *The first-order theory of dense linear order without endpoints is PSPACE-complete.*

Lecture 22

Complexity of the Theory of Real Addition

Among Alfred Tarski's many significant achievements, one of the most important was showing the decidability of $\text{Th}(\mathbb{R}, +, \cdot, =)$, the first-order theory of reals with addition and multiplication. This is often called the theory of *real closed fields*. This result stands in stark contrast to the undecidability of number theory $\text{Th}(\mathbb{N}, +, \cdot, =)$.

In the next two lectures, we show that the weaker theory of *real addition* $\text{Th}(\mathbb{R}, +, \leq)$, the first-order theory of the real numbers with addition $+$ and order $\leq$, is complete for the complexity class $STA(*, 2^{O(n)}, n)$.

We have not included a constant for the real number 0 in the language, but it is definable: it is the unique x such that $\forall y \; x + y = y$. The identity relation $=$ is also definable: $x = y$ iff $x \leq y \wedge y \leq x$. The element 1 is not definable, but any positive element is as good as 1 in this theory, because for any $a > 0$, the map $x \mapsto ax$ is an isomorphism of the structure $(\mathbb{R}, +, \leq)$. This says that $\mathbb{R} \vDash \varphi(1)$ iff $\mathbb{R} \vDash \exists z \; z > 0 \wedge \varphi(z)$, so we might as well assume that the constant 1 is in the language.

Ferrante and Rackoff [40] showed that the theory is decidable in exponential space, or $STA(*, 2^{O(n)}, *)$. Fischer and Rabin [42] showed that the theory is hard for $NEXPTIME$, or $STA(*, 2^{O(n)}, \Sigma 1)$. These arguments are given in [63]. Berman [14] improved both the upper and lower bounds and showed that the exact complexity is $STA(*, 2^{O(n)}, n)$. We prove the upper bound in this lecture and the lower bound in Lecture 23.

Extend the first-order language to allow multiplication by rational constants. Thus we now allow statements such as $\forall x \, \exists y \, \frac{2}{3}x + \frac{4}{5}y \leq \frac{8}{9}$. It suffices to show the upper bound for this extended language.

Definition 22.1 *An integer affine function is a function $f : \mathbb{R}^k \to \mathbb{R}$ defined by a linear polynomial with integer coefficients:*

$$f(x_1, \ldots, x_k) \;\; = \;\; c_0 + \sum_{i=1}^{k} c_i x_i$$

for some $c_0, \ldots, c_k \in \mathbb{Z}$. For such functions, define

$$\| f \| \;\; = \;\; \max_{i=0}^{k} |c_i|.$$

Let A^k be the set of all affine functions $f : \mathbb{R}^k \to \mathbb{R}$ with integer coefficients, and let $A_m^k \overset{\text{def}}{=} \{f \in A^k \mid \|f\| \leq m\}$.

As with the theory of dense linear order without endpoints, we partition the set of all k-tuples of reals into finitely many equivalence classes that determine the formulas that a given k-tuple satisfies.

Definition 22.2 *For $a, b \in \mathbb{R}^k$, define*

$$a \equiv_m^k b \;\; \overset{\text{def}}{\iff} \;\; \forall f \in A_m^k \; \mathrm{sign}(f(a)) = \mathrm{sign}(f(b)),$$

where

$$\mathrm{sign}(x) \;\; \overset{\text{def}}{=} \;\; \begin{cases} -1 & \text{if } x < 0, \\ 0 & \text{if } x = 0, \\ 1 & \text{if } x > 0. \end{cases}$$

Equivalently, $a \equiv_m^k b$ iff $a = a_1, \ldots, a_k$ and $b = b_1, \ldots, b_k$ satisfy the same linear inequalities $\sum_i c_i x_i \leq c_0$ with integer coefficients c_i of absolute value at most m.

Unlike the theory of dense linear order, the definition here depends on a parameter m, which will determine the size of constants and quantifier depth of formulas on which two $\equiv_m^k$-equivalent k-tuples are guaranteed to agree. For more complex formulas, we need to take larger m. Note that the equivalence relation $\equiv_m^k$ is finer (equates fewer elements) for larger values of m.

The following lemma is used inductively to eliminate one quantifier.

Lemma 22.3 *Let $a, b \in \mathbb{R}^k$. If $a \equiv_{2m^2}^k b$, then for all $a' \in \mathbb{R}$ there exists $b' \in \mathbb{R}$ such that $a, a' \equiv_m^{k+1} b, b'$.*

Proof. Suppose $a \equiv_{2m^2}^k b$. Let $a' \in \mathbb{R}$ be arbitrary. We would like to find $b' \in \mathbb{R}$ such that for any $f \in A_m^k$ and $|c| \leq m$,

$$\text{sign}(f(a) + ca') = \text{sign}(f(b) + cb'), \tag{22.1}$$

or equivalently,

$$\text{sign}(a' + f(a)/c) = \text{sign}(b' + f(b)/c), \tag{22.2}$$

assuming $c \neq 0$ (if $c = 0$ then (22.1) is immediate from the assumption $a \equiv_{2m^2}^k b$). Now (22.2) is equivalent to the assertion that

$$a' \leq -\frac{f(a)}{c} \quad \Leftrightarrow \quad b' \leq -\frac{f(b)}{c}.$$

In order to show the existence of such a b' for an arbitrarily chosen a', it suffices to show that the numbers $f(a)/c$ for all affine functions $f \in A_m^k$ and nonzero c such that $|c| \leq m$ lie in the same order on the real line as the corresponding numbers $f(b)/c$. This occurs if and only if for all $f, g \in A_m^k$ and nonzero c, d such that $|c|, |d| \leq m$,

$$\frac{f(a)}{c} \leq \frac{g(a)}{d} \quad \Leftrightarrow \quad \frac{f(b)}{c} \leq \frac{g(b)}{d},$$

or in other words,

$$\text{sign}(df(a) - cg(a)) = \text{sign}(df(b) - cg(b)). \tag{22.3}$$

But the affine function

$$h(x) = df(x) - cg(x)$$

satisfies $\|h\| \leq 2m^2$, so (22.3) follows from the assumption $a \equiv_{2m^2}^k b$. $\square$

Let σ be an $\equiv_{2m^2}^k$-equivalence class and and τ an $\equiv_m^{k+1}$-equivalence class. We say that τ is *consistent with* σ if there exist $a \in \mathbb{R}^k$ and $a' \in \mathbb{R}$ such that $a \in \sigma$ and $a, a' \in \tau$.

Lemma 22.4 *Let $a \in \mathbb{R}^k$ and σ the $\equiv_{2m^2}^k$-equivalence class of a. The set*

$$\{(a, f(a)/c) \mid f \in A_{2m^2}^k, \ |c| \leq 2m^2\}$$

contains a representative of every $\equiv_m^{k+1}$-equivalence class consistent with σ.

Proof. Let $a', b' \in \mathbb{R}$. Then $a, a' \equiv_m^{k+1} a, b'$ iff for all nonzero c such that $|c| \le m$ and $f \in A_m^k$,

$$\text{sign}(ca' + f(a)) = \text{sign}(cb' + f(a));$$

in other words, $a' \le -f(a)/c$ iff $b' \le -f(a)/c$. This says that all $\equiv_m^{k+1}$-equivalence classes are represented by a, a', where a' is either

(i)   a rational number $f(a)/c$ for $|c| \le m$ and $f \in A_m^k$;

(ii) a rational number contained in an interval strictly between adjacent rational numbers $f(a)/c$ and $g(a)/d$ for $|c|, |d| \le m$, $c, d \ne 0$, and $f, g \in A_m^k$; or

(iii) a rational number strictly less than the smallest rational number of the form $f(a)/c$ for $|c| \le m$ and $f \in A_m^k$, or strictly greater than the largest such rational number.

For (i), take $a' = f(a)/c$. For (ii), take the midpoint of the interval:

$$a' = \tfrac{1}{2}(f(a)/c + g(a)/d) = \frac{f(a)d + g(a)c}{2cd}.$$

Letting $h(a) = f(a)d + g(a)c$, we have $h \in A_{2m+1}^k$ and $|2cd| \le 2m^2$. For (iii), take $a' = f(a)/c - 1 = (f(a) - c)/c$ or $a' = f(a)/c + 1 = (f(a) + c)/c$.

$\square$

Let us say a formula with variables $x = x_1, \dots, x_k$ is in *reduced form* if it is in prenex form and all atomic formulas are linear inequalities of the form $f(x) \ge 0$ for some affine function f with integer coeffients. Any formula can be put into reduced form efficiently and without significant increase in size, so we assume henceforth that all formulas are of this form.

Define

$$r(0, \ell) \stackrel{\text{def}}{=} \ell \qquad r(n+1, \ell) \stackrel{\text{def}}{=} 2r(n, \ell)^2.$$

It follows inductively that $r(n, \ell) = 2^{2^n - 1} \ell^{2^n}$, which is $2^{2^{O(n + \log \log \ell)}}$. Thus a number whose absolute value is bounded by $r(n, \ell)$ can be written down in binary with exponentially many bits.

Lemma 22.5 *If $a, b \in \mathbb{R}^k$ and $a \equiv_{r(n, \ell)}^k b$, then for any formula $\varphi(x)$ in reduced form with free variables $x = x_1, \dots, x_k$ and at most n quantifiers such that all integer constants are of size at most ℓ,*

$$\mathbb{R} \vDash \varphi(a) \quad \Leftrightarrow \quad \mathbb{R} \vDash \varphi(b).$$

Proof. The proof is by induction on the structure of the formula. For atomic formulas, the result is immediate from the definition of $\equiv_\ell^k$. For formulas of the form $\varphi \wedge \psi$, the argument is by straightforward appeal to the induction hypothesis for the simpler formulas φ and ψ:

$$\mathbb{R} \vDash \varphi(a) \wedge \psi(a)$$
$$\Leftrightarrow \quad \mathbb{R} \vDash \varphi(a) \text{ and } \mathbb{R} \vDash \psi(a)$$
$$\Leftrightarrow \quad \mathbb{R} \vDash \varphi(b) \text{ and } \mathbb{R} \vDash \psi(b)$$
$$\Leftrightarrow \quad \mathbb{R} \vDash \varphi(b) \wedge \psi(b).$$

The argument for the other propositional operators is equally straightforward. Finally, for formulas $\exists x \; \varphi(a, x)$, suppose $\mathbb{R} \vDash \exists x \; \varphi(a, x)$, and let $a' \in \mathbb{R}$ be such that $\mathbb{R} \vDash \varphi(a, a')$. By Lemma 22.3, there exists b' such that

$$a, a' \quad \equiv_{r(n-1, \ell)}^{k+1} \quad b, b',$$

and by the induction hypothesis, $\mathbb{R} \vDash \varphi(b, b')$, thus

$$\mathbb{R} \quad \vDash \quad \exists x \; \varphi(b, x).$$

The case where the leading quantifier is $\forall$ is similar. $\qquad\square$

Theorem 22.6 $\mathrm{Th}(\mathbb{R}, +, \leq) \; \in \; STA(*, 2^{O(n)}, n).$

Proof. We describe an alternating Turing machine that accepts the set of true formulas of the theory of real addition. The machine M runs in time 2^n and makes n alternations on formulas of length n.

The first step is to put the formula into reduced form. This can be done in polynomial time and entails no significant increase in size. Let n and ℓ be the quantifier depth and maximum absolute value of a constant in the formula, respectively.

Suppose the first quantifier is $\exists x_1$. Using existential branching, choose a rational number $a_1 = d/c$, where $|c|, |d| \leq r(n, \ell)$. If the first quantifier is $\forall$, do the same thing with universal branching. By the bound on $|c|$ and $|d|$, this takes exponential time. By Lemma 22.4, the leaves of the computation tree have a representative of each $\equiv_{r(n, \ell)}^1$-equivalence class.

We eliminate the second quantifier $Q_2 x_2$ by choosing $f \in A_{r(n-1, \ell)}^1$ and c such that $|c| \leq r(n-1, \ell)$ either existentially or universally according as Q_2 is $\exists$ or $\forall$, respectively, and let $a_2 = f(a_1)/c$. By Lemma 22.4, the leaves of the computation tree have a representative a_1, a_2 of each $\equiv_{r(n-1, \ell)}^2$-equivalence class consistent with the $\equiv_{r(n, \ell)}^1$-equivalence class of a_1.

Continuing in this fashion, after eliminating all quantifiers, we have chosen a sequence $a = a_1, \ldots, a_n$ representing some $\equiv_\ell^n$-equivalence class.

We then evaluate the quantifier-free part of the formula on a, which takes exponential time.

The correctness of this alternating algorithm follows from Lemma 22.5.

$\square$

Lecture 23

Lower Bound for the Theory of Real Addition

Last time we proved that $\text{Th}(\mathbb{R}, +, \leq)$, the theory of real addition, is contained in the complexity class $STA(*, 2^{n^{O(1)}}, n)$. In this lecture we show that the theory is hard for this complexity class as well. We actually only show hardness for $STA(*, 2^{n^{O(1)}}, \Sigma 1) = NEXPTIME$, as most of the main ideas are already contained there. This result is originally due to Fischer and Rabin [42], and the improvement to $STA(*, 2^{n^{O(1)}}, n)$ is due to Berman [14].

The idea is to encode full arithmetic—addition and multiplication—on integers up to a certain magnitude with short formulas of $\text{Th}(\mathbb{R}, +, \leq)$.

The language does not contain a predicate that picks out the integers, and there is no way to define such a predicate. Quantifiers range over reals, and we have no way of saying "x is an integer" in general. However, we can say that "x is an integer in the range $0 \leq x \leq n$" for any large constant n. In fact, the formula

$$x = 0 \ \lor \ x = 1 \ \lor \ x = 1 + 1 \ \lor \ \cdots \ \lor \ x = \underbrace{1 + \cdots + 1}_{n} \tag{23.1}$$

does this, albeit very inefficiently. We can even encode multiplication on integers in this range in the same way, just by writing down the multiplication table. However, this is too inefficient an encoding to get a decent lower complexity bound.

To get the lower complexity bound we want, we show how to encode full arithmetic on integers in the range $0 \leq x < 2^{2^n}$ with formulas of length $O(n)$. This allows us to manipulate bit strings of length 2^n with short formulas and index into them to extract their bits, which in turn allows us to describe computation histories of exponential-time machines, thereby encoding the halting problem for such machines.

Constructing Short Formulas for Large Integers

We make use of a trick due to Fischer and Rabin [42] for constructing short formulas for full arithmetic on large integers. We illustrate the trick by constructing a formula of length $O(n)$ that says $x = 2^{2^n}$. We actually construct inductively a formula $\psi_n(x, z)$ that says $x = 2^{2^n} z$. Then $x = 2^{2^n}$ is expressed by $\psi_n(x, 1)$.

Here is an inductive construction that produces formulas that are exponentially smaller than (23.1), although still too big:

$$\psi_0(x, z) \quad \overset{\text{def}}{\Longleftrightarrow} \quad x = z + z, \tag{23.2}$$

$$\psi_{n+1}(x, z) \quad \overset{\text{def}}{\Longleftrightarrow} \quad \exists y \; \psi_n(x, y) \wedge \psi_n(y, z). \tag{23.3}$$

Note that we have used only addition $+$, thus the ψ_n are syntactically correct. Under this definition, the ψ_n express the desired property:

$$\psi_0(x, z) \quad \Leftrightarrow \quad x = z + z \quad \Leftrightarrow \quad x = 2^{2^0} z,$$

and by induction,

$$\begin{aligned}
\psi_{n+1}(x, z) \quad &\Leftrightarrow \quad \exists y \; \psi_n(x, y) \wedge \psi_n(y, z) \\
&\Leftrightarrow \quad \exists y \; x = 2^{2^n} y \wedge y = 2^{2^n} z \\
&\Leftrightarrow \quad x = 2^{2^n} 2^{2^n} z \\
&\Leftrightarrow \quad x = 2^{2^{n+1}} z.
\end{aligned}$$

The problem is that ψ_n appears twice in the inductive definition (23.3) of ψ_{n+1}, so the length of ψ_{n+1} is roughly twice the length of ψ_n. Thus the formulas ψ_n grow exponentially with n.

Here is where the trick comes in. We use universal quantification to write ψ_{n+1} with only one occurrence of ψ_n. Instead of the definition (23.3), we take

$$\psi_{n+1}(x, z) \quad \overset{\text{def}}{\Longleftrightarrow}$$
$$\exists y \; \forall u \; \forall v \; ((u = x \wedge v = y) \vee (u = y \wedge v = z)) \to \psi_n(u, v).$$

This says the same thing, but now ψ_{n+1} contains only one occurrence of ψ_n, so its length is that of ψ_n plus a constant. Thus the formulas ψ_n grow only linearly with n.

Encoding Multiplication

Let

$$I_n \overset{\text{def}}{=} \{z \mid z \text{ is an integer and } 0 \le z < 2^{2^n}\}.$$

We now show how to construct a formula of length $O(n)$ that says, "$z \in I_n$ and $x = yz$." The basis is given by:

$$\text{MULT}_0(x, y, z) \overset{\text{def}}{\Longleftrightarrow} (z = 0 \wedge x = 0) \vee (z = 1 \wedge x = y). \qquad (23.4)$$

Before we give the induction step, let us prove a motivating lemma.

Lemma 23.1 $I_{n+1} = \{z_1 z_2 + z_3 + z_4 \mid z_i \in I_n,\ 1 \le i \le 4\}.$

Proof. ($\supseteq$) The sum of four integers in the range $0 \le z_i < 2^{2^n}$ must be an integer and cannot be less than 0 nor more than

$$(2^{2^n} - 1)(2^{2^n} - 1) + (2^{2^n} - 1) + (2^{2^n} - 1) = 2^{2^{n+1}} - 1.$$

($\subseteq$) If $0 \le z < 2^{2^{n+1}} = 2^{2^n} 2^{2^n}$, and if we divide z by 2^{2^n} using integer division with remainder, we obtain $z = 2^{2^n} q + r$, $0 \le q < 2^{2^n}$, and $0 \le r < 2^{2^n}$. Thus

$$z = (2^{2^n} - 1)q + q + r.$$

$\square$

This gives us a way of defining MULT_{n+1} inductively in terms of MULT_n. By Lemma 23.1,

$$z \in I_{n+1} \wedge x = yz$$
$$\Leftrightarrow\ \exists z_1\ \exists z_2\ \exists z_3\ \exists z_4$$
$$\bigwedge_{i=1}^{4} z_i \in I_n\ \wedge\ z = z_1 z_2 + z_3 + z_4\ \wedge\ x = yz_1 z_2 + yz_3 + yz_4$$
$$\Leftrightarrow\ \exists z_1\ \exists z_2\ \exists z_3\ \exists z_4\ \exists s\ \exists t\ \exists u\ \exists v\ \exists w\ \ z = s + z_3 + z_4\ \wedge\ x = u + v + w$$
$$\wedge \bigwedge_{i=1}^{4} z_i \in I_n\ \wedge\ s = z_1 z_2\ \wedge\ t = yz_1\ \wedge\ u = tz_2$$
$$\wedge\ v = yz_3\ \wedge\ w = yz_4$$
$$\Leftrightarrow\ \exists z_1\ \exists z_2\ \exists z_3\ \exists z_4\ \exists s\ \exists t\ \exists u\ \exists v\ \exists w\ \ z = s + z_3 + z_4\ \wedge\ x = u + v + w$$
$$\wedge\ \text{MULT}_n(s, z_1, z_2)\ \wedge\ \text{MULT}_n(t, y, z_1)\ \wedge\ \text{MULT}_n(u, t, z_2)$$
$$\wedge\ \text{MULT}_n(v, y, z_3)\ \wedge\ \text{MULT}_n(w, y, z_4).$$

The last of these formulas is semantically correct but way too large, because it involves five occurrences of MULT_n; however, we can combine these into one occurrence using the Fischer–Rabin trick. We define $\text{MULT}_{n+1}(x, y, z)$ to be the resulting formula.

Once we have defined MULT_n, we can express membership in the set I_n with the formula

$$I_n(x) \stackrel{\text{def}}{\Longleftrightarrow} \text{MULT}_n(x, 1, x).$$

Bitstring Manipulation

Once we know how to define and manipulate large integers with short formulas, we are on our way to manipulating long bitstrings. Here are some useful formulas that will help us with this task.

- Integer division with remainder: "$y, q, r \in I_n$, q is the quotient and r the remainder obtained when dividing x by y" = "$y, q, r \in I_n \wedge x = yq + r \wedge 0 \le r < y$":

$$
\begin{aligned}
\text{INTDIV}_n(x, y, q, r) \stackrel{\text{def}}{\Longleftrightarrow} \ & I_n(q) \wedge I_n(r) \wedge \exists u \, \text{MULT}_n(u, q, y) \\
& \wedge \, x = u + r \wedge 0 \le r < y \\
\text{REM}_n(x, y, r) \stackrel{\text{def}}{\Longleftrightarrow} \ & \exists q \, \text{INTDIV}_n(x, y, q, r).
\end{aligned}
$$

- "$x, y \in I_n$ and y divides x":

$$\text{DIV}_n(y, x) \stackrel{\text{def}}{\Longleftrightarrow} \text{REM}_n(x, y, 0).$$

- "$x \in I_n$ and x is even":

$$\text{EVEN}_n(x) \stackrel{\text{def}}{\Longleftrightarrow} \text{DIV}_n(2, x).$$

Here 2 is an abbreviation for $1 + 1$.

- "$p \in I_n$ and p is prime":

$$\text{PRIME}_n(p) \stackrel{\text{def}}{\Longleftrightarrow} I_n(p) \wedge \forall z \, \text{DIV}_n(z, p) \rightarrow (z = 1 \vee z = p).$$

- "y has no odd factors less than 2^{2^n} except 1":

$$\text{POWER2}_n(y) \stackrel{\text{def}}{\Longleftrightarrow} \forall z \, \text{DIV}_n(z, y) \rightarrow (z = 1 \vee \text{EVEN}_n(z)).$$

For numbers $y \in I_n$, this says that y is a power of 2.

- "$x, y \in I_n$, y is a power of two, say 2^k, and the kth bit of the binary representation of x is 1":

$$\text{BIT}_n(x, y) \overset{\text{def}}{\Longleftrightarrow} I_n(x) \wedge I_n(y) \wedge \text{POWER2}_n(y)$$
$$\wedge \, \forall q \, \forall r \, (\text{INTDIV}_n(x, y, q, r) \rightarrow \neg\text{EVEN}_n(q)).$$

Here is an explanation of the formula $\text{BIT}_n(x, y)$. Suppose x and y are numbers satisfying $\text{BIT}_n(x, y)$. Because y is a power of two, its binary representation consists of a 1 followed by a string of zeros. The formula $\text{BIT}_n(x, y)$ is true precisely when x's bit in the same position as the 1 in y is 1. We get hold of this bit in x by dividing x by y using integer division; the quotient q and remainder r are the binary numbers illustrated. The bit we are interested in is 1 iff q is odd.

$$
\begin{aligned}
y &= \quad\quad\quad 1\,0\,0\,0\,0\,0\,0\,0\,0\,0\,0\,0 \\
x &= \underbrace{1\,1\,0\,1\,1\,0\,0\,1}_{q}\,\underbrace{0\,1\,0\,0\,0\,1\,0\,1\,1\,0\,1\,1}_{r}.
\end{aligned}
$$

This formula is useful for treating numbers as bit strings of exponential length and indexing into them with other numbers to extract bits.

Now we can use this capability to write a formula $\varphi_{M,x}(w)$ that expresses the fact that w is a bitstring of exponential length encoding an accepting computation history of a nondeterministic exponential-time-bounded Turing machine M on input x. Then M accepts x iff $\mathbb{R} \models \exists w \, \varphi_{M,x}(w)$. The accepting computation history is a sequence of exponentially many configurations of M beginning with the start configuration on input x and ending with an accepting configuration such that each successive configuration follows from the previous according to the transition rules of M. A string of exponential length is used as a yardstick to compare corresponding bits of adjacent configurations. The construction of $\varphi_{M,x}(w)$ is fairly standard using the tools we have provided; for a detailed account, see the proof of Gödel's incompleteness theorem as given in Lectures 38 and 39 of [76]. The only difference here is that because of the bound on the length of strings we can talk about, we can only encode accepting computation histories of Turing machines running in exponential time.

Lecture 24

Lower Bound for Integer Addition

The theory of integer addition, $\text{Th}(\mathbb{Z}, +, \leq)$, also known as *Presburger arithmetic*, is complete for the complexity class $STA(*, 2^{2^{n^{O(1)}}}, n)$—one exponential up from $\text{Th}(\mathbb{R}, +, \leq)$. In this lecture we describe how to obtain the lower bound.

The decidability of Presburger arithmetic is due to Presburger [98]. The precise upper and lower bounds are due to Berman [14], based on work of Fischer and Rabin [42], Cooper [33], Oppen [91], and Ferrante and Rackoff [41].

As with $\text{Th}(\mathbb{R}, +, \leq)$, the trick is to encode full arithmetic—addition and multiplication—on large integers with short formulas. Intepreting integers as bit strings, this enables us to describe accepting computation histories of time-bounded Turing machines.

The main difference between $\text{Th}(\mathbb{Z}, +, \leq)$ and $\text{Th}(\mathbb{R}, +, \leq)$ is that quantifiers in $\text{Th}(\mathbb{Z}, +, \leq)$ range over $\mathbb{Z}$, so we do not have to encode this as we did with $\text{Th}(\mathbb{R}, +, \leq)$. This allows us to encode full arithmetic on integers in the range $0 \leq x < 2^{2^{2^n}}$ with formulas of length $O(n)$, which in turn allows us to encode computation histories of Turing machines running in double-exponential time.

Encoding Huge Numbers

Recall from Lecture 23 the definition

$$I_n = \{z \mid z \text{ is an integer and } 0 \le z < 2^{2^n}\}.$$

Let

$$\ell_n = \prod_{\substack{p \in I_n \\ p \text{ prime}}} p,$$

the product of all the primes less than 2^{2^n}. By [51, Theorem 414, p. 341], $\ell_n \ge 2^{c2^{2^n}}$ for some constant $c > 0$; so for $n \ge 1 + \log \log \frac{1}{c}$, $\ell_n \ge 2^{2^{2^{n-1}}}$. We can define the number ℓ_n with a short formula:

$$L_n(x) \overset{\text{def}}{\iff} x \ge 1 \ \wedge \ \forall p \ (\text{PRIME}_n(p) \ \to \ \text{DIV}_n(p, x))$$
$$\wedge \ \forall y \ge 1 \ (\forall p \ (\text{PRIME}_n(p) \ \to \ \text{DIV}_n(p, y))) \ \to \ x \le y.$$

In other words, ℓ_n is the least positive integer divisible by all primes less than 2^{2^n}. Once we have defined ℓ_n, we can say that x is an integer in the range $0 \le x < \ell_n$ by just saying $\exists \ell \ L_n(\ell) \ \wedge \ 0 \le x < \ell$. Here we are also using the fact that variables range over integers, which was not true with $\text{Th}(\mathbb{R}, +, \le)$.

Chinese Remaindering

The Chinese remainder theorem (Theorem B.1) says that we can do arithmetic on numbers modulo ℓ_n by doing arithmetic on their remainders modulo primes less than 2^{2^n}.

We can express that r is the remainder of x modulo a prime $p < 2^{2^n}$ with the predicate $\text{REM}_n(x, p, r)$ defined in Lecture 23. However, to get the higher complexity bound, we must amend the definition slightly to avoid saying anything about the size of x. We therefore redefine

$$\text{INTDIV}_n(x, y, q, r)$$
$$\overset{\text{def}}{\iff} \exists u \ \text{MULT}_n(u, q, y) \ \wedge \ x = u + r \ \wedge \ 0 \le r < y. \tag{24.1}$$

Here we are taking advantage of the fact that quantifiers range over integers, so that we do not have to include the predicates $I_n(q)$ and $I_n(r)$. We also amend the definition of $\text{BIT}_n(x, y)$ to replace the subexpression $I_n(x) \wedge I_n(y)$ with $x < \ell_n \wedge y < \ell_n$. The definitions of the other predicates—$\text{REM}_n(x, y, r)$, $\text{DIV}_n(y, x)$, and so on—are the same. The inductive definition of $\text{MULT}_n(u, q, y)$ given in Lecture 23 constrains y to be in I_n, but does not say anything about u and q. Thus $\text{INTDIV}_n(x, y, q, r)$, defined as in (24.1), says nothing about the size of x or q. Consequently, $\text{DIV}_n(y, x)$, although constraining y to be in I_n, says nothing about x.

Write $x \equiv y \pmod{p}$ if x and y are congruent modulo p; that is, if p divides $x - y$. The Chinese remainder theorem says that $x \equiv yz \pmod{\ell_n}$ iff for all primes $p < 2^{2^n}$, $x \equiv yz \pmod{p}$. For $p < 2^{2^n}$, we can express $x \equiv yz \pmod{p}$ using MULT_n and DIV_n:

$$\mathrm{MULT}_n^{\mathrm{mod}}(x, y, z, p)$$

$$\overset{\mathrm{def}}{\Longleftrightarrow} \quad \exists u\; \exists v\; \mathrm{MULT}_n(u, y, v) \;\wedge\; \mathrm{DIV}_n(p, x - u) \;\wedge\; \mathrm{DIV}_n(p, z - v).$$

This does not place any size constraints on x, y, or z, because MULT_n does not care how big its first two arguments are (it does care how big its third argument is, which is why we need the v), and DIV_n does not care how big its second argument is. Also, we have used the subtraction operator in the above definition, but we can express subtraction because we can express addition:

$$z = x - y \quad \overset{\mathrm{def}}{\Longleftrightarrow} \quad x = y + z.$$

We can now define a multiplication predicate that expresses $x \equiv yz \pmod{\ell_n}$:

$$\mathrm{MULT}'_n(x, y, z) \quad \overset{\mathrm{def}}{\Longleftrightarrow} \quad \forall p\; \mathrm{PRIME}_n(p) \;\rightarrow\; \mathrm{MULT}_n^{\mathrm{mod}}(x, y, z, p).$$

This is all the machinery we need to do full arithmetic on numbers in the range $0 \leq x \leq 2^{2^{2^n}}$ with formulas of length $O(n)$. The rest of the construction is the same as in Lecture 23.

Lecture 25

Automata on Infinite Strings and S1S

Here is a logical theory that is decidable, but not in *elementary time*; that is, not in any time bounded by a stack of exponentials

$$2^{2^{2^{\cdots^{2^n}}}}$$

of any fixed height. It is the *monadic second-order theory of successor* (S1S). This is the set of true statements about the natural numbers $\omega = \{0, 1, 2, \ldots\}$ expressible in a language with a symbol s for the successor function and allowing quantification over elements and subsets of ω. The term *second-order* refers to the fact that we allow quantification over relations, not just individual elements. The term *monadic* means that we allow quantification only over sets (monadic or unary relations). If we allow quantification over dyadic (binary) relations, the theory becomes undecidable. The theory S1S was originally shown to be decidable by Büchi [23, 24]. The result was generalized to the monadic second-order theory of n successors (SnS) by Rabin [99].

The structure in question is $(\omega, s, \in)$, where ω is the set of natural numbers, s is the successor function $x \mapsto x+1$, and $\in \subseteq \omega \times 2^\omega$ is the usual membership relation between elements and sets. The variables $x, y, z, \ldots$ range over elements and $X, Y, Z, \ldots$ range over sets, and quantification is allowed over both types of variables.

Here are some predicates definable in this language.

- "$x = y$": $\forall X \; x \in X \leftrightarrow y \in X$.

- "$X \subseteq Y$": $\forall x \; x \in X \rightarrow x \in Y$.

- "$X = Y$": $X \subseteq Y \land Y \subseteq X$.

- "$x = 0$": $\forall y \; \neg(x = \mathsf{s}y)$; in other words, x has no predecessor.

- "$x = 1$": $x = \mathsf{s}0$. Because we can define 0, we can use 0 informally as if it were a symbol in the language.

- "$x \leq y$":

$$\forall X \; (x \in X \land (\forall z \; z \in X \rightarrow \mathsf{s}z \in X)) \rightarrow y \in X.$$

  In other words, any set X that contains x and is closed under successor also contains y.

- "X is finite": $\exists x \; \forall y \; y \in X \rightarrow y \leq x$.

Automata on Infinite Strings

We prove the decidability of S1S using automata on infinite strings. We define three types of such automata, namely *Büchi*, *Rabin*, and *Muller*. These automata are similar in most respects, but differ in the form of their acceptance conditions.

Let Σ be a finite alphabet, and let Σ^ω denote the set of infinite strings $a_0 a_1 a_2 \cdots$ over Σ. If $|\Sigma| \geq 2$, then Σ^ω is uncountable—it contains as many elements as there are real numbers. These are the inputs to our automata.

A *nondeterministic Büchi automaton* is a structure

$$M \;=\; (Q, \Sigma, \delta, s, F),$$

where Q is a finite set of states, $\delta \subseteq Q \times \Sigma \times Q$ is the transition relation, $s \in Q$ is the start state, and $F \subseteq Q$ is the set of accept states.

The transition $(p, a, q) \in \delta$ means that M can move from state p to state q under input symbol a. The machine M is *deterministic* if δ is single-valued; that is, if it is equivalent to a function $\delta : Q \times \Sigma \rightarrow Q$.

A *run* of M on input $a_0 a_1 a_2 \cdots \in \Sigma^\omega$ is a sequence $q_0 q_1 q_2 \cdots \in Q^\omega$ such that

- $q_0 = s$;

- $(q_i, a_i, q_{i+1}) \in \delta$, $i \geq 0$.

If M is deterministic, there is a unique run on each input string. If M is nondeterministic, there could be anywhere from zero to uncountably many runs on a given input string.

The *IO set* of a run σ, denoted $\mathrm{IO}(\sigma)$, is the set of states of Q that appear in σ infinitely often. Formally, if $\sigma = q_0 q_1 q_2 \cdots$, then

$$\mathrm{IO}(\sigma) \overset{\text{def}}{=} \{q \in Q \mid q = q_i \text{ for infinitely many } i\},$$

or equivalently,

$$\mathrm{IO}(\sigma) \overset{\text{def}}{=} \bigcap_{n \geq 0} \{q_i \mid i \geq n\}.$$

A run of a Büchi automaton is an *accepting run* if its IO set intersects F; that is, if $\mathrm{IO}(\sigma) \cap F \neq \varnothing$. A string is *accepted by M* if it has an accepting run. The set of strings in Σ^ω accepted by M, denoted $L(M)$, is the set of strings that have accepting runs:

$$L(M) \overset{\text{def}}{=} \{x \in \Sigma^\omega \mid \text{there is a run } \sigma \text{ of } M \text{ on } x \text{ such that } \\ \mathrm{IO}(\sigma) \cap F \neq \varnothing\}.$$

Example 25.1 Here is a deterministic Büchi automaton that accepts the set

$$\{x \in \{a,b,c\}^\omega \mid \text{every } a \text{ is followed sometime later by a } b\}$$
$$= c^\omega + (a+b+c)^* b (b+c)^\omega + ((a+c)^* b)^\omega.$$

(The short arrow indicates the start state and a circle around a state indicates that it is an accept state.)   □

Example 25.2 Here is a nondeterministic Büchi automaton that accepts the set $(0+1)^* 0^\omega$, the set of infinite strings over the alphabet $\{0,1\}$ with only finitely many occurrences of 1.

$$0,1 \quad \bigcirc \xrightarrow{ 0 } \circledcirc \quad 0$$

□

This was the original definition used by Büchi to prove that S1S is decidable. Unfortunately, the acceptance condition is not quite general enough for our needs, because it turns out that nondeterministic and deterministic Büchi automata are not equivalent. For example, the set of Example 25.2 is not accepted by any deterministic Büchi automaton (Homework 8, Exercise 1).

Muller Automata

To correct this inadequacy, we generalize the acceptance condition as follows. Instead of a *set* of accept states F, we designate a *set of sets* of states $\mathcal{F}$; that is, $\mathcal{F} \subseteq 2^Q$. We think of the sets in $\mathcal{F}$ as the "good" IO sets. We define a run σ to be *accepting* if $\mathrm{IO}(\sigma) \in \mathcal{F}$. The new acceptance condition is called *Muller acceptance*, and this type of automaton is called a *Muller automaton* [89].

Every Büchi automaton is a Muller automaton: take $\mathcal{F} = \{A \subseteq Q \mid A \cap F \neq \varnothing\}$. We show in Lectures 26 and 27 that nondeterministic Büchi automata, nondeterministic Muller automata, and deterministic Muller automata are all equivalent. For the remainder of this lecture, we assume this has been done. We use the equivalence between deterministic and nondeterministic Muller automata to eliminate existential quantifiers in S1S.

Encoding S1S with Automata

Büchi showed that S1S was decidable by reducing it to the emptiness problem for nondeterministic Büchi automata [23, 24]. In fact, S1S and nondeterministic Büchi automata are equivalent in expressive power.

To make sense of this statement, let us represent a set $A \subseteq \omega$ by its *characteristic string*, the infinite string over $\{0, 1\}$ with a 1 in position i iff $i \in A$. For example, the characteristic strings of the sets {multiples of 3} and {primes} are

$$100100100100100100100\cdots,$$
$$001101010001010001010\cdots,$$

respectively. We represent an element $a \in \omega$ as we would the singleton $\{a\}$. For example, we would represent the number 4 as the string

$$000010000000000000000\cdots.$$

For a tuple $a_1, \ldots, a_n, A_1, \ldots, A_m \in \omega^n \times (2^\omega)^m$, the characteristic string is a string over the alphabet $\{0, 1\}^{m+n}$. Think of the string as divided into $m + n$ tracks, each of which contains the characteristic string of one of the a_i or A_i. For example, the characteristic string of the 5-tuple

3, 5, {even numbers}, {multiples of 3}, {primes}

would be

$$
\begin{array}{l}
0\,0\,0\,1\,0\,0\,0\,0\,0\,0\,0\,0\,0\,0\,0\,0\,0\,0\,0\,0\,0\,0\,0\,0\,0\,0 \\
0\,0\,0\,0\,0\,1\,0\,0\,0\,0\,0\,0\,0\,0\,0\,0\,0\,0\,0\,0\,0\,0\,0\,0\,0\,0 \\
1\,0\,1\,0\,1\,0\,1\,0\,1\,0\,1\,0\,1\,0\,1\,0\,1\,0\,1\,0\,1\,0\,1\,0\,1\,0 \quad \cdots\ . \\
1\,0\,0\,1\,0\,0\,1\,0\,0\,1\,0\,0\,1\,0\,0\,1\,0\,0\,1\,0\,0\,1\,0\,0\,1 \\
0\,0\,1\,1\,0\,1\,0\,1\,0\,0\,0\,1\,0\,1\,0\,0\,0\,1\,0\,1\,0\,0\,0\,1\,0\,0\,0\,0
\end{array}
$$

Theorem 25.3 (i) *Let $\varphi(\overline{x}, \overline{X})$ be a formula of* S1S *with free individual variables among* $\overline{x} = x_1, \ldots, x_n$ *and free set variables among* $\overline{X} = X_1, \ldots, X_m$. *There exists a Muller automaton* M_φ *over alphabet* $\{0, 1\}^{m+n}$ *such that*

$$L(M_\varphi) = \{(\overline{a}, \overline{A}) \mid \varphi(\overline{a}, \overline{A}) \text{ is true}\},$$

where $\overline{a} = a_1, \ldots, a_n$ *and* $\overline{A} = A_1, \ldots, A_m$.

(ii) *For any nondeterministic Büchi automaton* M *over the alphabet* $\{0, 1\}$, *there is a formula* $\varphi_M(X)$ *of* S1S *such that* $\varphi_M(A)$ *is true iff* M *accepts the characteristic string of* A.

Proof. We only prove (i), leaving (ii) as an exercise.

To prove (i), we proceed by induction on the structure of the given formula φ. Any atomic formula is of the form

$$\underbrace{\mathsf{s}\,\mathsf{s}\cdots\mathsf{s}}_{k}\,x \in X$$

for some fixed $k \geq 0$. A deterministic Büchi automaton can be built to read the tracks corresponding to x and X only, checking that the track corresponding to x has a unique 1, say in position i, and that the track corresponding to X has a 1 in position $i + k$. Here is the automaton for $k = 3$:

In the label (b, c), b and c represent the input bits on the tracks corresponding to x and X, respectively, and $-$ indicates that the transition is enabled under either bit. Transitions not shown can be assumed to go to a dead state.

For a formula of the form

$$\varphi_1(\overline{x}, \overline{X}) \;\wedge\; \varphi_2(\overline{x}, \overline{X}), \tag{25.1}$$

assume that we have already constructed deterministic Muller automata $M_i = (Q_i, \Sigma, \delta_i, s_i, \mathcal{F}_i)$ for φ_i, $i \in \{1, 2\}$. To get an automaton for (25.1), we do a product construction. Let

$$M_3 = (Q_3, \Sigma, \delta_3, s_3, \mathcal{F}_3),$$

where

$$Q_3 \stackrel{\text{def}}{=} Q_1 \times Q_2,$$
$$s_3 \stackrel{\text{def}}{=} (s_1, s_2),$$
$$\delta_3((q_1, q_2), a) \stackrel{\text{def}}{=} (\delta_1(q_1, a), \delta_2(q_2, a)),$$
$$\mathcal{F}_3 \stackrel{\text{def}}{=} \{A \subseteq Q_3 \mid \pi_1(A) \in \mathcal{F}_1 \text{ and } \pi_2(A) \in \mathcal{F}_2\},$$

where π_1 and π_2 are the appropriate projections:

$$\pi_1(A) \quad \overset{\text{def}}{=} \quad \{p \in Q_1 \mid \exists q \in Q_2 \ (p, q) \in A\}$$
$$\pi_2(A) \quad \overset{\text{def}}{=} \quad \{q \in Q_2 \mid \exists p \in Q_1 \ (p, q) \in A\}.$$

For a formula of the form

$$\neg\varphi(\overline{x}, \overline{X}),$$

assume we have constructed a *deterministic* Muller automaton

$$M \ = \ (Q, \Sigma, \delta, s, \mathcal{F})$$

for φ. To get an automaton for the negation, take the complement of $\mathcal{F}$ in 2^Q.

The constructions for $\vee$, $\rightarrow$, and $\leftrightarrow$ are obtained by composing the constructions above.

For a formula of the form

$$\exists X_1 \ \varphi(\overline{x}, X_1, \ldots, X_m),$$

assume we have constructed the deterministic Muller automaton M_φ (we show how to do this in Lectures 26 and 27). Build a new automaton M that on input $a_1, \ldots, a_n, A_2, \ldots, A_m$ nondeterministically guesses A_1 and simulates M_φ on $a_1, \ldots, a_n, A_1, \ldots, A_m$.

The construction for a formula of the form

$$\exists x_1 \ \varphi(x_1, \ldots, x_n, \overline{X})$$

is similar, except that the automaton only guesses strings of the form 0^*10^ω; that is, strings containing a single 1.

For universal quantifiers $\forall$, we use the fact that $\exists x \ \varphi$ is equivalent to $\neg\forall x \ \neg\varphi$ and compose the constructions for $\neg$ and $\exists$ above.

Continuing inductively in this fashion, we can obtain a nondeterministic Büchi automaton M_φ such that $L(M_\varphi)$ is nonempty iff the sentence φ is true. For Büchi automata, nonemptiness can be tested by checking if there is an accessible loop containing a state of F.   □

In Lectures 26 and 27 we discuss the relationship between Büchi and Muller automata and give an efficient determinization construction due to Safra.

Undecidability of the Dyadic Theory

The monadic second-order theory of successor, which is decidable, allows second-order quantification only over sets (monadic predicates, unary relations). The *dyadic* theory, which allows quantification over dyadic predicates (binary relations), is undecidable. To show this we encode the Post correspondence problem (PCP), a well-known undecidable problem.

An instance of the PCP consists of a pair of monoid homomorphisms $f, g : \Sigma^* \to \Gamma^*$, where Σ and Γ are finite alphabets. Recall that a monoid homomorphism is a map that satisfies $f(xy) = f(x)f(y)$ and $f(\varepsilon) = \varepsilon$. A *solution* of the instance is a string $x \in \Sigma^*$ such that $f(x) = g(x)$. Of course, there is always the trivial solution ε; the interesting question is whether there exists a nonnull solution. This problem is undecidable, and we use this fact to show the undecidability of dyadic second-order theory of successor. See [61, §9.4] for more details and a proof of undecidability of PCP (or, if you are in a do-it-yourself mood, try Miscellaneous Exercise 96).

Let f, g be an instance of PCP with $\Gamma = \{0, 1\}$. We encode the problem as a formula of dyadic S1S as follows. Let $z \in \{0, 1\}^*$. Considering subsets $A \subseteq \omega$ as infinite-length strings $A \in \{0, 1\}^\omega$, let $\psi_z(A, i)$ be the predicate, "z is the substring of A of length $|z|$ beginning at position i." This predicate is easily expressible in our language: for example,

$$\psi_{10011}(A, i) \;=\; i \in A \wedge \mathsf{s}(i) \notin A \wedge \mathsf{s}^2(i) \notin A \wedge \mathsf{s}^3(i) \in A \wedge \mathsf{s}^4(i) \in A.$$

Now let R be a dyadic predicate (binary relation) variable, and consider the formula

$$\varphi(A, R) \;\overset{\text{def}}{=}\; R(0, 0) \;\wedge\; \forall i\, \forall j \bigwedge_{a \in \Sigma} (R(i, j) \wedge \psi_{f(a)}(A, i) \wedge \psi_{g(a)}(A, j) \\ \to\; R(\mathsf{s}^{|f(a)|}(i), \mathsf{s}^{|g(a)|}(j))).$$

If this formula is true of A, R, then $R(i, j)$ holds for any pair of numbers i, j such that for some $x \in \Sigma^*$, the prefix of A of length i is $f(x)$ and the prefix of A of length j is $g(x)$; in other words,

$$\{(|f(x)|, |g(x)|) \mid f(x), g(x) \text{ are prefixes of } A\} \;\subseteq\; R. \tag{25.2}$$

Note that for each A, the intersection of all relations R satisfying $\varphi(A, R)$ is also a relation satisfying $\varphi(A, R)$, and is the smallest such. For that relation, equality holds in (25.2). Then there is a solution iff the sentence

$$\exists A\, \forall R \;\; \varphi(A, R) \to \exists i > 0\; R(i, i)$$

is true.

Lecture 26

Determinization of ω-Automata

Rabin Automata

Last time we defined Büchi and Muller automata. Recall that in Büchi acceptance, we specify a subset $F \subseteq Q$ and call a run *accepting* if $\mathrm{IO}(\sigma) \cap F \neq \varnothing$. In Muller acceptance, we specify a set $\mathcal{F} \subseteq 2^Q$ and call a run *accepting* if $\mathrm{IO}(\sigma) \in \mathcal{F}$. Every Büchi automaton is a Muller automaton: take $\mathcal{F} = \{A \subseteq Q \mid A \cap F \neq \varnothing\}$.

A third type that lies somewhere in between is *Rabin automata*. In the Rabin acceptance condition, we specify a finite set of pairs (G_i, R_i), $1 \leq i \leq k$, where G_i and R_i are subsets of Q. Think of a green light flashing every time the machine enters a state in G_i and a red light flashing every time the machine enters a state in R_i. A run is defined to be *accepting* if for some i, the ith green light flashes infinitely often and the ith red light flashes only finitely often. Formally, a run σ is *accepting* if there exists an i such that $\mathrm{IO}(\sigma) \cap G_i \neq \varnothing$ and $\mathrm{IO}(\sigma) \cap R_i = \varnothing$.

Every Büchi automaton is a Rabin automaton: take $k = 1$, $G_1 = F$, and $R_1 = \varnothing$. In turn, every Rabin automaton is a Muller automaton: take

$$\mathcal{F} \;=\; \{A \subseteq Q \mid \bigvee_i (A \cap G_i \neq \varnothing \wedge A \cap R_i = \varnothing)\}.$$

In this lecture and the next we show that nondeterministic and deterministic Rabin automata, nondeterministic and deterministic Muller automata, and nondeterministic Büchi automata are all equivalent. As mentioned last time, deterministic Büchi automata are strictly weaker (Homework 8, Exercise 1).

We show the equivalence of all these automata by means of the inclusions in the following diagram, where the arrows mean "can be simulated by."

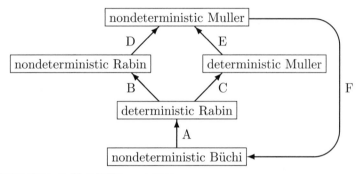

Inclusions B and E are immediate. Inclusions C and D are straightforward and have already been argued above.

For inclusion F, let M be a nondeterministic Muller automaton with acceptance set $\mathcal{F}$. Design a nondeterministic Büchi automaton N (a Rabin automaton with a single green light) that operates as follows. On any input x, N simulates the computation of M on x, guessing some run σ of M nondeterministically. At some point, N nondeterministically guesses that all states of M not in $\mathrm{IO}(\sigma)$ have already been seen for the last time. It also guesses the IO set A and verifies that $A \in \mathcal{F}$. For the remainder of the computation, it verifies that A is indeed $\mathrm{IO}(\sigma)$. It checks that from that point on,

- every state of σ is in A, and

- σ hits every state in A infinitely often.

To do this, it continues to simulate M, marking each state that M ever enters. If at any point M enters a state not in A, N just rejects. As soon as all the states of A become marked, N flashes its green light, erases all the marks, and begins the process anew.

The most difficult of the six inclusions is A. This was first shown by McNaughton in 1966 [84], but his construction was doubly exponential. A more efficient singly exponential construction was given by Safra in 1988 [107].

In order to motivate the construction, we start with an incorrect construction and show why it does not work. Our attempt to rectify the situation leads us to a second incorrect construction that errs too far in the opposite direction. Finally we give a correct construction.

First construction (incorrect)   Let $M = (Q, \Sigma, \Delta, s, F)$ be a nondeterministic Büchi automaton with n states. Recall that a nondeterministic transition function is of type $\Delta : Q \times \Sigma \to 2^Q$. Intuitively, if the machine is in state p and sees input symbol a, then it can move to any state in $\Delta(p, a)$.

The function Δ extends uniquely to

$$\widehat{\Delta} : 2^Q \times \Sigma^* \ \to \ 2^Q$$

by induction as follows: for $x \in \Sigma^*$ and $a \in \Sigma$,

$$\widehat{\Delta}(A, \varepsilon) \ \stackrel{\text{def}}{=} \ A \qquad \widehat{\Delta}(A, xa) \ \stackrel{\text{def}}{=} \ \bigcup_{q \in \widehat{\Delta}(A, x)} \Delta(q, a).$$

It follows that

$$\Delta(p, a) \ = \ \widehat{\Delta}(\{p\}, a) \qquad \widehat{\Delta}(A, xy) \ = \ \widehat{\Delta}(\widehat{\Delta}(A, x), y).$$

Build a deterministic Rabin automaton N that keeps track of all the states M could possibly be in. It starts with a single token on the start state of M. In each step, there is a set of tokens occupying the states of N, which mark all states M could possibly be in at that point according to the nondeterministic choices M has made so far. On each input symbol, N moves the tokens on M in all possible ways according to the transition relation of M. Thus N is deterministic.

Now N must somehow decide if there is a run of M that hits a state in F infinitely often. Let us make N flash a green light every time one of the tokens on M occupies a state of F. Thus N is actually a deterministic Büchi automaton.

Formally, take

$$N \ = \ (2^Q, \Sigma, \widehat{\Delta}, \{s\}, G),$$

where $\widehat{\Delta}$ is defined above, and

$$G \ \stackrel{\text{def}}{=} \ \{A \subseteq Q \mid A \cap F \neq \varnothing\}.$$

It is true that $L(M) \subseteq L(N)$, because if there is an accepting run of M (that is, a run that hits F infinitely often), then in N's simulation, a state of F is occupied by a token infinitely often, therefore N flashes green infinitely often, hence N accepts. Unfortunately, the reverse inclusion does not hold. Here is a counterexample:

(The short arrow indicates the start state and a circle indicates an accept state.) This automaton accepts no strings, because there is only one infinite run, and that run has IO set $\{s\}$, which does not contain a final state. However, the construction above gives the automaton

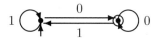

where the state on the left represents the set $\{s\}$ and the state on the right represents the set $\{s, t\}$. This automaton accepts the set of strings with infinitely many zeros.

Second construction (also incorrect)   The problem with the previous construction was that it is possible for M to hit F at infinitely many time instants, even though none of those hits occur on the same run. To remedy this, we modify the construction to make sure that when such a situation arises, we can always reconstruct an infinite run of M that hits F infinitely often.

We describe N in terms of blue and white tokens on the states of M. It starts with a single blue token on the start state of M. On each input symbol, it moves the tokens according to the transition rules of M, preserving colors; except that if a blue and white token both occupy a state M, then the blue token is removed, and if a blue token occupies a state of F, then it is replaced by a white token. Whenever all the tokens become white, N flashes green and replaces all the white tokens with blue tokens. Again, N is a deterministic Büchi automaton.

Formally, the set of states of N is $2^Q \times 2^Q$, where a state (B, W) of N represents the set of states B of M occupied by blue tokens and the set of states W of M occupied by white tokens. Define

$$N \;=\; (2^Q \times 2^Q, \Sigma, \delta, (\{s\}, \varnothing), G)$$

with

$$\delta : 2^Q \times 2^Q \times \Sigma \;\to\; 2^Q \times 2^Q$$
$$\delta((B, W), a) \stackrel{\text{def}}{=} (B', W'),$$

where

$$W' \stackrel{\text{def}}{=} \widehat{\Delta}(W, a) \cup (F \cap \widehat{\Delta}(B, a))$$
$$B' \stackrel{\text{def}}{=} \widehat{\Delta}(B, a) - W'$$

if $B \neq \varnothing$, and

$$W' \stackrel{\text{def}}{=} \widehat{\Delta}(W, a) \cap F$$
$$B' \stackrel{\text{def}}{=} \widehat{\Delta}(W, a) - F$$

if $B = \varnothing$; and

$$G \stackrel{\text{def}}{=} \{(\varnothing, W) \mid W \subseteq Q\}.$$

The idea here is that at any green flash, any token occupying any state must have gotten there via a path through a state of F since the last green flash, because that is the only way the token could have become white. If there are infinitely many green flashes, we can reconstruct an accepting run of M (one whose IO set intersects F infinitely often). We do this with the aid of *König's lemma*:

Lemma 26.1   **(König's lemma)**   *Every infinite finite-degree tree has an infinite path.*

Proof. If the root has infinitely many descendants but only finitely many children, then some child must have infinitely many descendants. Move down to that child and repeat the argument. In this way we can trace an infinite path down through the tree.   □

König's lemma is false without the degree restriction. A tree consisting of a root with countably many childen is an infinite tree, all of whose paths are finite.

Now suppose that N flashes green infinitely often on input x. At every green flash, every state occupied by a token is accessible from a state at the previous green flash by a path segment that goes through a state of F. Let these path segments be the edges of an infinite tree whose nodes are labeled with pairs (q, t), where t is a time instant at which N flashes green and q is a state of M occupied by a token at time t. By the construction, every node labeled $(q, t + 1)$ has a parent labeled (p, t), thus we have an infinite tree of finite degree. By König's lemma, there exists an infinite path in this tree, which represents an infinite run intersecting F infinitely often.

The problem here is that we have erred too far in the other direction. It is true that if N flashes green infinitely often on input x, then there is an accepting run of M on x, so $L(N) \subseteq L(M)$, but not vice versa.

Unfortunately, it is not necessarily the case that $L(M) \subseteq L(N)$. Here is a counterexample:

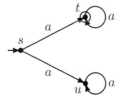

This automaton accepts a^ω, but the construction above gives

which does not. Here the state on the left represents the pair $(\{s\}, \varnothing)$ and the state on the right represents the pair $(\{u\}, \{t\})$.

Thus the first construction gives a deterministic automaton that accepts too many strings and the second too few. Next time we give a construction due to Safra [107] that achieves a happy compromise.

Lecture 27

Safra's Construction

In Lecture 26 we gave two incorrect constructions for the determinization of Büchi automata. The first constructed a deterministic automaton that accepted too many strings; the second erred in the opposite direction by accepting too few strings.

In this lecture we give a third construction due to Safra [107] that is a compromise between the first two. This construction correctly produces a deterministic Rabin automaton equivalent to a given nondeterministic Büchi automaton.

For a given nondeterministic Büchi automaton $B = (Q, \Sigma, \Delta, s, F)$ with n states, the equivalent deterministic Rabin automaton R will have $2^{O(n \log n)}$ states and n pairs in the acceptance condition.

Say we have a collection of *colors*, each with an associated *bell* and *buzzer*. There are several tokens of each color, which are placed on the states of B at various times, moved around, and sometimes removed. A *stack* is a pile of tokens on a state. The *height* of a stack σ is the number of tokens in σ and is denoted $|\sigma|$.

A token is *in play* at time t if it is in some stack on some state at time t. A color is *in play* at time t if it is the color of a token in play at time t. A color is *visible* at time t if it is the color of the top token of some stack at time t.

The colors in play at time t are ordered by *age*, which is the time they last came into play. All tokens of any one color in play will always come into play at the same time, therefore will have the same age. When we bring a new token into play, we always place it on top of a stack; when we remove a token from play, we always remove all the tokens above it; and when we move tokens around, we always move an entire stack at once. Thus it is an invariant of the simulation that the tokens on any stack are always ordered by age from the oldest on the bottom to the youngest on top.

The stacks are linearly ordered at time t as follows: $\sigma \ll_t \tau$ if either

- σ is a proper extension of τ (that is, τ can be obtained by removing tokens from the top of σ); or

- neither σ nor τ is an extension of the other and σ is lexicographically older than τ; that is, starting from the bottom and moving up, at the first position where σ and τ differ in color, the color of σ is older.

The ages of colors and the order $\ll_t$ are time-dependent, because colors can go in and out of play.

The simulation starts with a single white token on the start state s. Now assume we have stacks of colored tokens on the states of B at time t. To construct the configuration at time $t+1$, execute the following three steps. (In this construction, the intermediate configurations are transitory and do not count as states of the simulating automaton.)

Move   Suppose the next input symbol is a. For each state q, remove the stack currently on q. For each $p \in \Delta(q, a)$, clone the stack that was on q and try to put it on p. If we try to put more than one stack on p, resolve in favor of the $\ll_t$-least stack. If any color completely disappears from the board in this process, buzz its buzzer.

Cover   For each accept state $q \in F$, place a token of an unused color on top of q's stack. For this purpose we bring k unused colors into play, where k is the number of distinct visible colors on states $q \in F$. If two stacks on two different accept states have the same visible color, then we cover them with the same new color. We bring the new colors into play in some arbitrary order to determine their relative age, but the order is not important.

This is the only way new colors can come into play; thus it is an invariant of the simulation that if a token of color c in play is directly over a token of color d, then all tokens of color c in play are directly over a token of color d.

Audio Check   For every invisible color c in play, ring its bell and remove all tokens above any token of color c. Buzz the buzzers of the removed tokens. Note that if any token is removed in this process, then all tokens of that color are removed. There may be more than one invisible color in play, but the order in which we process them does not matter.

After these three steps are executed, there are at most n colors remaining in play; otherwise there must be an invisible one, contradicting the Audio Check step.

Claim 27.1 *B accepts x iff there is a color whose bell rings infinitely often but whose buzzer buzzes only finitely often.*

Proof. Suppose there is a color, say cyan, whose bell rings infinitely often but whose buzzer buzzes only finitely often. Let $t_0, t_1, \ldots$ be the times at which cyan's bell rings *after* its buzzer has already buzzed for the last time. From t_0 on, cyan is continuously in play, otherwise its buzzer would have buzzed. At the times t_i, all cyan tokens are visible. Between t_i and t_{i+1}, each cyan token gets covered with another token of a different color, because the only way cyan's bell can ring at time t_{i+1} is if cyan becomes invisible. Thus for every state q with a cyan token at time t_{i+1}, there is a segment of a run from some state with a visible cyan token at time t_i through a state of F to q. As in the previous lecture, König's lemma (Lemma 26.1) can be used to construct a run with infinitely many occurrences of states in F.

Conversely, suppose there is an accepting run ρ of B. Let σ_t be the stack on the state of ρ at time t. Let $m = \liminf |\sigma_t|$; that is, m is the maximum height such that from some point on, the stacks along ρ reach height m and then never go below height m again. Note that $m \geq 1$ because white (the oldest color) is always in play, and $m \leq n$ because there are at most n colors in play. From some point on, the stack is of height at least m, and infinitely often exactly m. After that point, the colors on the stack at height m and below may change due to being replaced by a $\ll_t$-lesser stack in the Move step, but this can happen only finitely often because lexicographic order on the age space ω^m is well-founded. So from some point on, the colors at height m and below do not change. Say the color at height m at this point is magenta. Then infinitely often after that point, the run ρ goes through a state of F, because it is an accepting run, at which point the stack acquires a new token; but sometime after that the stack must shrink back to height m again, and the only way that can happen is if magenta's bell rings. Thus magenta rings infinitely often and buzzes only finitely often. $\square$

A state of the simulating Rabin automaton R will consist of a stack of tokens (possibly empty) on each state of B, the current ordering $\ll_t$, and an indication of which bells rang. Each token configuration can be specified by a map $Q \rightarrow \{\text{colors}\} \cup \{\text{none}\}$ giving the top color of the stack on each state and a map $\{\text{colors}\} \rightarrow \{\text{colors}\} \cup \{\text{none}\}$ telling for each color in play the color immediately below it. There are at most $n!$ orderings $\ll_t$ and 2^n ways the bells can ring. Thus there are at most $n^n \cdot n^n \cdot n! \cdot 2^n = 2^{O(n \log n)}$ states in all. The acceptance condition consists of n pairs, one for each color.

We have shown

Theorem 27.2   (Safra [107])   *Every nondeterministic Büchi automaton with n states can be simulated by a deterministic Rabin automaton with at most $2^{O(n \log n)}$ states and n pairs in the acceptance condition.*

Lecture 28

Relativized Complexity

Many fundamental questions in complexity theory remain unanswered. Probably the most important of these is the $P = NP$ question. Although it is generally believed that $P \neq NP$—mainly because it is inconceivable how one might perform an exhaustive search through exponentially many candidates in polynomial time—we have up to now been unable to prove this formally.

In trying to understand this question and similar questions regarding containment and separation of complexity classes, researchers have looked at relativized versions, for which answers are somewhat easier to obtain. Fairly early on, Baker, Gill, and Solovay [10] constructed oracles A and B such that $P^A = NP^A$ and $P^B \neq NP^B$ (oracle Turing machines and relativized complexity classes were defined in Lecture 9). We present these results below.

Rise and Fall of the Random Oracle Hypothesis

Not long after the Baker–Gill–Solovay results, Bennett and Gill [12] showed that with respect to a random oracle C (every element is either in or out of C with probability $\frac{1}{2}$), $P \neq NP$ with probability 1. This and similar results for other classes agreed well with the generally accepted beliefs about what happens in the corresponding unrelativized cases. These observations led to the *random oracle hypothesis*: every containment or separation result that holds with probability 1 with respect to a random oracle also holds in the unrelativized case. Intuitively, as the argument went, one should be able to get no more information from a random oracle than from no oracle at all. This conjecture went through a series of qualifications and reformulations to rule out simple counterexamples that kept popping up, ultimately to be refuted by Chang et al. in 1994 [27]: with respect to a random oracle C, $IP^C \neq PSPACE^C$ with probability 1.

Relativized $P^A = NP^A$

Theorem 28.1   *There exists a recursive oracle A such that $P^A = NP^A$.*

Proof. This is the easier of the two relativized P versus NP results. For the oracle A we can take QBF, the set of true quantified Boolean formulas, a *PSPACE*-complete set. Then

$$PSPACE \quad \subseteq \quad P^{\text{QBF}} \quad \subseteq \quad NP^{\text{QBF}} \quad \subseteq \quad NPSPACE^{\text{QBF}}$$
$$\subseteq \quad NPSPACE \quad \subseteq \quad PSPACE.$$

The first inclusion holds because membership in a set in *PSPACE* can be determined by reducing to QBF, then consulting the oracle. The next-to-last inclusion holds because an *NPSPACE* machine has enough space to decide membership in QBF directly without consulting the oracle. The last inclusion is Savitch's theorem. $\square$

The following result is somewhat harder and requires diagonalization.

Theorem 28.2   *There exists a recursive oracle B such that $P^B \neq NP^B$.*

Proof. Let Σ contain at least two elements. Let $C \subseteq \Sigma^*$ be any oracle, and consider the set

$$E^C \quad \overset{\text{def}}{=} \quad \{x \in \Sigma^* \mid \exists y \in C \; |y| = |x|\}.$$

That is, $x \in E^C$ if C contains an element of the same length as x. The set E^C can be accepted by a nondeterministic oracle machine N with oracle C that operates as follows. On input x, guess a string y of the same length as x, then query the oracle C to determine whether $y \in C$. Regardless of the oracle, this is a nondeterministic polynomial time algorithm, therefore $E^C \in NP^C$ for any C.

Now we construct by diagonalization an oracle B such that $E^B \neq L(M^B)$ for any deterministic polynomial time oracle machine M. This says that $E^B \in NP^B - P^B$. The idea behind the construction is that a polynomial-time-bounded oracle machine M has only time enough to ask polynomially many questions of the oracle, but there are exponentially many strings of length n. This allows us to adjust the oracle by throwing in or leaving out elements that M never looks at, thereby causing M to have been wrong.

Let $M_0, M_1, \ldots$ be a list of all deterministic polynomial time bounded oracle machines. Assume without loss of generality that each M_i is equipped with a uniform clock and parameter c such that M_i shuts itself off after n^c steps; thus for each i, the time bound n^c of M_i is recognizable from the description of M_i.

We construct the oracle B as the limit of a sequence of finite approximations. Each approximation B_k is a partial function $B_k : \Sigma^* \to \{0, 1\}$ with finite domain. $B_k(x) = 1$ means $x \in B$, $B_k(x) = 0$ means $x \notin B$, and $B_k(x)$ undefined means it has not yet been determined whether $x \in B$. The approximations B_k become more defined at later stages of the construction; that is, $B_k \sqsubseteq B_{k+1}$, which means that B_{k+1} is defined wherever B_k is, and where both are defined, they take equal values.

In this construction, we maintain the invariant that, just before commencing stage k, we have successfully diagonalized away from $M_0, M_1, \ldots, M_{k-1}$; that is, we have built B_k such that for any total extension C of B_k and for any $i < k$, there exists an x such that

$$M_i^C \text{ accepts } x \quad \Leftrightarrow \quad x \notin E^C$$
$$\Leftrightarrow \quad C \text{ contains no elements of length } |x|. \qquad (28.1)$$

We start at stage 0 with B_0 completely undefined. Now say at stage k we have constructed B_k satisfying (28.1). We now look at M_k. Say the time bound of M_k is n^c. Pick n greater than the length of all elements in the domain of B_k—this is possible because the domain of B_k is finite—and large enough that $2^n > n^c$. Initially set $B_{k+1} := B_k$.

Now simulate M_k on some input of length n, say a^n. Whenever M_k attempts to query its oracle on some string y, if B_{k+1} is already defined on y, we supply the value of $B_{k+1}(y)$. If B_{k+1} is undefined on y, we set $B_{k+1}(y) := 0$ and supply the value 0. The machine runs to completion and either accepts or rejects.

Now we adjust the oracle so that M_k is guaranteed to have been wrong in its attempt to accept E^C. Note that because M_k runs in time $n^c < 2^n$, there must be at least one string of length n that was never the subject of an oracle query by M_k during the computation, therefore B_{k+1} is still undefined on those strings. If M_k accepted, we set $B_{k+1}(y) := 0$ for all remaining strings y of length n on which B_{k+1} is still undefined. In this case $B_{k+1}(y) = 0$ for all strings y of length n. If M_k rejected, we set $B_{k+1}(y) := 1$ for all remaining strings y of length n on which B_{k+1} is still undefined, and there is at least one such string. These adjustments do not affect the computation of M_k on input a^n, because they were never the subject of an oracle query.

Then for any total extension C of B_{k+1}, (28.1) holds, therefore $L(M_k^C) \neq E^C$. If we take B to be any total extension of all the B_k for $k \geq 0$, then we are guaranteed that $L(M_k^B) \neq E^B$ for any M_k. Thus $E^B \in NP^B - P^B$. $\qquad\square$

Lecture 29

Nonexistence of Sparse Complete Sets

Here are some results that lend a little insight into the $P = NP$ question. Let Σ be a finite alphabet of at least two letters. A subset $S \subseteq \Sigma^*$ is *sparse* if there is a polynomial bound on the number of elements of length n; that is, if there exists a constant $k \geq 0$ such that for all $n \geq 2$, $|S \cap \Sigma^n| \leq n^k$. A set is *dense* if it is not sparse.

Intuitively, if $P \neq NP$, we would not expect a sparse set to be NP-complete. If it were, then the few elements of the set would have to be well hidden so that they could not be found in polynomial time; but then this would make reductions difficult.

This intuition was confirmed in a series of results by Berman [15], Fortune [43], and Mahaney [82]. Berman showed that no set of strings over a single-letter alphabet can be NP-complete unless $P = NP$. Fortune showed that no sparse set can be co-NP-complete unless $P = NP$. Finally, Mahaney showed that no sparse set can be NP-complete unless $P = NP$. All known NP- and co-NP-complete sets are dense, so this confirms our expectations. Of course, if $P = NP$, then all nontrivial sets in NP, including sparse ones, are NP-complete. We prove the Fortune and Mahaney results below (Theorems 29.1 and 29.2, respectively).

The proof of Theorem 29.1 is a little easier and we start with that. The proof of Theorem 29.2 is built on Theorem 29.1 but requires an extra idea, namely *census functions*. We saw the use of census functions in the proof of the Immerman–Szelepcsényi theorem in Lecture 4, but this is where they were first introduced.

The proof of Theorem 29.1 is based on the idea of *self-reducibility*. This technique allows a polynomial-time decision procedure for Boolean satisfiability to be converted to an algorithm for *computing* a satisfying truth assignment for a given satisfiable Boolean formula. Given a satisfiable formula $\varphi(x_1, \ldots, x_n)$ of n variables, we instantiate x_1 with both truth values and use the decision procedure to determine which of $\varphi(0, x_2, \ldots, x_n)$ or $\varphi(1, x_2, \ldots, x_n)$ is satisfiable; at least one of them must be. Say it is the former. We then instantiate x_2 in both ways and ask which of $\varphi(0, 0, x_3, \ldots, x_n)$ or $\varphi(0, 1, x_3, \ldots, x_n)$ is satisfiable, and so on. Continuing in this fashion, we can instantiate all the variables while always maintaining satisfiability. The final result is a satisfying truth assignment.

Theorem 29.1   (Fortune [43])   *If there exists a sparse set that is $\leq_{\mathrm{m}}^{\mathrm{P}}$-hard for co-NP, then $P = NP$.*

Proof. Let Boolean formulas be coded as strings over a binary alphabet Γ. It is reasonable to assume of the coding that a formula of length n has at most n variables and that the length of any formula obtained by substituting Boolean values 0 or 1 for any of the variables is no greater than the length of the original formula.

Suppose $S \subseteq \Sigma^*$ is sparse with $|S \cap \Sigma^n| \leq n^k$. If in addition S is co-NP-hard, then $\sim$SAT $\leq_{\mathrm{m}}^{\mathrm{P}} S$, where $\sim$SAT is the set of unsatisfiable Boolean formulas. Let $\sigma : \Gamma^* \to \Sigma^*$ be a polynomial-time many–one reduction that reduces $\sim$SAT to S with time bound n^c. Thus for all formulas φ coded as a string in Γ^*, $\varphi \in \sim$SAT iff $\sigma(\varphi) \in S$, and if φ is of length at most n, then $\sigma(\varphi)$ is of length at most n^c.

Let S' be the set of elements of S of length n^c or less. An upper bound on the size of S' is $N = n^{c(k+1)}$. If a formula φ is of length n or less, and if φ is unsatisfiable, then $\sigma(\varphi)$ must be in S'.

We show that under these assumptions, there exists a deterministic polynomial-time algorithm to decide Boolean satisfiability.

Let T be the full binary tree of depth n. The nodes of T are binary strings of length n or less, the root is ε, and each string α of length less than n has exactly two children $\alpha 0$ and $\alpha 1$. Denote the set of nodes at depth m by T_m. We interpret the α as partial truth assignments to Boolean variables $x_1, \ldots, x_n$. If α is of length m, then the truth value of x_i is given by the ith symbol of α, $1 \leq i \leq m$; the values of $x_{m+1}, \ldots, x_n$ are not determined by α.

Say we are given a Boolean formula φ of length n with (at most) n Boolean variables $x_1, \ldots, x_n$. If α is a node of T, let $\alpha(\varphi)$ denote the formula obtained by substituting the values determined by α in φ. If $|\alpha| = m$, then $\alpha(\varphi)$ is a formula with variables $x_{m+1}, \ldots, x_n$.

The full tree T has exponentially many nodes, but we now show how to prune it to get a subtree T' satisfying the following properties at all levels m.

(i) T'_m has at most $N + 1$ elements.

(ii) φ is satisfiable iff at least one $\alpha(\varphi)$, $\alpha \in T'_m$, is satisfiable.

We build T'_m inductively from the top down. At stage 0, we take $T'_0 = \{\varepsilon\}$. Now suppose we have constructed T'_m satisfying properties (i) and (ii). For each $\alpha \in T'_m$, calculate $\sigma(\alpha 0(\varphi))$ and $\sigma(\alpha 1(\varphi))$, and let A be the set of values obtained, up to a maximum of $N + 1$ distinct values. If there are more than $N + 1$, just take A to be any subset of size $N + 1$ of the set of all values obtained. Now define T'_{m+1} to be any subset of $\{\alpha 0, \alpha 1 \mid \alpha \in T'_m\}$ of size $|A|$ representing all the values in A; that is, such that

$$\{\sigma(\beta(\varphi)) \mid \beta \in T'_{m+1}\} \;\; = \;\; A.$$

It is obvious from the construction that (i) holds for all m. Also, if $\alpha(\varphi)$ is satisfiable for some $\alpha \in T'_m$, then φ is satisfiable, because $\alpha(\varphi)$ is a substitution instance of φ. It therefore remains to show that if φ is satisfiable, then at least one $\alpha(\varphi)$, $\alpha \in T'_m$, is satisfiable. We prove this by induction on m. The basis $m = 0$ holds trivially. Now suppose it holds for m. By the induction hypothesis, $\alpha(\varphi)$ is satisfiable for some $\alpha \in T'_m$. Then at least one of $\alpha 0(\varphi)$ or $\alpha 1(\varphi)$ is satisfiable. In the construction of T'_{m+1} above, either $|A| \leq N$ or $|A| = N + 1$. If $|A| \leq N$, then $\sigma(\alpha 0(\varphi)) = \sigma(\beta(\varphi))$ and $\sigma(\alpha 1(\varphi)) = \sigma(\gamma(\varphi))$ for some $\beta, \gamma \in T'_{m+1}$, because A contains the values $\sigma(\alpha 0(\varphi))$ and $\sigma(\alpha 1(\varphi))$. Thus either $\beta(\varphi)$ or $\gamma(\varphi)$ is satisfiable. On the other hand, if $|A| = N + 1$, then $\sigma(\beta(\varphi)) \notin S'$ for some $\beta \in T'_{m+1}$ because $|S'| \leq N$, therefore $\beta(\varphi)$ is satisfiable.

This construction can be carried out in polynomial time. By (ii), we can determine whether φ is satisfiable by evaluating $\alpha(\varphi)$ for all $\alpha \in T'_n$. This gives a deterministic polynomial-time algorithm for Boolean satisfiability. $\square$

Mahaney's Theorem

Now we show how census functions can be combined with Theorem 29.1 to prove nonexistence of sparse NP-complete sets.

Theorem 29.2   (Mahaney [82])   *If there exists a sparse set that is $\leq^P_m$-complete for NP, then $P = NP$.*

Proof. Suppose that S is sparse and NP-complete. Let σ be a polynomial-time reduction from SAT to S with time bound n^c. Let $C(n)$

be the number of elements of S of length n^c or less:

$$C(n) \stackrel{\text{def}}{=} |S \cap \Sigma^{\leq n^c}|.$$

We know that $C(n)$ is bounded by a polynomial in n, say n^k. If we could only compute the exact value of $C(n)$ in polynomial time, then $\sim S$ would be in NP, because

$$\sim S = \{x \mid \exists y_1 \cdots \exists y_{C(|x|)} \bigwedge_i (|y_i| \leq |x|^c \wedge y_i \neq x \wedge y_i \in S)$$
$$\wedge \bigwedge_{i \neq j} y_i \neq y_j\}.$$

Because S is NP-complete, $\sim S \leq_m^P S$. Then $S \leq_m^P \sim S$ by the same reduction, so $S \equiv_n^p \sim S$, therefore S is also co-NP-complete. But then we could apply Theorem 29.1.

The trouble is that we do not know the true value of $C(|x|)$. Nevertheless, we can do the construction of Theorem 29.1 for all guesses $m = 0, 1, 2, \ldots, |x|^k$ for the value of $C(|x|)$. For all $m \neq C(|x|)$, we will get garbage results. But for $m = C(|x|)$, we are sure to get a satisfying truth assignment to the given formula φ if indeed φ is satisfiable. If φ is unsatisfiable, then no value of m will give a satisfying assignment.

Interestingly, we may never know when we hit the true value of $C(|x|)$. But we try them all, and we know one of them must be correct. Therefore if we try all values $m = 0, 1, 2, \ldots, |x|^k$ and fail to find a satisfying assignment, we know that φ must be unsatisfiable. Of course, if we ever find a satisfying assignment, even for the wrong value of m, then φ is satisfiable.

More formally, define

$$\widehat{S} \stackrel{\text{def}}{=} \{(x, m) \mid \exists y_1 \ldots \exists y_m \bigwedge_i (|y_i| \leq |x|^c \wedge y_i \neq x \wedge y_i \in S)$$
$$\wedge \bigwedge_{i \neq j} y_i \neq y_j\}.$$

This is like the characterization of $\sim S$ above, but parameterized by m. The number m is represented in unary. Note that

- if $m < C(|x|)$, then $(x, m) \in \widehat{S}$,

- if $m = C(|x|)$, then $(x, m) \in \widehat{S}$ iff $x \notin S$, and

- if $m > C(|x|)$, then $(x, m) \notin \widehat{S}$.

Thus

$$\widehat{S} = \{(x, m) \mid m < C(|x|) \vee (m = C(|x|) \wedge x \notin S)\}.$$

Also $\widehat{S} \in NP$, therefore $\widehat{S} \leq_{\mathrm{m}}^{\mathrm{p}} S$ via some polynomial-time reduction τ. Then $(x, m) \in \widehat{S}$ iff $\tau(x, m) \in S$.

Now consider the polynomial-time maps $\varphi \mapsto \tau(\sigma(\varphi), m)$. We have

$$
\begin{aligned}
\tau(\sigma(\varphi), C(|\sigma(\varphi)|)) \in S \quad &\Leftrightarrow \quad (\sigma(\varphi), C(|\sigma(\varphi)|)) \in \widehat{S} \\
&\Leftrightarrow \quad \sigma(\varphi) \notin S \\
&\Leftrightarrow \quad \varphi \in {\sim}\mathrm{SAT}.
\end{aligned}
$$

Also, if $m < C(|\sigma(\varphi)|)$, then $\tau(\sigma(\varphi), m) \in S$, and if $m > C(|\sigma(\varphi)|)$, then $\tau(\sigma(\varphi), m) \notin S$, but this is not really relevant.

Thus if we attempt to perform the construction of Theorem 29.1 for all functions $\tau(\cdot, m)$ for m up to n^{kc}, we are sure to hit the correct value of $C(|\sigma(\varphi)|)$, and the construction of Theorem 29.1 will give us a satisfying assignment if there is one.   □

An alternative proof of Mahaney's theorem is given in Miscellaneous Exercise 59.

Supplementary Lecture F

Unique Satisfiability

In this lecture and the next, we study some interesting reducibility relations involving randomness and counting and prove two notable results in structural complexity theory. The first is a result of Valiant and Vazirani [125] that Boolean satisfiability is no easier for formulas that are guaranteed to have at most one solution than for arbitrary formulas. The second is a result of Toda [120] that the ability to count solutions to instances of *NP*-complete problems allows one to compute any set in the polynomial-time hierarchy. In this lecture we prove the Valiant–Vazirani result, on which Toda's theorem is partly based.

As shown in Theorem 10.2, problems in *NP* can be defined in terms of the existence of polynomial-size *witnesses* that can be recognized easily. For instance, Boolean satisfiability (SAT) is characterized by the existence of satisfying truth assignments. The number of witnesses for different instances of the problem can vary over an exponential range, and one might ask whether this variation is at least partly responsible for the difficulty of *NP*-complete problems.

Valiant and Vazirani [125] answered this question in the negative. They showed that the general satisfiability problem SAT reduces by an efficient randomized reduction to USAT, the satisfiability problem for Boolean formulas that are promised to have at most one satisfying assignment.

The proof is based on the following idea. Consider the vector space $\mathbb{F}^n$ of dimension n over a finite field $\mathbb{F}$. One can construct a random tower of linear subspaces

$$\{0\} \;=\; E_0 \;\subset\; E_1 \;\subset\; \cdots \;\subset\; E_n \;=\; \mathbb{F}^n \tag{F.1}$$

such that E_i has dimension i, all such towers equally likely. This can be done by choosing a random basis $x_1,\ldots,x_n$ of $\mathbb{F}^n$ and defining $E_i = \{x_1,\ldots,x_{n-i}\}^{\perp}$, where

$$A^{\perp} \;\overset{\text{def}}{=}\; \{y \mid \forall x \in A \;\; x \bullet y = 0\},$$

the *orthogonal complement* of A, and where $\bullet$ denotes inner product. A random basis $x_1,\ldots,x_n$ can be obtained efficiently (Miscellaneous Exercise 60(b)).

For $\mathbb{F} = \mathbb{GF}_2$, the field on two elements, Valiant and Vazirani [125, Theorem 4(ii)] show that for any nonempty set $S \subseteq \mathbb{GF}_2^n$, for a randomly chosen tower (F.1), some $S \cap E_i$ contains exactly one vector with probability at least $1/2$. This leads to a randomized reduction from the general satisfiability problem to the unique satisfiability problem.

The proof of this result as given in [125] is inductive on dimension. We give an alternative proof here that achieves a slightly better lower bound of $3/4$.

Lemma F.1   *Let S be a nonempty subset of $\mathbb{GF}_2^n$. Let $E_0,\ldots,E_n$ be a random tower of linear subspaces of $\mathbb{GF}_2^n$ with $\dim E_i = i$. Then*

$$\Pr(\exists i \;\; |S \cap E_i| = 1) \;\geq\; \tfrac{3}{4}. \tag{F.2}$$

Proof. If $0 \in S$, then the probability is 1, because $S \cap E_0 = \{0\}$. Similarly, if $|S| = 1$, then the probability is 1, because $|S| = |S \cap E_n| = 1$. Ruling out these two cases, we can assume that $|S \cap E_n| \geq 2$ and $|S \cap E_0| = 0$.

Because $S \cap E_i \subseteq S \cap E_{i+1}$ for $0 \leq i \leq n-1$, we have that $|S \cap E_i| \leq |S \cap E_{i+1}|$ for $0 \leq i \leq n-1$. Thus there exists a least $k \geq 1$ such that $|S \cap E_k| \geq 2$. Here k is a random variable whose value depends on the choice of the random tower E_i as well as S.

If A is a set of vectors, let $<A>$ denote the linear span of A in $\mathbb{GF}_2^n$. Over $\mathbb{GF}_2$, any pair of distinct nonzero vectors x,y are linearly independent, because there are only three nonzero linear combinations, namely x, y, and $x + y$, and the last is 0 iff $x = y$. Thus $\dim <S \cap E_k>$ is at least 2.

Now (F.2) is equivalent to the statement

$$\Pr(S \cap E_{k-1} = \varnothing) \;\leq\; \tfrac{1}{4}, \tag{F.3}$$

and this is what we would like to show. Let $H = \{x_{n-k+1}\}^{\perp}$, the hyperplane consisting of all vectors orthogonal to x_{n-k+1}, and let T be a maximal linearly independent subset of $S \cap E_k$. Then $E_{k-1} = E_k \cap H$ and

$$T \cap H \;\subseteq\; S \cap E_k \cap H \;=\; S \cap E_{k-1},$$

so it suffices to show

$$\Pr(T \cap H = \varnothing) \;\leq\; \tfrac{1}{4}.$$

Because $\dim \langle T \rangle = \dim \langle S \cap E_k \rangle \geq 2$, by Miscellaneous Exercise 73,

$$\Pr(T \cap H = \varnothing) \;\leq\; \max_{d \geq 2} \Pr(T \cap H = \varnothing \mid \dim \langle T \rangle = d), \tag{F.4}$$

thus it suffices to bound

$$\Pr(T \cap H = \varnothing \mid \dim \langle T \rangle = d) \tag{F.5}$$

for $d \geq 2$. Using the fact that for subspaces A and B of a finite-dimensional vector space, $\operatorname{codim} A \cap B \leq \operatorname{codim} A + \operatorname{codim} B$ (Miscellaneous Exercise 61), we have that if $\dim \langle T \rangle = d$, then $\dim(\langle T \rangle \cap H)$ is either d or $d - 1$. Again by Miscellaneous Exercise 73, (F.5) is bounded by the maximum of

$$\Pr(T \cap H = \varnothing \mid \dim \langle T \rangle = d \wedge \dim(\langle T \rangle \cap H) = d) \tag{F.6}$$

$$\Pr(T \cap H = \varnothing \mid \dim \langle T \rangle = d \wedge \dim(\langle T \rangle \cap H) = d - 1). \tag{F.7}$$

But (F.6) is 0, because $\dim \langle T \rangle = \dim(\langle T \rangle \cap H) = d$ implies $T \subseteq H$, therefore $T \cap H$ cannot be empty. Thus we must bound (F.7). But this is the same as the probability, given a set Q of $d \geq 2$ linearly independent vectors and a random hyperplane G in $\langle Q \rangle$, that none of the vectors of Q lie in G:

$$\Pr(Q \cap G = \varnothing). \tag{F.8}$$

Let x be a vector in $\langle Q \rangle$ such that $G = \{x\}^{\perp}$. Then (F.8) becomes

$$\Pr(\forall y \in Q \;\; y \bullet x \neq 0). \tag{F.9}$$

To bound this expression, we use the inclusion–exclusion principle (see Lecture 13). Over the field $\mathbb{GF}_q$ of q elements,

$$
\begin{aligned}
\Pr(\forall y \in Q \;\; y \bullet x \neq 0) &= 1 - \Pr(\exists y \in Q \;\; y \bullet x = 0) \\[2mm]
&= 1 - \left(\sum_{m=1}^{|Q|} (-1)^{m+1} \sum_{\substack{A \subseteq Q \\ |A| = m}} \Pr(x \in A^{\perp}) \right) \\[2mm]
&= 1 - \left(\sum_{m=1}^{d} (-1)^{m+1} \binom{d}{m} q^{-m} \right) \\[2mm]
&= \sum_{m=0}^{d} \binom{d}{m} (-1)^m q^{-m} \\[2mm]
&= \left(1 - \frac{1}{q} \right)^{d}.
\end{aligned}
$$

This is maximized at $d = 2$, which for $q = 2$ gives $1/4$. $\square$

It can be shown that the lower bound $3/4$ in Lemma F.1 is the best possible for all n (Miscellaneous Exercise 62).

Unique Satisfiability and General Satisfiability

The class RP (for *random polynomial time*) was defined in Lecture 13. This is the class of sets accepted by polynomial-time probabilistic computations with one-sided error. Recall that RP is defined in terms of probabilistic machines, which are exactly like nondeterministic Turing machines except that they make choices probabilistically rather than nondeterministically. At each choice point, the machine chooses one of its possible next configurations randomly with uniform probability. Equivalently, we can supply a deterministic polynomial-time machine with a polynomial-length string of random bits that it can consult during its computation.

Valiant and Vazirani's result says that an RP computation supplied with an oracle USAT for unique satisfiability can determine general satisfiability with arbitrarily small one-sided error. The oracle USAT may be any oracle that answers affirmatively when queried on a Boolean formula that has exactly one satisfying assignment, negatively when queried on a formula that has no satisfying assignments, and arbitrarily on any other query.

Theorem F.2   **(Valiant and Vazirani [125])** $NP \subseteq RP^{\text{USAT}}$.

Proof. We show how to decide Boolean satisfiability by an RP computation with a USAT oracle. That is, we show that there is a polynomial-time-bounded probabilistic oracle Turing machine M with oracle USAT such that

$$\varphi \text{ is satisfiable} \;\Rightarrow\; \Pr(M \text{ accepts } \varphi) \geq \tfrac{3}{4},$$
$$\varphi \text{ is unsatisfiable} \;\Rightarrow\; \Pr(M \text{ accepts } \varphi) = 0.$$

Equivalently, encoding the random choices as part of the input, we show that there is a deterministic polynomial-time oracle machine N such that for w a string of random bits of sufficient polynomial length chosen with uniform probability among all strings of that length,

$$\varphi \text{ is satisfiable} \;\Rightarrow\; \Pr_w(N \text{ accepts } \varphi \# w) \geq \tfrac{3}{4},$$
$$\varphi \text{ is unsatisfiable} \;\Rightarrow\; \Pr_w(N \text{ accepts } \varphi \# w) = 0.$$

The machine N uses its random bits w to construct a random tower of linear subspaces $H_i \subseteq \mathbb{GF}_2^n$ as in Lemma F.1, where n is the number of

Boolean variables of φ. This requires $O(n^2)$ random bits. The tower can be represented as a sequence of n linearly independent vectors, as described in the proof of Lemma F.1.

Say the variables of φ are $x_1, \ldots, x_n$. For each $0 \leq i \leq n$, N constructs a formula ψ_i stating that $(x_1, \ldots, x_n)$, regarded as an n-vector in $\mathbb{GF}_2^n$, lies in H_i. This construction is a straightforward encoding of the inner product of $(x_1, \ldots, x_n)$ with the random vectors representing H_i. The machine N then queries the oracle on the conjunctions $\varphi \wedge \psi_i$ and accepts if the oracle responds "yes" to any of these queries.

Let S be the set of truth assignments satisfying φ. Then

$$\mathrm{Pr}_w(N \text{ accepts } \varphi \# w) \;=\; \mathrm{Pr}(\exists i \; \varphi \wedge \psi_i \in \mathrm{USAT}),$$

which by Lemma F.1 is at least

$$\mathrm{Pr}(\exists i \; |S \cap H_i| = 1) \;\geq\; \tfrac{3}{4}$$

if $S \neq \varnothing$ and 0 if $S = \varnothing$. □

The error can be made arbitrarily small by amplification (Lemma 14.1).

Supplementary Lecture G

Toda's Theorem

In this lecture we present Toda's theorem (Theorem G.1), which states that the ability to count solutions to instances of NP-complete problems allows one to solve any decision problem in the polynomial-time hierarchy PH. This theorem came as quite a surprise when it first appeared in 1989 [120, 121], because it showed that the power to count solutions was much stronger than previously thought. This result is a convergence of several important ideas and constructions involving various complexity classes and is a fine example of structural complexity theory at its best.

Counting Classes

The power to count solutions is captured in the class $P^{\#P}$. We can define the class $\#P$ to be the class of all functions $f_M : \{0,1\}^* \to \mathbb{N}$ such that M is a nondeterministic polynomial-time Turing machine and $f_M(x)$ gives the number of accepting computation paths of M on input x. Equivalently, $f \in \#P$ iff there is a set $A \in P$ and $k \geq 0$ such that for all $x \in \{0,1\}^*$,

$$f(x) \quad = \quad |\{y \in \{0,1\}^{|x|^k} \mid x\#y \in A\}| \tag{G.1}$$

(Miscellaneous Exercise 64). For example, the function that returns the number of satisfying truth assignments to a given Boolean formula is in $\#P$. Note that unlike other complexity classes we have studied, $\#P$ is not a class of decision problems, but a class of integer-valued functions. The class $P^{\#P}$ is the class of decision problems solvable in polynomial time with an oracle for some $f \in \#P$. The classes $\#P$ and $P^{\#P}$ were introduced by Valiant [124].

The formulation (G.1) suggests that we can view $\#P$ as the set of functions that give the cardinality of a set of *witnesses* to an existential formula. If $p : \mathbb{N} \to \mathbb{N}$ and A is a set of strings, a *witness* for x with respect to p and A is a binary string y of length $p(|x|)$ such that $x \# y \in A$. Let us denote the set of all witnesses for x with respect to p and A by $W(p, A, x)$. Thus

$$W(p, A, x) \overset{\text{def}}{=} \{y \in \{0,1\}^{p(|x|)} \mid x \# y \in A\}.$$

Many of the complexity classes we have considered in this course can be defined in terms of the cardinality of witness sets $W(p, A, x)$ for various parameters A and p. The defining conditions often do not require full knowledge of $|W(p, A, x)|$, but only bounds or certain bits. For example,

$$L \in NP \overset{\text{def}}{\Longleftrightarrow} \exists A \in P \; \exists c \geq 0 \; \forall x \; \; x \in L \Leftrightarrow |W(n^c, A, x)| > 0$$

$$L \in \Sigma_{k+1}^{\mathrm{p}} \overset{\text{def}}{\Longleftrightarrow} \exists A \in \Pi_k^{\mathrm{p}} \; \exists c \geq 0 \; \forall x \; \; x \in L \Leftrightarrow |W(n^c, A, x)| > 0$$

$$L \in RP \overset{\text{def}}{\Longleftrightarrow} \exists A \in P \; \exists c \geq 0 \; \forall x$$
$$x \in L \;\Rightarrow\; |W(n^c, A, x)| \geq \tfrac{3}{4} \cdot 2^{|x|^c}$$
$$x \notin L \;\Rightarrow\; |W(n^c, A, x)| = 0$$

$$L \in BPP \overset{\text{def}}{\Longleftrightarrow} \exists A \in P \; \exists c \geq 0 \; \forall x$$
$$x \in L \;\Rightarrow\; |W(n^c, A, x)| \geq \tfrac{3}{4} \cdot 2^{|x|^c}$$
$$x \notin L \;\Rightarrow\; |W(n^c, A, x)| \leq \tfrac{1}{4} \cdot 2^{|x|^c}$$

$$L \in PP \overset{\text{def}}{\Longleftrightarrow} \exists A \in P \; \exists c \geq 0 \; \forall x \; \; x \in L \Leftrightarrow |W(n^c, A, x)| \geq 2^{|x|^c - 1}$$

$$L \in \oplus P \overset{\text{def}}{\Longleftrightarrow} \exists A \in P \; \exists c \geq 0 \; \forall x \; \; x \in L \Leftrightarrow |W(n^c, A, x)| \text{ is odd}$$

$$L \in \#P \overset{\text{def}}{\Longleftrightarrow} \exists A \in P \; \exists c \geq 0 \; \forall x \; \; L(x) = |W(n^c, A, x)|.$$

Note that PP and $\oplus P$ are defined in terms of the first and last bit, respectively, of $|W(n^c, A, x)|$.

We can generalize further to define a set of operators $BP, R, \#$, and so on, on complexity classes $\mathcal{C}$. These operators can be viewed as reducibility relations. The definitions are assembled in the following table, which should be read as follows. The complexity class whose name appears in the left-hand column is defined to be the class of sets or functions L for which there exist $A \in \mathcal{C}$ and $k \geq 0$ such that for all x, the condition in the right-hand column holds.

Class	Defining Condition
$R \cdot \mathcal{C}$	$x \in L \implies |W(n^k, A, x)| \geq \frac{3}{4} \cdot 2^{|x|^k}$ $x \notin L \implies |W(n^k, A, x)| = 0$
$BP \cdot \mathcal{C}$	$x \in L \implies |W(n^k, A, x)| \geq \frac{3}{4} \cdot 2^{|x|^k}$ $x \notin L \implies |W(n^k, A, x)| \leq \frac{1}{4} \cdot 2^{|x|^k}$
$P \cdot \mathcal{C}$	$x \in L \iff |W(n^k, A, x)| \geq \frac{1}{2} \cdot 2^{|x|^k}$
$\oplus \cdot \mathcal{C}$	$x \in L \iff |W(n^k, A, x)| \equiv 1 \pmod 2$
$\Sigma^{\mathrm{p}} \cdot \mathcal{C}$	$x \in L \iff |W(n^k, A, x)| > 0$
$\Pi^{\mathrm{p}} \cdot \mathcal{C}$	$x \in L \iff |W(n^k, A, x)| = 2^{|x|^k}$
$\Sigma^{\log} \cdot \mathcal{C}$	$x \in L \iff |W(\lceil k \log n \rceil, A, x)| > 0$
$\Pi^{\log} \cdot \mathcal{C}$	$x \in L \iff |W(\lceil k \log n \rceil, A, x)| = 2^{\lceil k \log |x| \rceil}$
$\# \cdot \mathcal{C}$	$L(x) = |W(n^k, A, x)|$

Thus $RP = R \cdot P$, $BPP = BP \cdot P$, $PP = P \cdot P$, $\oplus P = \oplus \cdot P$, $NP = \Sigma^{\mathrm{p}} \cdot P$, co-$NP = \Pi^{\mathrm{p}} \cdot P$, $\Sigma^{\mathrm{p}}_{k+1} = \Sigma^{\mathrm{p}} \cdot \Pi^{\mathrm{p}}_k$, and $\# \cdot P = \#P$. Note that all of these operators are monotone with respect to set inclusion.

The characterization of complexity classes in terms of operators and witnesses reveals an underlying similarity among the classes that is quite striking. It also provides a common framework in which to carry out the constructions in the proof of Toda's theorem. Many of the closure properties we need are naturally expressed as algebraic properties of these operators, such as commutativity and idempotence.

Toda's Theorem

Theorem G.1   (Toda [120]) $PH \subseteq P^{\#P}$.

Much of the proof of Toda's theorem can be broken down into a series of inclusions (Lemma G.2(i)–(vi)) that establish basic algebraic properties of the operators on complexity classes defined in the previous section. These are combined to show that $PH \subseteq BP \cdot \oplus P$. In addition, there is a final ingenious argument showing that $BP \cdot \oplus P \subseteq P^{\#P}$.

Lemma G.2   *Let $\mathcal{C}$ be a complexity class closed downward under the polynomial-time Turing reducibility relation $\leq^{\mathrm{p}}_{\mathrm{T}}$. Then*

(i) $\Sigma^{\mathrm{p}} \cdot \mathcal{C} \subseteq R \cdot \Sigma^{\log} \cdot \oplus \cdot \mathcal{C}$;

(ii) $\Pi^{\log} \cdot \oplus \cdot \mathcal{C} \subseteq \oplus \cdot \mathcal{C}$;

(iii) $\oplus \cdot BP \cdot \mathcal{C} \subseteq BP \cdot \oplus \cdot \mathcal{C}$;

(iv) $BP \cdot BP \cdot \mathcal{C} \subseteq BP \cdot \mathcal{C}$;

(v) $\oplus \cdot \oplus \cdot \mathcal{C} \subseteq \oplus \cdot \mathcal{C}$;

(vi) $BP \cdot \mathcal{C}$ and $\oplus \cdot \mathcal{C}$ are closed downward under $\leq_{\mathrm{T}}^{\mathrm{p}}$.

Proof. Believe it or not, clause (i) is a minor variant of the result of Valiant and Vazirani presented in Lecture F (Lemma F.1). We prove this clause explicitly and leave the remaining clauses (ii)–(vi) as exercises (Miscellaneous Exercise 66).

By definition, $L \in \Sigma^{\mathrm{p}} \cdot \mathcal{C}$ iff there exist $A \in \mathcal{C}$ and $k \geq 0$ such that for all x,

$$x \in L \quad \Leftrightarrow \quad W(n^k, A, x) \neq \varnothing.$$

Let $N = n^k$. As in Lemma F.1, let H_i^w, $0 \leq i \leq N$, be a random tower of linear subspaces of $\mathbb{GF}_2^N$ with $\dim H_i^w = i$. Recall that the tower is specified by a random nonsingular matrix over $\mathbb{GF}_2^N$, where H_i^w is the orthogonal complement of the first $N-i$ columns of the matrix. The random nonsingular matrix in turn is determined by a string w of random bits of length $O(N^2)$. By Lemma F.1,

$$
\begin{aligned}
x \in L \quad &\Rightarrow \quad W(N, A, x) \neq \varnothing \\
&\Rightarrow \quad \Pr_w(\exists i \leq N \, | W(N, A, x) \cap H_i^w | = 1) \;\geq\; \tfrac{3}{4} \\
&\Rightarrow \quad \Pr_w(\exists i \leq N \, | W(N, A, x) \cap H_i^w | \text{ is odd}) \;\geq\; \tfrac{3}{4}, \\
x \notin L \quad &\Rightarrow \quad W(N, A, x) = \varnothing \\
&\Rightarrow \quad \Pr_w(\exists i \leq N \, | W(N, A, x) \cap H_i^w | \text{ is odd}) \;=\; 0.
\end{aligned}
$$

Let p be such that $p(|x\#w\#i|) = |x|^k$. The function p can be made to depend only on the length of $x\#w\#i$ provided we represent i in binary with $\lceil k \log |x| \rceil$ bits. Similarly, let q be such that $q(|x\#w|) = \lceil k \log |x| \rceil$. Define

$$
\begin{aligned}
A' \;&\overset{\text{def}}{=}\; \{x\#w\#i\#y \mid x\#y \in A \wedge y \in H_i^w\} \\
A'' \;&\overset{\text{def}}{=}\; \{z \mid |W(p, A', z)| \text{ is odd}\} \\
A''' \;&\overset{\text{def}}{=}\; \{u \mid |W(q, A'', u)| > 0\}.
\end{aligned}
$$

Because $A \in \mathcal{C}$ and $A' \leq_{\mathrm{T}}^{\mathrm{p}} A$, we have $A' \in \mathcal{C}$, $A'' \in \oplus \cdot \mathcal{C}$, and $A''' \in \Sigma^{\log} \cdot \oplus \cdot \mathcal{C}$ by definition of these classes. Then

$$
\begin{aligned}
W(p, A', x\#w\#i) \;&=\; \{y \in \{0,1\}^{|x|^k} \mid x\#w\#i\#y \in A'\} \\
&=\; \{y \in \{0,1\}^{|x|^k} \mid x\#y \in A\} \cap H_i^w \\
&=\; W(N, A, x) \cap H_i^w,
\end{aligned}
$$

therefore

$$\exists i \leq N \; |W(N, A, x) \cap H_i^w| = 1$$
$$\Rightarrow \quad \exists i \leq N \; |W(N, A, x) \cap H_i^w| \text{ is odd}$$
$$\Leftrightarrow \quad \exists i \leq N \; |W(p, A', x\#w\#i)| \text{ is odd}$$
$$\Leftrightarrow \quad \exists i \leq N \; x\#w\#i \in A''$$
$$\Leftrightarrow \quad |\{i \in \{0,1\}^{\lceil k \log |x| \rceil} \mid x\#w\#i \in A''\}| > 0$$
$$\Leftrightarrow \quad |W(q, A'', x\#w)| > 0$$
$$\Leftrightarrow \quad x\#w \in A'''.$$

It follows that

$$x \in L \quad \Rightarrow \quad W(N, A, x) \neq \varnothing$$
$$\Rightarrow \quad \Pr_w(\exists i \leq N \; |W(N, A, x) \cap H_i^w| = 1) \geq \tfrac{3}{4}$$
$$\Rightarrow \quad \Pr_w(x\#w \in A''') \geq \tfrac{3}{4}$$
$$x \notin L \quad \Rightarrow \quad W(N, A, x) = \varnothing$$
$$\Rightarrow \quad \forall i \leq N \; |W(N, A, x) \cap H_i^w| \text{ is even}$$
$$\Rightarrow \quad \Pr_w(x\#w \in A''') = 0.$$

Thus $L \in R \cdot \Sigma^{\log} \cdot \oplus \cdot \mathcal{C}$. $\qquad\qquad\qquad\qquad\qquad\qquad\qquad$ $\square$

Proof of Theorem G.1. We first argue that the various inclusions of Lemma G.2 combine to imply that $PH \subseteq BP \cdot \oplus P$. The proof is by induction on the level of the polynomial-time hierarchy. Certainly $\Sigma_0^p = P \subseteq BP \cdot \oplus P$. Now suppose that $\Sigma_k^p \subseteq BP \cdot \oplus P$. By Lemma G.2(vi), $BP \cdot \oplus P$ is closed under complement, therefore $\Pi_k^p \subseteq BP \cdot \oplus P$. Then

$$
\begin{array}{rll}
\Sigma_{k+1}^p & = \quad \Sigma^p \cdot \Pi_k^p & \\
& \subseteq \quad \Sigma^p \cdot BP \cdot \oplus P & \text{by monotonicity} \\
& \subseteq \quad R \cdot \Sigma^{\log} \cdot \oplus \cdot BP \cdot \oplus P & \text{by Lemma G.2(i)} \\
& \subseteq \quad R \cdot \oplus \cdot BP \cdot \oplus P & \text{by Lemma G.2(ii) and (vi)} \\
& \subseteq \quad BP \cdot \oplus \cdot BP \cdot \oplus P & \text{because } R \cdot \mathcal{C} \subseteq BP \cdot \mathcal{C} \text{ trivially} \\
& \subseteq \quad BP \cdot BP \cdot \oplus \cdot \oplus P & \text{by Lemma G.2(iii)} \\
& \subseteq \quad BP \cdot \oplus P & \text{by Lemma G.2(iv) and (v).}
\end{array}
$$

We have shown that $\Sigma_k^p \subseteq BP \cdot \oplus P$ for all k, therefore

$$PH \quad = \quad \bigcup_k \Sigma_k^p \quad \subseteq \quad BP \cdot \oplus P.$$

It remains to show that $BP \cdot \oplus P \subseteq P^{\#P}$. Let $L \in BP \cdot \oplus P$. Then there exist $A \in \oplus P$ and $k \geq 0$ such that for all x,

$$x \in L \quad \Rightarrow \quad \mathrm{Pr}_w(x \# w \in A) \geq \tfrac{3}{4}$$
$$x \notin L \quad \Rightarrow \quad \mathrm{Pr}_w(x \# w \in A) \leq \tfrac{1}{4},$$

where the w are chosen uniformly at random among all binary strings of length $|x|^k$. Equivalently,

$$
\begin{aligned}
x \in L \quad &\Rightarrow \quad |W(n^k, A, x)| \geq \tfrac{3}{4} \cdot 2^{|x|^k} \\
x \notin L \quad &\Rightarrow \quad |W(n^k, A, x)| \leq \tfrac{1}{4} \cdot 2^{|x|^k}.
\end{aligned}
\tag{G.2}
$$

Furthermore, because $A \in \oplus P$, there exists a polynomial-time nondeterministic TM M such that $x \# w \in A$ iff $f(x \# w)$ is odd, where $f(x \# w)$ is the number of accepting computation paths of M on input $x \# w$.[1]

Now we modify M to obtain a new machine N that on input $x \# w$ has $p(f(x \# w))$ accepting computation paths instead of $f(x \# w)$, where p is a polynomial function represented by a certain polynomial in $\mathbb{N}[z]$. Specifically, the polynomial p we are interested in is $h^m(z)$, where

$$h(z) \quad \overset{\text{def}}{=} \quad 4z^3 + 3z^4 \quad \in \quad \mathbb{N}[z],$$

$h^i(z)$ is the i-fold composition of h with itself, that is,

$$
\begin{aligned}
h^0(z) \quad &\overset{\text{def}}{=} \quad z \\
h^{i+1}(z) \quad &\overset{\text{def}}{=} \quad h(h^i(z)),
\end{aligned}
$$

and $m = \log(n^k + 1)$. One can show inductively that $h^m(z)$ is of degree $4^m = (n^k + 1)^2$ and has coefficients that can be represented by at most $\tfrac{2}{3}(4^m - 1) = \tfrac{2}{3}((n^k + 1)^2 - 1)$ bits. Moreover, the polynomial $h^m(z)$ can be constructed from $x \# w$ in polynomial time. The construction of N from M is detailed in Miscellaneous Exercise 67(d).

One can show by induction that $h^m(z)$ has the following agreeable property:

$$
\begin{aligned}
z \text{ is odd} \quad &\Rightarrow \quad h^m(z) \equiv -1 \ (\mathrm{mod}\ 2^{2^m}) \\
z \text{ is even} \quad &\Rightarrow \quad h^m(z) \equiv 0 \ (\mathrm{mod}\ 2^{2^m})
\end{aligned}
$$

(Miscellaneous Exercise 65), thus

$$
\begin{aligned}
z \text{ is odd} \quad &\Rightarrow \quad p(z) \equiv -1 \ (\mathrm{mod}\ 2^{n^k+1}) \\
z \text{ is even} \quad &\Rightarrow \quad p(z) \equiv 0 \ (\mathrm{mod}\ 2^{n^k+1}).
\end{aligned}
\tag{G.3}
$$

---

[1]This characterization of $\oplus P$ and the characterization involving witness sets are equivalent; this follows immediately from the corresponding equivalence for $\#P$ (Miscellaneous Exercise 64).

Now we can determine membership in L by a $P^{\#P}$ computation as follows. Build a new machine K that on input x of length n

(i)   generates all possible strings $x\#w$ with $|w| = n^k$ by nondeterministic branching, one computation path for each such string;

(ii)   for each $x\#w$, runs N on $x\#w$.

The number of accepting computation paths of K on input x is then

$$\sum_{|w|=n^k} p(f(x\#w)).$$

Modulo 2^{n^k+1}, this quantity is

$$
\begin{aligned}
\sum_{|w|=n^k} p(f(x\#w)) \;\; &\equiv \sum_{\substack{|w|=n^k \\ f(x\#w)\ \text{odd}}} -1 \quad \text{by (G.3)} \\
&\equiv 2^{n^k+1} - |\{w \mid |w| = n^k \wedge f(x\#w)\ \text{is odd}\}| \\
&= 2^{n^k+1} - |\{w \mid |w| = n^k \wedge x\#w \in A\}| \\
&= 2^{n^k+1} - |W(n^k, A, x)|.
\end{aligned}
$$

But by (G.2),

$$
\begin{aligned}
x \in L \;\; &\Rightarrow \;\; \tfrac{3}{4}\cdot 2^{n^k} \leq |W(n^k, A, x)| \leq 2^{n^k} \\
&\Rightarrow \;\; 2^{n^k} \leq 2^{n^k+1} - |W(n^k, A, x)| \leq \tfrac{5}{4}\cdot 2^{n^k} \\
x \notin L \;\; &\Rightarrow \;\; 0 \leq |W(n^k, A, x)| \leq \tfrac{1}{4}\cdot 2^{n^k} \\
&\Rightarrow \;\; \tfrac{7}{4}\cdot 2^{n^k} \leq 2^{n^k+1} - |W(n^k, A, x)| \leq 2^{n^k+1},
\end{aligned}
$$

and these are disjoint intervals. Thus the number of accepting computation paths of K, reduced modulo 2^{n^k+1}, determines membership in L. This says that $L \in P^{\#P}$. $\qquad\square$

Lecture 30

Circuit Lower Bounds and Relativized $PSPACE = PH$

In this lecture and the next we study lower bounds for families of Boolean circuits of constant depth and their application to relativized complexity. These results are interesting not only from the point of view of the design and analysis of algorithms, but also for their consequences regarding the structure of the polynomial-time hierarchy.

The first lower bound results for constant-depth circuits were obtained by Furst, Saxe, and Sipser [46] and Ajtai [5] in the mid-1980s. They showed the nonexistence of circuit families of constant depth and polynomial size for the PARITY function and other Boolean functions. A series of later papers established explicit superpolynomial lower bounds on the size of constant-depth circuits for some of these functions, culminating in a nearly optimal result of Håstad [56]. A detailed history and exposition of this work can be found in the survey article of Boppana and Sipser [20].

In this lecture we introduce the complexity class AC^0 of Boolean functions that can be computed by a family of constant-depth, polynomial-size circuits of unbounded indegree. We then show that a sufficiently strong superpolynomial lower bound on the size of constant-depth circuits for PARITY implies the existence of an oracle separating $PSPACE$ from PH. This relationship was first observed by Furst, Saxe, and Sipser [46] and was one of the primary motivations for the study of the complexity of constant-depth circuits.

At the end of this lecture, we set up the machinery needed to prove that PARITY is not in AC^0, then dive into the details in Lecture 31. Our proof is a minor variant of the original proof of Furst, Saxe, and Sipser [46], which introduced the idea of a *random partial valuation*. This was one of the first applications of probability to lower bound proofs, although the conclusion is not probabilistic at all.

Unfortunately, this result is not strong enough to obtain an oracle separating $PSPACE$ from PH. Stronger bounds achieving separation were first obtained by Yao [126] and Håstad [56]. A complete proof of Håstad's result is given in Supplementary Lecture H.

Parity and the Class AC^0

A *parity function* is a Boolean function $\{0,1\}^n \to \{0,1\}$ whose value changes whenever any one of its inputs changes. There are exactly two such functions for any n, namely the function that computes the mod-2 sum of its inputs and the function that computes the complement of this value. We refer to these functions collectively as PARITY.

We have studied uniform families of Boolean circuits previously in Lectures 11 and 12. Recall that the complexity class NC discussed in those lectures was defined in terms of logspace-uniform families of Boolean circuits of polylog depth and polynomial size. There was also a degree restriction: the gates of the circuits were binary $\wedge$ and $\vee$ or unary $\neg$ gates.

To define the class AC^0, we restrict to constant depth, but we relax the degree restriction to allow $\wedge$ and $\vee$ gates to compute the conjunction and disjunction, respectively, of an unbounded number of Boolean inputs in one step. (It is easy to show that no function of a single output that depends on all its inputs can be computed by a circuit of bounded degree in $o(\log n)$ depth.)

For example, for fixed k, the Boolean function that returns 1 if at least k of its inputs are 1 can be computed by an AC^0 circuit of size $\binom{n}{k} + 1$ and depth 2 as

$$\bigvee_{\substack{S \subseteq \{x_1,\dots,x_n\} \\ |S|=k}} \bigwedge S.$$

As with NC, the official definition of AC^0 includes a uniformity condition. However, we are concerned here only with lower bounds, and any lower bound that we can derive for nonuniform circuits holds *a fortiori* for uniform circuits, so we can conveniently ignore this issue.

We can also assume without loss of generality that all negations are applied to inputs only, and that the gates are arranged in levels with inputs and their negations at level 0 and conjunctions and disjunctions strictly alternating level by level. The assumption regarding negations can be enforced by constructing the dual circuit as in Lecture 12. The assumption regarding the alternation of $\wedge$ and $\vee$ can be enforced by the addition of dummy gates if necessary without significantly increasing the depth or size.

Separating PH from $PSPACE$

The main result of this lecture is that a sufficiently strong lower bound on the size of constant-depth circuits for PARITY implies the existence of an oracle separating $PSPACE$ from the polynomial-time hierarchy.

Theorem 30.1   (Furst, Saxe, and Sisper [46])   *Suppose that there exists no family of circuits for* PARITY *of depth d and size $2^{(\log n)^c}$ for any constants c and d. Then there exists an oracle $A \subseteq \{0,1\}^*$ such that $PH^A \neq PSPACE^A$.*

Proof. Let $A \subseteq \{0,1\}^*$. The *characteristic function* of A, also denoted by A, is the function

$$A(x) \quad \stackrel{\text{def}}{=} \quad \begin{cases} 1, & \text{if } x \in A \\ 0, & \text{otherwise.} \end{cases}$$

Define the set

$$P(A) \quad \stackrel{\text{def}}{=} \quad \{x \in \{0,1\}^* \mid \sum_{y \in A, \, |y|=|x|} A(y) \equiv 1 \,(\text{mod } 2)\}$$

$$= \quad \{x \in \{0,1\}^* \mid \text{the number of strings in } A \text{ of length } |x| \text{ is odd}\}.$$

For the set A shown in the diagram below, $P(A)$ would contain all strings of length 4, as the number of strings of length 4 in A is 9, which is odd.

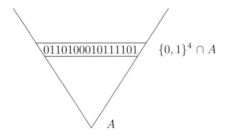

There is an oracle Turing machine that for any oracle A accepts the set $P(A)$ and runs in linear space. The machine on input x simply steps through all the strings of length $|x|$ in some order, querying the oracle on each one and keeping track of the number of positive responses mod 2. Thus $P(A) \in PSPACE^A$ for any A.

Now for any oracle machine M, let us say that M is *correct on n* if, supplied with any oracle A and input x of length n, M accepts x iff $x \in P(A)$; that is, for any A and x of length n, M correctly computes the parity of the number of elements of A of length n. We argue that there can be no polynomial-time Σ_d-machine that is correct on all but finitely many n, otherwise we could build a family of circuits violating the assumption of the theorem.

Suppose for a contradiction that there were such a machine M. On inputs of length n, M runs in time n^c and makes at most d alternations of universal and existential states along any computation path, where c and d are constants. The computation tree of M on any input can be divided into d levels such that the configurations on any level are either all universal configurations or all existential configurations, and the first level contains only existential configurations.

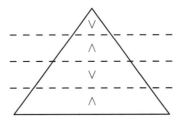

We can assume without loss of generality that on input x, M never queries its oracle on any string that is not the same length as x, because these strings are irrelevant in determining whether $x \in P(A)$. If M does so, we can build another machine N that simulates M, supplying an arbitrary truth value whenever M would attempt to query the oracle on a string of the wrong length. The machine N must be correct on n if M is.

Now we build a parity circuit C_n for each n on which M is correct. The circuit C_n is built from the computation tree of N on input 0^n and all possible oracles A. The gates of the circuit are the configurations of N on input 0^n; there are at most 2^{n^c} of these for some constant c. The inputs are Boolean variables representing the truth values of $A(y)$, $|y| = n$. There are 2^n of these. The output gate is the start configuration of N. An oracle query just reads the input gate corresponding to the query string. An existential configuration α becomes an $\vee$-gate of unbounded indegree, taking its inputs from all the universal or halting configurations β for which there exists a computation path $\alpha \to \beta$ in which all configurations, except possibly the last, are existential. A universal configuration becomes an $\wedge$-gate of unbounded indegree in a similar fashion. Thus we have flattened each existential and each universal level of the computation tree into a circuit of depth 1:

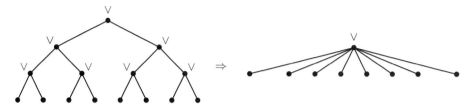

The resulting circuits C_n are of depth d and size 2^{n^c}, have 2^n inputs, and compute the parity of their inputs. In terms of the number of inputs $m = 2^n$, the size is $2^{(\log m)^c}$. We only have circuits whose input sizes are powers of 2, but we can obtain parity circuits for other input sizes m by taking the circuit for the next larger power of 2 and setting all but m inputs to 0. Thus we have a family of circuits of constant depth d and size $2^{(\log m)^c}$ for PARITY. This contradicts the assumption of the theorem.

Now we use this to construct an oracle A separating PH from $PSPACE$. We know from the contradiction that the premise that M was correct on all but finitely many n was erroneous. Thus there must exist arbitrarily large n on which M is incorrect. Proceeding by diagonalization, let $M_0, M_1, \ldots$ be a list of all polynomial-time Σ_d machines for all constants d. We construct A in stages. At the ith stage, let n_i be a number larger than any number chosen at any earlier stage such that M_i is incorrect on n_i. Let B_i be an oracle on which M_i gives the incorrect answer on input 0^{n_i}, let $A_i = B_i \cap \{0,1\}^{n_i}$, and let $A = \bigcup_i A_i$. Then M_i with oracle A is also incorrect on n_i, because it receives the same responses from A as from B_i. Thus no M_i computes $P(A)$. □

Lower Bounds for PARITY

In Lecture 31, we present the result of Furst, Saxe, and Sipser [46] that PARITY is not in AC^0. The proof is based on the observation that if we set some of the inputs of a parity circuit to 0 or 1, the resulting circuit still computes a parity function, albeit on fewer inputs. If we have a circuit of constant depth d, we set each input randomly and independently to 0 or 1, each with a certain probability $(1-p)/2$ close to but strictly less than $1/2$, or leave it unset with the remaining probability p. This is called a *random partial valuation*. By choosing p carefully, we can to show that with high likelihood, the resulting circuit is equivalent to one of depth strictly less than d while still retaining a good percentage of unassigned variables. By repeating this process, we can reduce the depth to 2 with bounded indegree at the first level; but we can show independently that no such family can compute PARITY.

Note that the degree of the gates and uniformity of the family of circuits are not issues here. The lower bounds hold even for nonuniform circuits containing gates of unbounded indegree and outdegree.

Lecture 31

Lower Bounds for Constant Depth Circuits

In this lecture we present the details of the result of Furst, Saxe, and Sipser [46] that PARITY is not in AC^0.

Recall that a formula or circuit is in *t-conjunctive normal form* (*t-CNF*) if it is a conjunction of clauses, each clause a disjunction of at most t literals, each literal a variable or its negation. Dually, a formula or circuit is in *t-disjunctive normal form* (*t-DNF*) if it is a disjunction of terms, each term a conjunction of at most t literals. We can assume without loss of generality that no term or clause contains a pair of complementary literals. By convention, the empty conjunction is equivalent to 1 and the empty disjunction is equivalent to 0.

Given a parity circuit of constant depth d, we can assume without loss of generality that the gates are arranged in levels with variables and their negations at level 0 and other levels alternating between disjunctions and conjunctions. We can also assume that the gates at level 1 are disjunctions; if not, consider the dual circuit instead, which is also a parity circuit.

The *length* of a term or clause M, denoted $|M|$, is the number of literals in M. We often omit the symbol $\wedge$ in terms and write $\overline{x}$ for $\neg x$. Thus $\overline{x}y\overline{z}$ means the same as $\neg x \wedge y \wedge \neg z$.

A *partial valuation* of a set X of Boolean variables is an assignment of constants to some of the variables of X, possibly leaving some variables unassigned. Formally, a *partial valuation* on X is a map $\rho : X \to X \cup \{0, 1\}$

such that for each $x \in X$, $\rho(x) \in \{x, 0, 1\}$. We say that x is *unassigned* under ρ if $\rho(x) = x$.

Any partial valuation ρ on X extends to a function on Boolean formulas over X in a natural way, replacing each variable x with $\rho(x)$ and then simplifying wherever possible using the Boolean algebra axioms $0 \vee x = x$, $0 \wedge x = 0$, $1 \vee x = 1$, $1 \wedge x = x$. For example, if $\rho(x) = 1$, $\rho(y) = 0$, and $\rho(z) = z$, then

$$\rho((x \vee y) \wedge (\overline{x} \vee z)) \;\; = \;\; z.$$

For the remainder of this lecture, we consider randomly chosen partial valuations in which each variable is independently assigned 0 or 1, each with probability $(1 - 1/\sqrt{n})/2$, or left unassigned with probability $1/\sqrt{n}$.

The central idea of the proof is that all the CNF subcircuits at level 2 will very likely become equivalent to DNF circuits after applying a finite number of partial valuations chosen randomly according to this distribution. Thus the chances are good that we will be able to replace all the CNF gates at level 2 with DNF gates, then absorb the disjunctions at level 2 into the disjunctions at level 3, thereby reducing the depth by one level. Continuing in this fashion, we will be able to get rid of all levels except two.

Lemma 31.1   *After a random partial valuation, the probability that there are fewer than $\sqrt{n}/2$ unassigned variables is at most $(2/e)^{\sqrt{n}/2}$.*

Proof. Each variable remains unassigned with probability $1/\sqrt{n}$. There are n variables in all, so the mean number of unassigned variables is $n/\sqrt{n} = \sqrt{n}$. The result now follows immediately from the Chernoff bound (I.7) with $\mu = \sqrt{n}$ and $\delta = 1/2$. $\square$

Lemma 31.1 is important, because it says that after the application of a random partial valuation, it is very likely that there are still enough unassigned variables left that the size of the circuit is still polynomial in the number of inputs.

The following are two technical lemmas that are used in our main development.

Lemma 31.2   *Let c be a constant, and let A be any set of variables of size at least c but $o(n^{1/c})$. The probability that a random partial valuation leaves more than c variables of A unassigned is bounded above by $n^{1-c/2}$ for sufficiently large n.*

Proof. Let X be a random variable representing the number of variables of A left unassigned by the random partial valuation. We wish to estimate $\Pr(X \geq c)$. If $s = |A|$, then the expected number of unassigned variables is $\mu = sn^{-1/2}$. Plugging this into the Chernoff bound (I.6), for sufficiently large n,

$$
\begin{aligned}
\Pr(X \geq c) &< e^{c-\mu}(\mu/c)^c \leq (e\mu/c)^c \\
&= (esn^{-1/2}/c)^c = n^{-c/2}(es/c)^c \leq n^{1-c/2},
\end{aligned}
$$

the last inequality by the assumption that s is $o(n^{1/c})$.   □

Lemma 31.3   *Let a and b be constants. Let S be any set of pairwise disjoint sets of variables such that S has size at least $b \log n$ and all elements of S have size less than a. For $A \in S$, let $E(A)$ be the event that a random partial valuation assigns 0 to all variables in A. For sufficiently large n, the probability that $E(A)$ does not occur for any $A \in S$ is bounded above by $n^{b \log(1-2^{-a})}$.*

Proof. Let $A \in S$ and let $s = |A|$. For sufficiently large n, the probability of $E(A)$ is $2^{-s}(1-n^{-1/2})^s \geq 2^{-s}/2 \geq 2^{-a}$. The probability that $E(A)$ does not occur is thus bounded by $1-2^{-a}$. Because the elements of S are pairwise disjoint, the events $E(A)$ are independent, therefore the probability that $E(A)$ fails for all $A \in S$ is the product of the probabilities that it fails for each of them, which is bounded by $(1 - 2^{-a})^{|S|}$. But

$$
(1 - 2^{-a})^{|S|} \leq (1 - 2^{-a})^{b \log n} = n^{b \log(1-2^{-a})}.
$$

□

Now we are ready to give the main part of the argument, which we have split into three lemmas. Let t be a constant. Call a circuit a *t-circuit* if every level-1 gate is of degree at most t; that is, if every level-2 gate is a t-CNF circuit.

Lemma 31.4   *If PARITY has polynomial-size circuits of depth d, then PARITY has polynomial-size t-circuits of depth d for some constant t.*

Lemma 31.5   *If PARITY has polynomial-size t-circuits of depth $d \geq 3$ and $t \geq 1$, then PARITY has polynomial-size $(t-1)$-circuits of depth d.*

Lemma 31.6   *If PARITY has polynomial-size 1-circuits of depth $d \geq 1$, then PARITY has polynomial-size circuits of depth $d - 1$.*

Proof of Lemma 31.4. Suppose PARITY has circuits of depth d and size n^k. Consider a random partial valuation ρ on the variables of the nth circuit. Let $t = 2k + 4$ and $b = (k + 1)/\log(3/2)$. For some level-1 clause C, consider the event $|\rho(C)| > t$; that is, more than t literals of C remain unassigned. We show that for sufficiently large n, this event occurs with probability at most $n^{-(k+1)}$.

There are three cases, depending on the size of C.

Case 1 If $|C| \leq t$, then the probability is already 0, so we are done.

Case 2 If $t \leq |C| \leq b \log n$, then by Lemma 31.2, for sufficiently large n, the probability that a random partial valuation leaves more than t literals in C unassigned is bounded above by $n^{1-t/2} = n^{-(k+1)}$.

Case 3 The last case is $|C| \geq b \log n$. If any literal of C is assigned 1, then $\rho(C) = 1$, so $|\rho(C)| = 0$. Thus the probability that there are at least t literals remaining unassigned is bounded by the probability that no literal in C is assigned 1. To calculate this probability, note that for sufficiently large n, the probability that any particular literal is not assigned 1 is

$$1 - \frac{1}{2}(1 - \frac{1}{\sqrt{n}}) \;=\; \frac{1}{2}(1 + \frac{1}{\sqrt{n}}) \;\leq\; \frac{2}{3}.$$

These events are independent, thus the probability that no literal in C is assigned 1 is at most

$$\left(\frac{2}{3}\right)^{|C|} \;\leq\; \left(\frac{2}{3}\right)^{b \log n} \;\leq\; n^{b \log(2/3)} \;=\; n^{-(k+1)}.$$

We have shown that for sufficiently large n, for each level-1 clause C, the probability that $|\rho(C)| > t$ is at most $n^{-(k+1)}$. By the law of sum, the probability that there exists a level-1 clause with more than t literals is bounded by the sum of these probabilities. Because there are at most n^k level-1 clauses in all,

$$\Pr\left(\exists C \; |\rho(C)| > t\right) \;\leq\; \sum_C \Pr\left(|\rho(C)| > t\right) \;\leq\; n^k n^{-(k+1)} \;=\; n^{-1},$$

which is vanishingly small.

Now by Lemma 31.1, the probability that there are fewer than $\sqrt{n}/2$ unassigned variables is also vanishingly small. Again by the law of sum, the probability that either event occurs in a single random partial valuation is vanishingly small. That is, with probability tending to 1, the random partial valuation leaves at least $\sqrt{n}/2$ variables unassigned and knocks all level-1 gates down to degree at most t. As the probability of this event is nonzero, there must be a partial valuation that realizes it. By making this partial valuation (and by assigning a few other inputs if necessary), we obtain a circuit for PARITY on $\sqrt{n}/2$ variables and all level-1 gates of degree at most t. These circuits are still polynomial-size, because polynomial in n is still polynomial in $\sqrt{n}/2$. $\square$

Proof of Lemma 31.5. Suppose PARITY has t-circuits of depth $d \geq 3$, $t \geq 1$, and size n^k. Let S be the set of clauses in some level-2 conjunction.

Let T be a maximal subset of S such that no two clauses of T contain the same variable, either positively or negatively. By *maximal*, we mean that there is no proper superset of T satisfying this property. Such a set T can be constructed by considering all the clauses in S in some order, taking the next clause into T if it does not have a variable in common with any of the clauses previously taken.

Let $a = 2k + 4$ and $b = -(k+1)/\log(1 - 2^{-a})$. Consider the effect of a random partial valuation on the variables of T. We consider two cases.

Case 1   If T contains at least $b \log n$ elements, then we have at least $b \log n$ clauses, no two of which share a variable. By Lemma 31.3, for sufficiently large n, some clause in T receives all 0 with probability at least $1 - n^{b \log(1 - 2^{-a})} = 1 - n^{-(k+1)}$, in which case the entire conjunction at level 2 disappears.

Case 2   If T contains at most $b \log n$ elements, then $\bigcup T$ contains at most $bt \log n$ elements and shares a variable with all clauses in S (otherwise T was not maximal). By Lemma 31.2, for sufficiently large n, the probability that a random partial valuation leaves more than a variables of $\bigcup T$ unassigned is bounded above by $n^{1-a/2} = n^{-(k+1)}$. Thus with very high probability, there are at most $2a$ literals $\ell_1, \ldots, \ell_{2a}$ of $\bigcup T$ unassigned, and every clause of $\rho(S)$ that is still of size t must contain one of these literals. Let φ_0 be the conjunction of all clauses of $\rho(S)$ of size at most $t - 1$. Of the remaining clauses of $\rho(S)$, let φ_j be the conjunction of those containing the literal ℓ_j and no literal ℓ_i for $i < j$, and let φ'_j be φ_j with all occurrences of ℓ_j deleted. Then φ_j is equivalent to $\ell_j \vee \varphi'_j$, and the original conjunction after the partial evaluation is equivalent to

$$\bigwedge_{j=0}^{2a} \varphi_j \;\equiv\; \varphi_0 \wedge \bigwedge_{j=1}^{2a} (\ell_j \vee \varphi'_j).$$

Using the distributive laws of Boolean algebra, this can be expressed as a disjunction of at most 2^{2a} $(t-1)$-CNF circuits of the form $\varphi_0 \wedge \psi_1 \wedge \cdots \wedge \psi_{2a}$, where each ψ_i is either ℓ_i or φ'_i.

We have shown that for sufficiently large n, with probability at least $1 - n^{-(k+1)}$, any t-CNF gate at level 2 becomes equivalent to a disjunction of a constant number of $(t-1)$-CNF gates under a random partial valuation. This disjunction can be merged with the disjunctions at level 3 to give a $(t-1)$-circuit.

The remainder of the proof proceeds as in Lemma 31.4. Briefly, the probabilities of each of these events and the event that there remain at least $\sqrt{n}/2$ unassigned variables tend to 1 sufficiently fast that they all occur simultaneously with nonzero probability. $\square$

Proof of Lemma 31.6. This is the easy one. A polynomial-size 1-circuit of depth $d \geq 1$ is equivalent to a circuit of depth $d-1$ simply by bypassing the singleton gates at level 1. (In fact, this is just a special case of Lemma 31.5 for $t = 1$.)   □

Lemma 31.7   *There is no $(n-1)$-CNF or $(n-1)$-DNF parity circuit on n inputs.*

Proof. An $(n-1)$-CNF circuit on n inputs is a conjunction of clauses with at most $n-1$ literals per clause. Any partial valuation of a parity circuit is a parity circuit on the remaining variables. But by setting at most $n-1$ variables, we can make all the literals in some clause 0, thus the whole circuit has constant value 0 regardless of the values of the remaining variables, so it cannot be a parity circuit. The argument for $(n-1)$-DNF circuits is similar.   □

Combining Lemmas 31.4–31.7, we have

Theorem 31.8   (Furst, Saxe, and Sipser [46])   PARITY $\notin AC^0$.

Proof. By repeating the constructions of Lemmas 31.4–31.6, we could start with any family of circuits for PARITY of constant depth and polynomial size and reduce them to a family of circuits for PARITY of depth 2, polynomial size, and constant indegree at level 1. These circuits would be t-DNF or t-CNF circuits for some constant t. But this is impossible by Lemma 31.7.   □

Supplementary Lecture H

The Switching Lemma

In this lecture we prove the Håstad switching lemma [56], which can be used to derive the lower bound on circuit size needed to achieve separation of PH^A and $PSPACE^A$ as described in Theorem 30.1.

Any set of variables X generates a free Boolean algebra $\mathcal{F}_X$, which is essentially the set of Boolean formulas over X modulo the axioms of Boolean algebra. If $|X| = n$, the algebra $\mathcal{F}_X$ has 2^{2^n} elements. There is a natural partial order $\leq$ on $\mathcal{F}_X$ corresponding to propositional implication; thus $\varphi \leq \psi$ iff $\varphi \to \psi$ is a propositional tautology.

Recall that a *term* of X is a conjunction of literals, no two of which form a complementary pair. For terms M and N, $M \leq N$ if every literal appearing in N also appears in M. The empty term 1 is the $\leq$-maximum element of $\mathcal{F}_X$. For a term M and CNF formula φ, $M \leq \varphi$ if M contains at least one literal from each clause of φ.

A *minterm* of a formula φ is a $\leq$-maximal term M such that $M \leq \varphi$. If φ is a CNF formula, then M is a minterm of φ iff M contains at least one literal from each clause of φ and no proper subterm of M has this property (Miscellaneous Exercise 71). One can show using the distributive laws of Boolean algebra that the disjunction of all minterms of φ is a DNF formula equivalent to φ (Miscellaneous Exercise 72). The set of minterms of φ is denoted $\mathrm{m}(\varphi)$.

A partial valuation $\rho : X \to X \cup \{0, 1\}$ gives rise to a Boolean algebra homomorphism $\mathcal{F}_X \to \mathcal{F}_Y$, where $Y \subseteq X$ is the set of variables not assigned by ρ. If $A \subseteq X$, we write $\rho(A) = A$ if no variable of A is assigned by ρ, that is, if $\rho(x) = x$ for all $x \in A$. For any term M, either $\rho(M) = 0$ or $M \leq \rho(M)$.

Because ρ is a homomorphism, it preserves $\leq$. Thus if $M \leq \varphi$, then $\rho(M) \leq \rho(\varphi)$. One can show that any minterm N of $\rho(\varphi)$ is $\rho(M)$ for some minterm M of φ (Miscellaneous Exercise 80). However, it is not true that $\rho(M)$ is a minterm of $\rho(\varphi)$ whenever M is a minterm of φ. For example, if $\psi = (x \vee z) \wedge (y \vee \overline{z})$ and $\rho(x) = x$, $\rho(y) = y$, and $\rho(z) = 0$, then xy is a minterm of ψ and $\rho(xy) = xy$, but xy is not a minterm of $\rho(\psi) = x$. However, we have the following special case.

Lemma H.1   *Let φ be a formula, C a clause, and $M \in \mathrm{m}(\varphi \wedge C)$. Let $\mathrm{var}(M, C)$ be the set of variables that occur in both M and C (not necessarily with the same polarity). Let $\sigma : \mathrm{var}(M, C) \to \{0, 1\}$ be the unique valuation such that $\sigma(M) \neq 0$. Then $\sigma(\varphi \wedge C) = \sigma(\varphi)$ and $\sigma(M)$ is a minterm of $\sigma(\varphi)$.*

Proof. The valuation σ is unique, because there is only one way to assign values to variables in $\mathrm{var}(M, C)$ so that the corresponding literals in M get the value 1. Because M must contain a literal of C, the set $\mathrm{var}(M, C)$ is nonempty and $\sigma(C) = 1$, thus $\sigma(\varphi \wedge C) = \sigma(\varphi) \wedge \sigma(C) = \sigma(\varphi)$.

The term M can be written as a conjunction of terms $M = \sigma(M)M'$, where M' contains all the variables in $\mathrm{var}(M, C)$. Note that $\sigma(M)$ contains no variables in $\mathrm{var}(M, C)$. Similarly, φ can be written as a conjunction $\varphi_0 \wedge \varphi_1$, where φ_1 is a CNF formula each of whose clauses contains a literal of M' and φ_0 is a CNF formula none of whose clauses contains a literal of M'. Then $\sigma(\varphi) \leq \varphi_0$ and $M' \leq \varphi_1 \wedge C$.

We now show that $\sigma(M)$ is a minterm of $\sigma(\varphi)$. Surely $\sigma(M) \leq \sigma(\varphi)$, because $M \leq \varphi$. If N were any other term such that $\sigma(M) \leq N \leq \sigma(\varphi)$, then

$$M \;=\; \sigma(M)M' \;\leq\; NM' \;\leq\; \sigma(\varphi) \wedge \varphi_1 \wedge C \;\leq\; \varphi \wedge C.$$

But M is a minterm of $\varphi \wedge C$, therefore $M = \sigma(M)M' = NM'$. Because $\sigma(M)$ and M' are on disjoint sets of variables, $\sigma(M) = N$. $\square$

Lemma H.2   *Let φ be a formula and W a set of variables. Let*

$$\varphi^{-W} \overset{\mathrm{def}}{=} \bigwedge_{\tau : W \to \{0,1\}} \tau(\varphi).$$

(i)   *If φ is written in CNF, then φ^{-W} is equivalent to a CNF formula with no more clauses than φ.*

(ii)   *If $M \in \mathrm{m}(\varphi)$ such that $\mathrm{var}(M, W) = \varnothing$, then $M \in \mathrm{m}(\varphi^{-W})$.*

(iii)   *If $\rho(W) = W$, then $\rho(\varphi)^{-W} = \rho(\varphi^{-W})$.*

(iv)   *If $\rho(W) = W$, then $\rho(\varphi) = 1$ iff $\rho(\varphi^{-W}) = 1$.*

Proof. For (i), let φ be written in CNF. We show that φ^{-W} is equivalent to φ with all literals involving variables in W erased. It suffices to show this for each clause C individually. For each $\tau : W \to \{0, 1\}$, $\tau(C)$ is either 1 or the clause with all literals involving variables in W erased. The former occurs if τ sets one of these literals to 1. The latter occurs if τ sets all these literals to 0, and there is at least one τ. Thus the conjunction C^{-W} is equivalent to this clause.

For (ii), we use (i). Let φ be written in CNF. Because $\mathrm{var}(M, W) = \varnothing$, M still has a literal in common with all clauses even after the variables in W are erased, thus $M \leq \varphi^{-W}$. Also, because each clause C is a disjunction of literals, $C^{-W} \leq C$, therefore $\varphi^{-W} \leq \varphi$. Because $M \leq \varphi^{-W} \leq \varphi$ and M is a minterm of φ, it is also a minterm of φ^{-W} (Miscellaneous Exercise 78).

We leave (iii) and (iv) as exercises (Miscellaneous Exercise 79).   □

Lemma H.3   (Switching Lemma [56])   *Let φ be a t-CNF formula over variables X. Let ρ be a random partial valuation in which every variable is independently assigned 0 or 1 each with probability $(1 - p)/2$ or left unassigned with probability p. Then*

$$\Pr(\rho(\varphi) \text{ is } not \text{ equivalent to an } s\text{-DNF formula}) \quad \leq \quad \alpha^s,$$

where $\alpha = 4pt/\ln 2 \sim 5.77pt$.

Proof. Every formula is equivalent to the disjunction of its minterms, so it suffices to show

$$\Pr(\exists M \in \mathrm{m}(\rho(\varphi)) \; |M| \geq s) \quad \leq \quad \alpha^s.$$

The proof is by induction on the number of clauses of φ. We actually need a stronger induction hypothesis, namely that the bound holds even when conditioned on an arbitrary formula ψ becoming 1 under ρ:

$$\Pr(\exists M \in \mathrm{m}(\rho(\varphi)) \; |M| \geq s \mid \rho(\psi) = 1) \quad \leq \quad \alpha^s. \tag{H.1}$$

For the basis $\varphi = 1$, the only minterm is 1, so the probability on the left-hand side is 0 unless $s = 0$, in which case the right-hand side is 1.

For the induction step, consider a formula $\varphi \wedge C$ of at least one nonempty clause. If the G_i are a family of mutually exclusive and exhaustive events, then to show that $\Pr(E \mid F) \leq a$, it suffices to show separately that $\Pr(E \mid G_i \wedge F) \leq a$ for all i (Miscellaneous Exercise 73). Thus to show (H.1), it suffices to show separately that

$$\Pr(\exists M \in \mathrm{m}(\rho(\varphi \wedge C)) \; |M| \geq s \mid \rho(C) = 1 \wedge \rho(\psi) = 1) \quad \leq \quad \alpha^s \tag{H.2}$$
$$\Pr(\exists M \in \mathrm{m}(\rho(\varphi \wedge C)) \; |M| \geq s \mid \rho(C) \neq 1 \wedge \rho(\psi) = 1) \quad \leq \quad \alpha^s. \tag{H.3}$$

The inequality (H.2) is the easier of the two. This is where we need the stronger induction hypothesis. Under the conditioning $\rho(C) = 1$, we have $\rho(\varphi \wedge C) = \rho(\varphi) \wedge \rho(C) = \rho(\varphi)$, and $\rho(C \wedge \psi) = 1$ iff $\rho(C) = 1$ and $\rho(\psi) = 1$, thus (H.2) is equivalent to

$$\Pr(\exists M \in \mathrm{m}(\rho(\varphi)) \ \ |M| \geq s \ | \ \rho(C \wedge \psi) = 1) \ \leq \ \alpha^s.$$

This follows immediately from the induction hypothesis.

Now for (H.3). For a partial valuation ρ, suppose

$$M \in \mathrm{m}(\rho(\varphi \wedge C)) \ \ \wedge \ \ |M| \geq s.$$

Let $A = \mathrm{var}(M, \rho(C)) \neq \varnothing$. By Lemma H.1, there exist $\sigma : A \to \{0, 1\}$ and $N \in \mathrm{m}(\sigma(\rho(\varphi)))$ (namely $N = \sigma(M)$) such that

- $|N| = |M| - |A| \geq s - |A|$
- $\mathrm{var}(N, C) = \varnothing$
- $\sigma(C) = 1$.

In addition, $\rho(A) = A$ because $A \subseteq \mathrm{var}(\rho(C))$, and σ and ρ commute, because they assign disjoint sets of variables.

Similarly, we can express ρ as a composition $\rho_1 \circ \rho_0$, where ρ_0 is a partial valuation on $\mathrm{var}(C)$ and ρ_1 is a partial valuation on $X - \mathrm{var}(C)$. Thus $\rho_1(C) = C$, $\rho_0(X - \mathrm{var}(C)) = X - \mathrm{var}(C)$, and ρ_1 and ρ_0 commute.

Now we have

$$N \in \mathrm{m}(\sigma(\rho(\varphi)))$$
$$\Rightarrow \quad N \in \mathrm{m}(\rho(\sigma(\varphi))) \qquad\qquad \text{because } \sigma \text{ and } \rho \text{ commute}$$
$$\Rightarrow \quad N \in \mathrm{m}(\rho_1(\rho_0(\sigma(\varphi))))$$
$$\Rightarrow \quad N \in \mathrm{m}(\rho_1(\rho_0(\sigma(\varphi)))^{-C}) \qquad \text{Lemma H.2(ii), } \mathrm{var}(N, C) = \varnothing$$
$$\Rightarrow \quad N \in \mathrm{m}(\rho_1(\rho_0(\sigma(\varphi))^{-C})) \qquad \text{Lemma H.2(iii), } \rho_1(C) = C.$$

We have shown that under the premise

$$\exists M \in \mathrm{m}(\rho(\varphi \wedge C)) \ \ |M| \geq s \ \wedge \ \mathrm{var}(M, \rho(C)) = A, \tag{H.4}$$

where $A \subseteq \mathrm{var}(C)$ and $A \neq \varnothing$, one can derive the conclusion

$$\exists \sigma : A \to \{0, 1\} \ \ \exists N \in \mathrm{m}(\rho_1(\rho_0(\sigma(\varphi))^{-C}))$$
$$|N| \geq s - |A| \ \wedge \ \rho(A) = A \ \wedge \ \sigma(C) = 1. \tag{H.5}$$

Now we are ready to prove the inequality (H.3). Using the law of sum, we can rewrite the left-hand side of (H.3) as

$$\sum_{\substack{A \subseteq \mathrm{var}(C) \\ A \neq \varnothing}} \Pr(E(A) \mid \rho(C) \neq 1 \ \wedge \ \rho(\psi) = 1), \tag{H.6}$$

where $E(A)$ is the event (H.4). Because (H.4) implies (H.5), this is at most

$$\sum_{\substack{A \subseteq \mathrm{var}(C) \\ A \neq \varnothing}} \Pr(F(A) \mid \rho(C) \neq 1 \ \wedge \ \rho(\psi) = 1), \tag{H.7}$$

where $F(A)$ is the event (H.5). By the law of sum, this is at most

$$\sum_{\substack{A \subseteq \mathrm{var}(C) \\ A \neq \varnothing}} \sum_{\substack{\sigma:A \to \{0,1\} \\ \sigma(C)=1}} \Pr(G(A) \mid \rho(C) \neq 1 \ \wedge \ \rho(\psi) = 1), \tag{H.8}$$

where $G(A)$ is the event

$$G(A) \ \overset{\mathrm{def}}{=} \ \exists N \in \mathrm{m}(\rho_1(\rho_0(\sigma(\varphi))^{-C})) \ |N| \geq s - |A| \ \wedge \ \rho(A) = A.$$

To bound $\Pr(G(A) \mid \rho(C) \neq 1 \wedge \rho(\psi) = 1)$, by Miscellaneous Exercise 73 it suffices to bound

$$\Pr(G(A) \mid \rho(C) \neq 1 \ \wedge \ \rho(\psi) = 1 \ \wedge \ \rho_0 = \tau)$$
$$= \ \Pr(G(A) \mid \rho_0(C) \neq 1 \ \wedge \ \rho_1(\rho_0(\psi)) = 1 \ \wedge \ \rho_0 = \tau) \tag{H.9}$$

for all partial valuations τ of $\mathrm{var}(C)$ such that $\tau(C) \neq 1$. But under the new condition $\rho_0 = \tau$, $\rho_1(\rho_0(\psi))$ becomes $\rho_1(\tau(\psi))$, $\rho_1(\rho_0(\sigma(\varphi))^{-C})$ becomes $\rho_1(\tau(\sigma(\varphi))^{-C})$, and the condition $\rho_0(C) \neq 1$ is redundant. Moreover, using Lemma H.2(iv), we can replace the condition $\rho_1(\tau(\psi)) = 1$ by $\rho_1(\tau(\psi)^{-C}) = 1$. After all these changes, (H.9) becomes

$$\Pr\left(\exists N \in \mathrm{m}(\rho_1(\tau(\sigma(\varphi))^{-C})) \ |N| \geq s - |A| \ \wedge \ \rho_0(A) = A \right.$$
$$\left. \mid \ \rho_1(\tau(\psi)^{-C}) = 1 \ \wedge \ \rho_0 = \tau\right). \tag{H.10}$$

Now the conditions involving ρ_0 and those involving ρ_1 are independent, because they refer to the action of ρ on disjoint sets of variables. Thus by Miscellaneous Exercise 75, (H.10) can be rewritten as a product

$$\Pr(\exists N \in \mathrm{m}(\rho_1(\tau(\sigma(\varphi))^{-C})) \ |N| \geq s - |A| \ \mid \ \rho_1(\tau(\psi)^{-C}) = 1)$$
$$\cdot \Pr(\rho_0(A) = A \mid \rho_0 = \tau).$$

The first factor is bounded by $\alpha^{s-|A|}$ by the induction hypothesis, and the second is bounded by $(2p/(1+p))^{|A|}$ by direct calculation. Thus (H.8) is at most

$$\sum_{\substack{A \subseteq \mathrm{var}(C) \\ A \neq \varnothing}} \sum_{\substack{\sigma: A \to \{0,1\} \\ \sigma(C)=1}} \left(\frac{2p}{1+p}\right)^{|A|} \alpha^{s-|A|}$$

$$= \sum_{k=1}^{|C|} \binom{|C|}{k}(2^k - 1)\left(\frac{2p}{1+p}\right)^k \alpha^{s-k}$$

$$\leq \sum_{k=0}^{t} \binom{t}{k}(2^k - 1)\left(\frac{2p}{1+p}\right)^k \alpha^{s-k}$$

$$= \alpha^s \left(\left(1 + \frac{4p}{(1+p)\alpha}\right)^t - \left(1 + \frac{2p}{(1+p)\alpha}\right)^t \right)$$

$$\leq \alpha^s \left(\left(1 + \frac{4p}{\alpha}\right)^t - 1 \right). \tag{H.11}$$

Substituting $4pt/\ln 2$ for α and using the fact that $(1 + 1/x)^x \leq e$ for all positive x, it follows that the larger parenthesized expression in (H.11) is at most 1, therefore (H.11) is at most α^s. $\qquad\square$

Theorem H.4 *There are no circuit families for PARITY of depth d and size $2^{(\log n)^c}$ for any constants c and d.*

Proof. Suppose there were. Let $t = (\log n)^{c+1}$ and $p = (\ln 2)/(8t)$. Without loss of generality, assume that the level-2 circuits are all t-CNF (add another level if necessary). Applying a random partial valuation, the probability that any given level-2 circuit does not become equivalent to a t-DNF circuit is bounded by α^t. By the law of sum, the probability that there exists a level-2 circuit that does not become equivalent to a t-DNF circuit is bounded by

$$2^{(\log n)^c} \cdot \alpha^t \;=\; 2^{(\log n)^c} \cdot 2^{-(\log n)^{c+1}},$$

which is vanishingly small as a function of n.

Afterwards, the expected number of unassigned variables is still quite large:

$$np \;=\; \frac{n \ln 2}{8(\log n)^{c+1}} \;\geq\; 2n(\log n)^{-(c+2)}$$

for sufficiently large n. The probability that the actual number of unassigned variables is less than half that is also vanishingly small (Miscellaneous Exercise 86).

Thus for sufficiently large n, there is nonzero probability that all t-CNF level-2 circuits can be replaced by equivalent t-DNF circuits and that there are still at least $n(\log n)^{-(c+2)}$ unassigned variables. The size of the new circuit in terms of the input size $m = n(\log n)^{-(c+2)}$ is still at most $2^{(\log m)^{c+3}}$. Because there is nonzero probability, there must exist a partial valuation making it true.

Iterating this construction $d - 2$ times, we obtain a family of circuits for PARITY of depth 2 and size $2^{(\log m)^k}$ for some k such that all level-1 circuits are of degree at most t. But for sufficiently large n, this contradicts Lemma 31.7. □

Corollary H.5   *There exists an oracle A such that $PH^A \neq PSPACE^A$.*

Proof. Theorems H.4 and 30.1. □

Supplementary Lecture I

Tail Bounds

In probabilistic analysis, we often need to bound the probability that a random variable deviates far from its mean. There are various formulas for this purpose. These are called *tail bounds*. The weakest of these is the *Markov bound*, which states that for any nonnegative random variable X with mean $\mu = \mathcal{E}X$,

$$\Pr(X \geq k) \;\; \leq \;\; \mu/k \tag{I.1}$$

(Miscellaneous Exercise 83). A better bound is the *Chebyshev bound*, which states that for a random variable X with mean $\mu = \mathcal{E}X$ and standard deviation $\sigma = \sqrt{\mathcal{E}((X - \mu)^2)}$, for any $\delta \geq 1$,

$$\Pr(|X - \mu| \geq \delta\sigma) \;\; \leq \;\; \delta^{-2} \tag{I.2}$$

(Miscellaneous Exercise 84).

The Markov and Chebyshev bounds converge linearly and quadratically, respectively, and are often too weak to achieve desired estimates. In particular, for the special case of Bernoulli trials (sum of independent, identically distributed 0,1-valued random variables) or more generally Poisson trials (sum of independent 0,1-valued random variables, not necessarily identically distributed), the convergence is exponential.

Consider Poisson trials X_i, $1 \leq i \leq n$, with sum $X = \sum_i X_i$ and $\Pr(X_i = 1) = p_i$. An exact expression for the upper tail is

$$\Pr(X \geq k) \;=\; \sum_{\substack{A \subseteq \{1,\dots,n\} \\ |A| \geq k}} \prod_{i \in A} p_i \prod_{i \notin A} (1 - p_i).$$

In the special case of Bernoulli trials with success probability p, this simplifies to the *binomial distribution*

$$\Pr(X \geq k) \;=\; \sum_{i \geq k} \binom{n}{i} p^i (1 - p)^{n-i}.$$

However, these expressions are algebraically unwieldy. A more convenient formula is provided by the *Chernoff bound*.

The Chernoff bound comes in several forms. One form states that for Poisson trials X_i with sum $X = \sum_i X_i$ and $\mu = \mathcal{E}X$, for any $\delta > 0$,

$$\Pr(X \geq (1 + \delta)\mu) \;<\; \left(\frac{e^\delta}{(1 + \delta)^{1+\delta}} \right)^{\mu}. \tag{I.3}$$

Equivalently,

$$\Pr(X \geq (1 + \delta)\mu) \;<\; \left(e \left(1 - \frac{\delta}{1 + \delta} \right)^{(1+\delta)/\delta} \right)^{\delta\mu}. \tag{I.4}$$

In (I.4), the subexpression

$$\left(1 - \frac{\delta}{1 + \delta} \right)^{(1+\delta)/\delta} \tag{I.5}$$

is a special case of the function $(1 - 1/x)^x$, which arises frequently in asymptotic analysis. It is worth remembering that this function is bounded above by e^{-1} for all positive x and tends to that value in the limit as x approaches infinity. Similarly, the function $(1 + 1/x)^x$ is bounded above by e for all positive x and tends to that limit as x approaches infinity (Miscellaneous Exercise 57(a)).

A third form equivalent to (I.3) and (I.4) is: for all $k > \mu$,

$$\Pr(X \geq k) \;<\; e^{k-\mu}(\mu/k)^k. \tag{I.6}$$

One can see clearly from (I.4) and (I.6) that the convergence is exponential with distance from the mean.

These formulas bound the upper tail of the distribution. There are also symmetric versions for the lower tail: for any δ such that $0 \leq \delta < 1$ and k

such that $0 < k \leq \mu$,

$$\Pr\left(X \leq (1-\delta)\mu\right) \quad < \quad \left(\frac{e^{-\delta}}{(1-\delta)^{1-\delta}}\right)^{\mu} \tag{I.7}$$

$$= \quad \left(e^{-1}\left(1 + \frac{\delta}{1-\delta}\right)^{(1-\delta)/\delta}\right)^{\delta\mu} \tag{I.8}$$

$$\Pr\left(X \leq k\right) \quad < \quad e^{k-\mu}(\mu/k)^{k}. \tag{I.9}$$

In the case of the lower tail, we also have a fourth version given by

$$\Pr\left(X \leq (1-\delta)\mu\right) \quad < \quad e^{-\delta^{2}\mu/2}. \tag{I.10}$$

This bound is slightly weaker than (I.7)–(I.9), but is nevertheless very useful because of its simple form.

Proof of the Chernoff Bound

We now prove the Chernoff bound (I.3) for Poisson trials X_i. It is easy to show that the other forms (I.4) and (I.6) are equivalent, and these are left as exercises (Miscellaneous Exercise 87). The proofs of the corresponding bounds (I.7)–(I.9) for the lower tail are similar and are also left as exercises (Miscellaneous Exercise 88). The weaker bound (I.10) requires a separate argument involving the Taylor expansion of $\ln(1-\delta)$, but is not difficult (Miscellaneous Exercise 89).

Although the success probabilities of the X_i may differ, it is important that the trials be independent. At a crucial step of the proof, we use the fact that the expected value of the product of independent trials is the product of their expectations (Miscellaneous Exercise 82).

Let X_i be Poisson trials with success probabilities p_i, sum $X = \sum_i X_i$, and mean $\mu = \mathcal{E}X = \sum_i p_i$. Fix $a > 0$. By the monotonicity of the exponential function and the Markov bound (I.1), we have

$$\Pr\left(X \geq (1+\delta)\mu\right) \quad = \quad \Pr\left(e^{aX} \geq e^{a(1+\delta)\mu}\right)$$

$$\leq \quad \mathcal{E}(e^{aX}) \cdot e^{-a(1+\delta)\mu}. \tag{I.11}$$

Because the expected value of the product of independent trials is the product of their expectations (Miscellaneous Exercise 82), and because the e^{aX_i} are independent if the X_i are, we can write

$$\mathcal{E}(e^{aX}) \quad = \quad \mathcal{E}(e^{\sum_i aX_i}) \quad = \quad \mathcal{E}(\prod_i e^{aX_i}) \quad = \quad \prod_i \mathcal{E}(e^{aX_i})$$

$$= \quad \prod_i (p_i e^a + (1-p_i)) \quad = \quad \prod_i (1 + p_i(e^a - 1)). \tag{I.12}$$

It follows from $(1 + 1/x)^x < e$ that $1 + y < e^y$ for all positive y. Applying this with $y = p_i(e^a - 1)$, we have $1 + p_i(e^a - 1) < e^{p_i(e^a - 1)}$, thus (I.12) is strictly bounded by

$$\prod_i e^{p_i(e^a - 1)} \;=\; e^{\sum_i p_i(e^a - 1)} \;=\; e^{(e^a - 1)\mu}.$$

Combining this with the expression $e^{-a(1+\delta)\mu}$ gives a strict bound

$$e^{(e^a - 1)\mu} \cdot e^{-a(1+\delta)\mu} \;=\; e^{(e^a - 1 - a - a\delta)\mu} \tag{I.13}$$

on (I.11). Now we wish to choose a minimizing $e^a - 1 - a - a\delta$. The derivative vanishes at $a = \ln(1 + \delta)$, and substituting this value for a in (I.13) and simplifying yields (I.3).

Lecture 32

The Gap Theorem and Other Pathology

One might get the impression from the structure of the complexity hierarchies we have studied that all problems have a natural inherent complexity, and that allowing slightly more time or space always allows more to be computed. Both these statements seem to be true for most natural problems and complexity bounds, but neither is true in general. One can construct pathological examples for which they provably fail.

For example, one can exhibit a computable function f with no asymptotically best algorithm, in the sense that for any algorithm for f running in time $T(n)$, there is another algorithm for f running in time $\log T(n)$. Thus f can be endlessly sped up. Also, there is nothing special about the log function—the result holds for any total recursive function.

For another example, one can show that there is a space bound $S(n)$ such that any function computable in space $S(n)$ is also computable in space $\log S(n)$. At first this might seem to contradict Theorem 3.1, but that theorem has a constructibility condition that is not satisfied by $S(n)$. Again, this holds for any recursive improvement, not just log.

Most of the examples of this lecture are constructed by intricate diagonalizations. They do not correspond to anything natural and would never arise in real applications. Nevertheless, they are worth studying as a way to better understand the power and limitations of complexity theory. We prove these results in terms of Turing machine time and space in this lecture; however, most of them are independent of the particular measure. A more abstract treatment is given in Supplementary Lecture J.

The first example we look at is the gap theorem, which states that there are arbitrarily large recursive gaps in the complexity hierarchy. This result is due independently to Borodin [21] and Trakhtenbrot [122].

Theorem 32.1   **(Gap Theorem [21, 122])**   *For any total recursive function $f : \omega \rightarrow \omega$ such that $f(x) \geq x$, there exists a time bound $T(n)$ such that $DTIME(f(T(n))) = DTIME(T(n))$; in other words, there is no set accepted by a deterministic TM in time $f(T(n))$ that is not accepted by a deterministic TM in time $T(n)$.*

Proof. Let $T_i(x)$ denote the running time of TM M_i on input x. For each n, define $T(n)$ to be the least m such that for all $i \leq n$, if $T_i(n) \leq f(m)$, then $T_i(n) \leq m$. To compute $T(n)$, start by setting $m := 0$. As long as there exists an $i \leq n$ such that $m < T_i(n) \leq f(m)$, set $m := T_i(n)$. This process must terminate, because there are only finitely many $i \leq n$. The value of $T(n)$ is the final value of m.

Now we claim that $T(n)$ satisfies the requirements of the theorem. Suppose M_i runs in time $f(T(n))$. Thus $T_i(n) \leq f(T(n))$ a.e.[1] By construction of T, for sufficiently large $n \geq i$, $T_i(n) \leq T(n)$. $\square$

What we have actually proved is stronger than the statement of the theorem. The theorem states that for any deterministic TM M_i running in time $f(T(n))$, there is an equivalent deterministic TM M_j running in time $T(n)$. But what we have actually shown is that any deterministic TM running in time $f(T(n))$ also runs in time $T(n)$.

Of course, all these bounds hold a.e., but we can make them hold everywhere by encoding the values on small inputs in the finite control and computing them by table lookup.

The next example gives a set for which any algorithm can be sped up arbitrarily many times by an arbitrary preselected recursive amount. This result is due to Blum [17].

Theorem 32.2   **(Speedup Theorem [17])**   *Let $T_i(x)$ denote the running time of TM M_i on input x. Let $f : \omega \rightarrow \omega$ be any monotone total recursive function such that $f(n) \geq n^2$. There exists a recursive set A such that for any TM M_i accepting A, there is another TM M_j accepting A with $f(T_j(x)) < T_i(x)$ a.e.*

---

[1] "a.e." means "almost everywhere" or "for all but finitely many n". Also, "i.o." means, "infinitely often" = "for infinitely many n".

Proof. Let f^n denote the n-fold composition of f with itself:

$$f^n \quad \overset{\text{def}}{=} \quad \underbrace{f \circ f \circ \cdots \circ f}_{n}.$$

Thus f^0 is the identity function, $f^1 = f$, and $f^{m+n} = f^m \circ f^n$. For example, if $f(m) = m^2$, then $f^n(m) = m^{2^n}$, and if $f(m) = 2^m$, then $f^n(m)$ is an iterated exponential involving a stack of 2's of height n.

We construct by diagonalization a set $A \subseteq 0^*$ such that

(i) for any machine M_i accepting A, $T_i(0^n) > f^{n-i}(2)$ a.e.,[2] and

(ii) for all k, there exists a machine M_j accepting A such that $T_j(0^n) \leq f^{n-k}(2)$ a.e.

This achieves our goal, because for any machine M_i accepting A, (ii) guarantees the existence of a machine M_j accepting A such that $T_j(0^n) \leq f^{n-i-1}(2)$ a.e.; but then

$$
\begin{aligned}
f(T_j(0^n)) \quad &\leq \quad f(f^{n-i-1}(2)) \text{ a.e.} \quad \text{by monotonicity of } f \\
&= \quad f^{n-i}(2) \\
&< \quad T_i(0^n) \text{ a.e.} \qquad \text{by (i).}
\end{aligned}
$$

Now we turn to the construction of the set A. Let $M_0, M_1, \ldots$ be a list of all one-tape Turing machines with input alphabet $\{0\}$. Let N be an enumeration machine that carries out the following simulation. It maintains a finite *active list* of descriptions of machines currently being simulated. We assume that a description of M_i suitable for universal simulation is easily obtained from the index i.

The computation of N proceeds in stages. Initially, the active list is empty. At stage n, N puts the next machine M_n at the end of the active list. It then simulates the machines on the active list in order, smallest index first. For each such M_i, it simulates M_i on input 0^n for $f^{n-i}(2)$ steps. It picks the first one that halts within its allotted time and does the opposite: if M_i rejects 0^n, N declares $0^n \in A$, and if M_i accepts 0^n, N declares $0^n \notin A$. This ensures that $L(M_i) \neq A$. It then deletes M_i from the active list. If no machine on the active list halts within its allotted time, then N just declares $0^n \notin A$.

This construction ensures that any machine M_i that runs in time $f^{n-i}(2)$ i.o. does not accept A. The machine M_i is put on the active list at stage i. Thereafter, if M_i halts within time $f^{n-i}(2)$ on 0^n but is not

---

[2]We are regarding $f^{n-i}(2)$ as a function of n with i a fixed constant. Thus "i.o." and "a.e." in this context are meant to be interpreted as "for infinitely many n" and "for all but finitely many n", respectively.

chosen for deletion, then some higher priority machine on the active list must have been chosen; but this can happen only finitely many times. So if M_i halts within time $f^{n-i}(2)$ on 0^n i.o., then eventually M_i will be the highest priority machine on the list and will be chosen for deletion, say at stage n. At that point, 0^n will be put into A iff $0^n \notin L(M_i)$, ensuring $L(M_i) \neq A$. This establishes condition (i) above.

For condition (ii), we need to show that for all k, A is accepted by a one-tape TM N_k running in time $f^{n-k}(2)$ a.e. The key idea is to hard-code the first m stages of the computation of N in the finite control of N_k for some sufficiently large m. Note that for each M_i, either

(A) $T_i(0^n) \leq f^{n-i}(2)$ i.o., in which case there is a stage $m(i)$ at which N deletes M_i from the active list; or

(B) $T_i(0^n) > f^{n-i}(2)$ a.e., in which case there is a stage $m(i)$ after which M_i always exceeds its allotted time.

Let $m = \max_{i \leq k} m(i)$. We cannot determine the $m(i)$ or m effectively (Miscellaneous Exercise 105), but we do know that they exist. The machine N_k has a list of elements $0^n \in A$ for $n \leq m$ hard-coded in its finite control. On such inputs, it simply does a table lookup to determine whether $0^n \in A$ and accepts or rejects accordingly. On inputs 0^n for $n > m$, it simulates the action of N on stages $m+1, m+2, \ldots, n$ starting with a certain active list, which it also has hard-coded in its finite control. The active list it starts with is N's active list at stage m with all machines M_i for $i \leq k$ deleted. This does not change the status of $0^n \in A$: for each M_i with $i \leq k$, in case A it has already been deleted from the active list by stage m, and in case B it will always exceed its allotted time after stage m, so it will never be a candidate for deletion. The simulation will therefore behave exactly as N would at stage m and beyond. The machine N_k can thus determine whether $0^n \in A$ and accept or reject accordingly.

It remains to estimate the running time of N_k on input 0^n. If $n \leq m$, N_k takes linear time, enough time to read the input and do the table lookup. If $n > m$, N_k must simulate at most $n - k$ machines on the active list on $n - m$ inputs, each for at most $f^{n-k-1}(2)$ steps. Under mild assumptions on the encoding scheme, interpreting the binary representation of the index i as a description of M_i, M_i has at most $\log i$ states, at most $\log i$ tape symbols, and at most $\log i$ transitions in its finite control, and one step of M_i can be simulated in roughly $c(\log i)^2$ steps of N_k. Thus the total time needed for all the simulations is at most $cn^2(\log n)^2 f^{n-k-1}(2)$. But

$$
\begin{aligned}
cn^2(\log n)^2 &\leq 2^{2^{n-k-1}} \quad \text{a.e.} \\
&\leq f^{n-k-1}(2) \quad \text{because } f(m) \geq m^2,
\end{aligned}
$$

therefore

$$
\begin{aligned}
cn^2 (\log n)^2 f^{n-k-1}(2) &\leq (f^{n-k-1}(2))^2 \text{ a.e.} \\
&\leq f(f^{n-k-1}(2)) \\
&= f^{n-k}(2).
\end{aligned}
$$

$\square$

There are a few interesting observations we can make about the proof of Theorem 32.2.

First, the "mild assumptions" on the encoding scheme are inconsequential. If they are not satisfied, the condition $f(m) \geq m^2$ can be strengthened accordingly. We only need to know that the overhead for universal simulation of Turing machines is bounded by a total recursive function.

The value $m = \max_{i \leq k} m(i)$ in the proof of Theorem 32.2 cannot be obtained effectively. We know that for each M_i there exists such an m, but it is undecidable whether M_i falls in case A or case B, so we do not know whether to delete M_i from the active list. Indeed, it is impossible to obtain a machine for A running in time $f^{n-k}(2)$ effectively from k (Miscellaneous Exercise 105).

Lecture 33

Partial Recursive Functions and Gödel Numberings

Partial and Total Recursive Functions

The next few lectures are an introduction to classical recursive function theory. For a more comprehensive treatment of this subject, see [104, 114].

A *partial recursive function* is a computable partial function $f : \omega \to \omega$. *Partial* means that it need not be defined on all inputs. *Computable* can be defined in several equivalent ways: by Turing machines, by Gödel's μ-recursive functions, by the λ-calculus, or by C programs, to name a few. A partial recursive function is *total* if it is everywhere defined.

For example, we can define the partial recursive function computed by a deterministic Turing machine M to be the partial function f such that

- if M does not halt on input x, then $f(x)$ is undefined, and

- if M halts on input x, then $f(x)$ is the value written on M's tape when it enters its halt state.

Pairing

We consider only unary functions $\omega \to \omega$. Functions of higher arity can be encoded using the one-to-one pairing function $\texttt{<} \ \texttt{>} : \omega^2 \to \omega$:

$$\texttt{<}i,j\texttt{>} \quad \stackrel{\text{def}}{=} \quad \binom{i+j+1}{2} + i. \tag{33.1}$$

$$
\begin{array}{c|cccccc}
 & \multicolumn{6}{c}{j} \\
 & 0 & 1 & 2 & 3 & 4 & 5 \\
\hline
0 & 0 & 1 & 3 & 6 & 10 & 15 \\
1 & 2 & 4 & 7 & 11 & 16 \\
2 & 5 & 8 & 12 & 17 \\
3 & 9 & 13 & 18 & & \ddots \\
4 & 14 & 19 \\
5 & 20 \\
\end{array}
$$

The corresponding projections are denoted π_1^2 and π_2^2:

$$\pi_1^2(\texttt{<}x,y\texttt{>}) \quad \stackrel{\text{def}}{=} \quad x \qquad\qquad \pi_2^2(\texttt{<}x,y\texttt{>}) \quad \stackrel{\text{def}}{=} \quad y.$$

For $n > 2$, we take

$$\texttt{<}x_1, \dots, x_n\texttt{>} \quad \stackrel{\text{def}}{=} \quad \texttt{<}x_1, \texttt{<}x_2, \dots, \texttt{<}x_{n-1}, x_n\texttt{>>>},$$

and for $m \le n$,

$$\pi_1^n \quad \stackrel{\text{def}}{=} \quad \pi_1^2 \qquad\qquad \pi_m^n \quad \stackrel{\text{def}}{=} \quad \pi_{m-1}^{n-1} \circ \pi_2^2.$$

To save notation, we write $f(x, y)$ instead of $f(\texttt{<}x,y\texttt{>})$ and $f(x_1, \dots, x_n)$ instead of $f(\texttt{<}x_1, \dots, x_n\texttt{>})$. But keep in mind that officially, all partial recursive functions are unary.

We also overload the symbol $\texttt{<} \ \texttt{>}$ by defining the following pairing operator on functions:

$$\texttt{<}f, g\texttt{>} \quad \stackrel{\text{def}}{=} \quad \lambda x.\texttt{<}f(x), g(x)\texttt{>}.$$

Basic Closure Properties

The partial recursive functions contain the constant functions $\kappa_c \stackrel{\text{def}}{=} \lambda x.c$ and projections π_1^2, π_2^2 and are closed under composition and pairing (among other closure properties). That is, if f and g are partial recursive functions, then so are $f \circ g \stackrel{\text{def}}{=} \lambda x.f(g(x))$ and $\texttt{<}f, g\texttt{>} \stackrel{\text{def}}{=} \lambda x.\texttt{<}f(x), g(x)\texttt{>}$.

The constant functions and projections are total. The composition $f \circ g$ is defined on x iff g is defined on x and f is defined on $g(x)$. The pair $\texttt{<}f, g\texttt{>}$ is defined on x iff both f and g are defined on x.

Gödel Numberings

There are only countably many partial recursive functions, because there are only countably many Turing machines (or C programs, or λ-terms, or μ-recursive functions, ...). An enumeration

$$\varphi_0, \ \varphi_1, \ \varphi_2, \ \ldots$$

of the partial recursive functions is called a *Gödel numbering* or *acceptable indexing* if it satisfies three properties:

(i)   Every partial recursive function is φ_i for some i.

(ii) The *universal function property*:   There is a partial recursive function U such that for all i and x,

$$U(i, x) \ = \ \varphi_i(x).$$

(iii) The s_n^m *property*:   There exist total recursive functions s_n^m such that for all i, $x_1, \ldots, x_n$, $y_1, \ldots, y_m$,

$$\varphi_{s_n^m(i,x_1,\ldots,x_n)}(y_1, \ldots, y_m) \ = \ \varphi_i(x_1, \ldots, x_n, y_1, \ldots, y_m).$$

The number i is called a *Gödel number* or *index* for the function φ_i.

Although the index i is just a number, it is convenient to think of i as a description of an algorithm or machine to compute the function φ_i. For example, i might encode some Turing machine to compute φ_i, or a C program, or something similar. The exact form depends on the particular indexing, and we are not so concerned with the exact form as we are with the properties (i)–(iii). All we really care about is that each partial recursive function has at least one index (property (i)), that it is possible to simulate functions uniformly given their indices (property (ii), the universal function property), and that it is possible to hard-code part of the input into the program (property (iii), the s_n^m property).

For example, let us take the indexing provided by Turing machines. Writing i as a binary string, we might interpret i as an encoded description of a Turing machine; see [61, 76] for such an encoding of a very specific form. Any number whose binary representation does not encode a Turing machine according to this scheme can be taken as an index of a trivial one-state Turing machine. Because every partial recursive function is computed by a Turing machine, and because every Turing machine has a description that can be encoded as a binary number, property (i) holds. Because there is a universal Turing machine that can take a description of another machine i and an input x and simulate the machine with description i on x, property (ii) holds. Finally, because it is possible to code parts of the input to a machine in the finite control, so that they can be accessed by table lookup, this indexing scheme satisfies property (iii).

Property (iii) of acceptable indexings assumes the existence of s_n^m functions. Actually, we only need to assume the existence of s_1^1; all the others are definable. For example, we can take

$$s_2^3 \stackrel{\text{def}}{=} s_1^1 \circ <s_1^1 \circ <\pi_1^3, \pi_2^3>, \pi_3^3>,$$

because then

$$
\begin{aligned}
\varphi_i(x_1, x_2, y_1, y_2, y_3) &= \varphi_i(<x_1, <x_2, <y_1, y_2, y_3>>>) \\
&= \varphi_{s_1^1(s_1^1(i,x_1),x_2)}(<y_1, y_2, y_3>) \\
&= \varphi_{s_1^1 \circ <s_1^1 \circ <\pi_1^3, \pi_2^3>, \pi_3^3>(i,x_1,x_2)}(<y_1, y_2, y_3>) \\
&= \varphi_{s_2^3(i,x_1,x_2)}(y_1, y_2, y_3).
\end{aligned}
$$

Comp, Const, and Pair

We can obtain an index for the composition $f \circ g$ of two partial recursive functions effectively[1] from indices for f and g. In other words, there exists a total recursive function **comp** such that

$$\varphi_{\text{comp}(i,j)} = \varphi_i \circ \varphi_j.$$

Here is the construction of **comp**. Let m be an index for the partial recursive function $U \circ <\pi_1^3, U \circ <\pi_2^3, \pi_3^3>>$. Then

$$
\begin{aligned}
(\varphi_i \circ \varphi_j)(x) &= \varphi_i(\varphi_j(x)) \\
&= U(i, U(j, x)) \\
&= U \circ <\pi_1^3, U \circ <\pi_2^3, \pi_3^3>>(i, j, x) \\
&= \varphi_m(i, j, x) \\
&= \varphi_{s_2^1(m,i,j)}(x),
\end{aligned}
$$

so we can take **comp** to be the total recursive function

$$
\begin{aligned}
\text{comp} \ &\stackrel{\text{def}}{=} \ \lambda <i,j>.s_2^1(m,i,j) \\
&= \ \lambda x.s_2^1(\kappa_m(x), \pi_1^2(x), \pi_2^2(x)) \\
&= \ s_2^1 \circ <\kappa_m, \pi_1^2, \pi_2^2>.
\end{aligned}
$$

This is total because s_2^1, κ_m, π_1^2, and π_2^2 are.

We can even get an index for **comp** if we like. Let ℓ be an index for s_2^1. Then

$$\text{comp}(i,j) = s_2^1(m,i,j) = \varphi_\ell(m,i,j) = \varphi_{s_1^2(\ell,m)}(i,j),$$

---

[1] *effectively* = by a recursive function.

so $s_1^2(\ell, m)$ is an index for comp.

We can also obtain an index for the pair $<f, g>$ of two partial recursive functions effectively from indices for f and g and an index for the constant function κ_c effectively from c. In other words, there exist total recursive functions pair and const such that

$$\varphi_{\mathsf{pair}(i,j)} \;=\; <\varphi_i, \varphi_j> \qquad\qquad \varphi_{\mathsf{const}(i)} \;=\; \kappa_i.$$

The construction of pair and const is similar to the construction of comp given above and is left as an exercise (Homework 10, Exercise 1).

The Recursion Theorem

One of the most intriguing aspects of recursive function theory is the power of self-reference. There is a general theorem called the *recursion theorem* that is a kind of a fixpoint theorem. It says that any total recursive functional (function that acts on indices) has a fixpoint. Formally,

Theorem 33.1 **(Recursion Theorem)**   *For any total recursive function σ, there exists an index i such that $\varphi_i = \varphi_{\sigma(i)}$. Moreover, such an i can be obtained effectively from an index for σ.*

We give several applications of this theorem in Lecture 34. For now we point out its similarity to Gödel's fixpoint lemma, which is used in the proof of the incompleteness theorem (see [76]). Recall that the fixpoint lemma states that for any formula $\Phi(x)$ of the language of number theory with one free variable x, there is a sentence Ψ such that

$$\mathbb{N} \;\vDash\; \Psi \leftrightarrow \Phi(\ulcorner\Psi\urcorner),$$

where $\ulcorner\Psi\urcorner$ is the numeric code of the sentence Ψ in some reasonable coding scheme. (Think: "code" = "Gödel number".) In fact, the fixpoint lemma and the recursion theorem are essentially the same phenomenon in different formalisms, and their proofs are very similar. There is also a strong connection to the fixpoint combinator $\lambda f.(\lambda x.f(xx)\ \lambda x.f(xx))$ of the λ-calculus.

The recursion theorem is due to Kleene [72] (see also [73]).

Proof of Theorem 33.1. Let h be a total recursive function that on input v produces the index of a function that on input x

(i)   computes $\varphi_v(v)$;

(ii) if $\varphi_v(v)$ is defined, applies σ to $\varphi_v(v)$; and

(iii) interprets $\sigma(\varphi_v(v))$ as an index and applies the function with that index to x.

Thus

$$\varphi_{h(v)}(x) \;\; = \;\; \varphi_{\sigma(\varphi_v(v))}(x)$$

if $\varphi_v(v)$ is defined, undefined otherwise. It is important to note that h itself is a total recursive function; it does not do any of the steps (i)–(iii) above, it only computes the index of a function that does them.

Now let u be an index for h. Then $h(u) = \varphi_u(u)$ is the desired fixpoint of σ: for all x,

$$\varphi_{h(u)}(x) \;\; = \;\; \varphi_{\sigma(\varphi_u(u))}(x) \;\; = \;\; \varphi_{\sigma(h(u))}(x).$$

We leave the second statement of the theorem, the fact that the fixpoint can be obtained effectively from an index for σ, as an exercise (Homework 10, Exercise 2). $\qquad\square$

We give several applications of the recursion theorem next time.

Lecture 34

Applications of the Recursion Theorem

Let $\varphi_0, \varphi_1, \ldots$ be a Gödel numbering of the partial recursive functions. Recall from last time the statement of the recursion theorem: any total recursive functional σ has a fixpoint i, that is, an index i such that $\varphi_i = \varphi_{\sigma(i)}$. Moreover, we can find i effectively from an index for σ.

The recursion theorem was originally conceived as a way to prove the existence of functions defined by recursion (hence the name). For example, the factorial function is a fixpoint of the total recursive transformation

$$P \;\mapsto\; \lambda x. \begin{cases} 1, & \text{if } x = 0 \\ x \cdot P(x-1), & \text{otherwise.} \end{cases}$$

We explore this application in Homework 12, Exercise 2. However, the recursion theorem has many other far-reaching consequences. It captures in a concise way the fundamental idea of *self-reference*.

A Self-Printing Program

As an example of self-reference, here is a C program that prints itself out:

```
char *s="char *s=%c%s%c;%cmain(){printf(s,34,s,34,10,10);}%c";
main(){printf(s,34,s,34,10,10);}
```

Here 34 and 10 are the ASCII codes for double quote and newline, respectively. In essence, the `printf` statement says that we should take the string `s` and print it after inserting a quoted copy of itself.

The same principle is illustrated in the following UNIX shell script. It is written with extra spaces and line breaks for readability; the self-printing program is actually the output of this program.

```
x='y=`echo .|tr . "\47"`;echo "x=$y$x$y;$x"'
y=`echo .|tr . "\47"`;echo "x=$y$x$y;$x"
```

Such programs are sometimes called *quines* after the philosopher Willard van Orman Quine.

Any general-purpose programming language has the power to construct quines. In any Gödel numbering of the partial recursive functions, the "program that prints itself out" would be an index i such that for all x, $\varphi_i(x) = i$. To obtain such an i, take the fixpoint of the functional const constructed in Lecture 33.

Rice's Theorem

Rice's theorem [102, 103] states that every nontrivial property of the recursively enumerable (r.e.) sets is undecidable. Here is a proof of this using the recursion theorem. Intuitively, if a nontrivial property were decidable, then one could construct a recursive functional with no fixpoint, contradicting the recursion theorem.

We show that every nontrivial property of the partial recursive functions is undecidable, where a *property of the partial recursive functions* is a map $P : \omega \to \{0,1\}$ such that if $\varphi_i = \varphi_j$, then $P(i) = P(j)$ (thus it is a property of functions, not of indices), and P is *nontrivial* if it is neither universally false nor universally true.

Suppose P is such a property. Because P is nontrivial, there exist indices i_0 and i_1 such that $P(i_0) = 0$ and $P(i_1) = 1$. Suppose P were decidable. Then the function

$$\sigma(j) \stackrel{\text{def}}{=} \begin{cases} i_0, & \text{if } P(j) = 1, \\ i_1, & \text{if } P(j) = 0 \end{cases}$$

would be a total recursive function. But σ has no fixpoint: for all j,

$$P(\sigma(j)) = \begin{cases} P(i_0) = 0, & \text{if } P(j) = 1, \\ P(i_1) = 1, & \text{if } P(j) = 0, \end{cases}$$

therefore $P(j) \neq P(\sigma(j))$. Because P is a property of functions, $\varphi_j \neq \varphi_{\sigma(j)}$. This contradicts the recursion theorem.

Minimal Programs

There is no algorithm to find a smallest program equivalent to a given one, for any reasonable definition of "smallest". This is true regardless of the programming language. For example, for Turing machines, there is no algorithm to find a Turing machine with the fewest states equivalent to a given one.

In general terms, for any Gödel numbering, there does not exist a total recursive function σ such that for all j, $\sigma(j)$ is a minimal index for φ_j. Here "minimal" means with respect to the natural order $\leq$ on ω. We prove an even stronger result:

Theorem 34.1 *In any Gödel numbering, there does not exist an infinite r.e. list of minimal indices.*

Proof. Suppose such a list did exist. Consider the total recursive function σ that on input x enumerates the list until encountering an index greater than x and takes that as its value. Then σ has no fixpoint: for all x, $\varphi_x \neq \varphi_{\sigma(x)}$, because $x < \sigma(x)$ and $\sigma(x)$ is a minimal index. This contradicts the recursion theorem.   □

Effective Padding

Even though you cannot effectively find a smaller index equivalent to a given one, you can always find a larger one. This is called *effective padding*. With Turing machines and Java programs, it is easy to pad the machine or program to get a larger one that is equivalent: for Turing machines, just throw in some dummy inaccessible states, and for programs, just include some dummy inaccessible statements. In general, any Gödel numbering has this padding property:

Lemma 34.2 *In any Gödel numbering, there exists a total recursive function σ such that for all x, $\sigma(x) > x$ and $\varphi_x = \varphi_{\sigma(x)}$.*

Proof. Say we are given x. To compute $\sigma(x)$, we go though a number of stages. We start at stage 0 with $B := \{x\}$. Now suppose that at some stage we have constructed $B \subseteq \{0, 1, 2, \ldots, x - 1, x\}$ such that for all $y \in B$, $\varphi_y = \varphi_x$. Consider the total recursive function

$$f(z) \;=\; \begin{cases} x + 1, & \text{if } z \in B, \\ x, & \text{if } z \notin B, \end{cases}$$

and let y be a fixpoint of f; that is, an index such that $\varphi_y = \varphi_{f(y)}$. We can get our hands on y by the effective version of the recursion theorem. If $y > x$, we are done:

$$\varphi_y \;=\; \varphi_{f(y)} \;=\; \varphi_x,$$

so we can set $\sigma(x) = y$. If $y \in B$, take $\sigma(x) = x+1$, and again we are done:

$$\varphi_{\sigma(x)} \;=\; \varphi_{x+1} \;=\; \varphi_{f(y)} \;=\; \varphi_y \;=\; \varphi_x.$$

Finally, if $y < x$ and $y \notin B$, we can set $B := B \cup \{y\}$ and repeat the process. The invariant is maintained, because

$$\varphi_y \;=\; \varphi_{f(y)} \;=\; \varphi_x.$$

This can go on for at most finitely many stages, because B can contain no more than $x+1$ elements. Eventually we find a fixpoint greater than x. $\quad\square$

The Isomorphism Theorem

We end this lecture by showing that all Gödel numberings are essentially the same up to recursive isomorphism. This result is due to Rogers (see [104]).

Theorem 34.3 *Let $\varphi_0, \varphi_1, \varphi_2, \ldots$ and $\psi_0, \psi_1, \psi_2, \ldots$ be two Gödel numberings of the partial recursive functions. There exists a one-to-one and onto total recursive function $\rho : \omega \to \omega$ such that for all i, $\varphi_i = \psi_{\rho(i)}$.*

Proof. Let U be the universal function for the φ numbering, and let ℓ be an index for U in the ψ numbering. Thus

$$\psi_\ell(i,x) \;=\; U(i,x) \;=\; \varphi_i(x).$$

Applying the s_1^1 function in the ψ numbering,

$$\psi_{s_1^1(\ell,i)}(x) \;=\; \psi_\ell(i,x) \;=\; \varphi_i(x).$$

Thus the total recursive function $\sigma \overset{\text{def}}{=} \lambda i.s_1^1(\ell,i)$ maps an index in the φ numbering to an equivalent index in the ψ numbering; that is, for all i, $\varphi_i = \psi_{\sigma(i)}$. The same construction in the other direction using the universal function of the ψ numbering and the s_1^1 function of the φ numbering yields a total recursive function τ such that for all j, $\psi_j = \varphi_{\tau(j)}$.

Now we combine σ and τ into a single total recursive one-to-one and onto function $\rho : \omega \to \omega$ such that for all i, $\varphi_i = \psi_{\rho(i)}$. We construct ρ in stages using a back-and-forth argument. After the nth stage, say we have constructed a finite matching $\rho : A_n \to B_n$, where A_n and B_n are n-element subsets of ω, ρ is one-to-one on A_n, and for all $i \in A_n$, $\varphi_i = \psi_{\rho(i)}$. If n is even, let m be the least element of $\sim A_n$. Starting with $\sigma(m)$, apply the effective padding function (Lemma 34.2) of the ψ numbering until encountering the first index $k \in \sim B_n$. This must happen eventually, because B_n is finite. Similarly, if n is odd, let k be the least element of $\sim B_n$. Starting with $\tau(k)$, apply the effective padding function of the φ numbering until encountering the first index $m \in \sim A_n$. In either case, we have $\varphi_m = \psi_k$, $m \notin A_n$, and $k \notin B_n$. Set $\rho(m) = k$, $A_{n+1} = A_n \cup \{m\}$, and $B_{n+1} = B_n \cup \{k\}$. We have increased the domain of definition of ρ by one and maintained the invariant.

Because we alternate and always process the least unmatched element on either side, eventually every element is matched.   □

An alternative proof of the Rogers isomorphism theorem is given in Miscellaneous Exercise 108.

There is another isomorphism theorem due to Myhill [90] (see [104]) known as the *Myhill isomorphism theorem*. It is an effective version of the *Cantor–Schröder–Bernstein theorem* of set theory, which says that if there are one-to-one functions $A \rightarrow B$ and $B \rightarrow A$, then A and B are of the same cardinality. The Myhill isomorphism theorem says that any two sets that are reducible to each other via one-to-one reductions are recursively isomorphic (Miscellaneous Exercise 109).

Both these isomorphism theorems, Rogers and Myhill, can be obtained as special cases of an even more general theorem (Miscellaneous Exercise 107).

Supplementary Lecture J

Abstract Complexity

There are many different ways to measure complexity of computations, time and space being the two most common. Among other possibilities are the number of times a Turing machine writes to a tape cell ("ink"), or the size and depth of Boolean circuits. These complexity measures share some common general properties that are independent of the particular measure. For example, the speedup theorem (Theorem 32.2) and gap theorem (Theorem 32.1) hold for both time and space, and indeed for any complexity measure satisfying a few easily stated axioms.

Quite early in the development of complexity theory, Blum [16] observed this phenomenon and attempted to formalize an abstract notion of *complexity measure* in order to derive such properties purely axiomatically. In their simplest form, the Blum axioms postulate a collection of functions Φ_i, one for each partial recursive function φ_i, such that

(i)   for all x, $\Phi_i(x)\downarrow$ iff $\varphi_i(x)\downarrow$; and

(ii)   it is uniformly decidable in i, x, and m whether $\Phi_i(x) = m$.

By (ii) we mean that there exists a total recursive function f of three variables such that

$$f(i, x, m) \;=\; \begin{cases} 1, & \text{if } \Phi_i(x) = m \\ 0, & \text{otherwise.} \end{cases}$$

The collection Φ is called an *abstract complexity measure*. Time complexity of Turing machines certainly satisfies these axioms, as does space, provided we consider the space usage on some input undefined if the machine does not halt on that input.

The functions Φ_i are partial recursive functions; moreover, one can obtain an index for Φ_i effectively from i (Miscellaneous Exercise 116).

Given an abstract complexity measure Φ, any total recursive function f defines a *complexity class*

$$\mathcal{C}_f^\Phi \stackrel{\text{def}}{=} \{\varphi_i \mid \Phi_i(n) \leq f(n) \text{ a.e.}\}.$$

Note that this is a class of functions, not of the programs that compute them; thus φ_i may be in $\mathcal{C}_f^\Phi$, even though $\Phi_i(n)$ exceeds $f(n)$ i.o.

The gap and speedup theorems presented in Lecture 32 can be reformulated in this more abstract setting. The proof of the gap theorem in this more abstract setting is a fairly straightforward generalization of the proof of Theorem 32.1, and we leave it as an exercise (Miscellaneous Exercise 120), but we redo the proof of the speedup theorem explicitly to illustrate how the Blum axioms capture the essential properties of complexity measures.

Theorem J.1   **(Gap Theorem [21])**   *Let Φ be an abstract complexity measure. For any total recursive function $f(x) \geq x$, there exists a total recursive function t such that $\mathcal{C}_t^\Phi = \mathcal{C}_{f \circ t}^\Phi$. In other words, for any total recursive $f(x) \geq x$, there is a total recursive t such that if $\Phi_i(x) \leq f(t(x))$ a.e., then there is an index j such that $\varphi_j = \varphi_i$ and $\Phi_j(x) \leq t(x)$ a.e.*

Proof. Miscellaneous Exercise 120.   □

Theorem J.2   **(Speedup Theorem [17])**   *Let Φ be an abstract complexity measure. For all total recursive f, there exists a total recursive g such that for all indices i for g, there exists another index j for g with $f(n, \Phi_j(n)) < \Phi_i(n)$ a.e.*

Proof. The proof mimics the proof of the speedup theorem for time (Theorem 32.2) to a large extent, except we rely only on the axioms for abstract complexity measures. It will help to understand that proof thoroughly before attempting to read this one.

We first diagonalize to get a g_r for each φ_r such that, if φ_r is total and i is an index for g_r, then

$$\Phi_i(n) > \varphi_r(n - i) \text{ a.e.}^1 \tag{J.1}$$

---

[1] As in Theorem 32.2, "a.e." and "i.o." refer to the variable n. Other variables, such as i in this expression, are regarded as fixed constants.

Stage 0   Let $g_r(0) = 0$ and $D_0 = \varnothing$. (Here D corresponds to those machines that have been deleted from the active list in the proof of Theorem 32.2.)

Stage $n \geq 1$ Choose the least i, if it exists, such that

(i) $i \leq n$

(ii) $i \notin D_{n-1}$

(iii) $\Phi_i(n) \leq \varphi_r(n - i)$.

If such an i exists, let $g_r(n) = \varphi_i(n) + 1$ and $D_n = D_{n-1} \cup \{i\}$. If no such i exists, just let $g_r(n) = 0$ and $D_n = D_{n-1}$.

Denote by $\varphi_{h(r,0,0)}$ the above program for g_r. Note that if $\varphi_r(i) \downarrow$ for all $0 \leq i \leq n$, then $\varphi_{h(r,0,0)}(n) \downarrow$. Thus if φ_r is total, then $\varphi_{h(r,0,0)}$ is total, and the function g_r computed by $\varphi_{h(r,0,0)}$ satisfies (J.1).

Next we construct an r.e. set of programs such that if φ_r is total, then all programs in the set are total, and g_r is represented among them i.o.

For $1 \leq k \leq m$, define $\varphi_{h(r,k,m)}$ by:

Stages $0, \ldots, m - 1$ Construct $\varphi_{h(r,k,m)}$ exactly as $\varphi_{h(r,0,0)}$.

Stage $n \geq m$ Choose the least i, if it exists, such that

(i) $k \leq i \leq n$

(ii) $i \notin D_{n-1}$

(iii) $\Phi_i(n) \leq \varphi_r(n - i)$.

If such an i exists, let $\varphi_{h(r,k,m)}(n) = \varphi_i(n) + 1$ and $D_n = D_{n-1} \cup \{i\}$. If no such i exists, just let $\varphi_{h(r,k,m)}(n) = 0$ and $D_n = D_{n-1}$.

Again, note that if $\varphi_r(i) \downarrow$ for $0 \leq i \leq n - k$, then $\varphi_{h(r,k,m)}(n) \downarrow$. Thus if φ_r is total, then $\varphi_{h(r,k,m)}$ is total. Moreover, we claim that

$$\forall k \overset{\infty}{\forall} m \; \varphi_{h(r,k,m)} \;=\; g_r.^2$$

This is true because during the computation of $\varphi_{h(r,0,0)}$, at some stage m, all $i \leq k$ that will ever be in some D_j are already in D_m. Thereafter, the same candidates for D_j are chosen by $\varphi_{h(r,k,m)}$ as by $\varphi_{h(r,0,0)}$, so the same function g_r is constructed.

---

[2]Here $\overset{\infty}{\forall}$ means "for all but finitely many ... " or "for all sufficiently large ... ". There is also $\overset{\infty}{\exists}$, which means "there exist infinitely many ... ".

Finally, we choose an appropriate φ_r. Define the recursive operator σ by

$$\varphi_{\sigma(r)}(0) \overset{\text{def}}{=} 0$$

$$\varphi_{\sigma(r)}(n+1) \overset{\text{def}}{=} 1 + \max_{k,m \le n} f(n+k, \Phi_{h(r,k,m)}(n+k)).$$

By the recursion theorem, σ has a fixpoint r, so that

$$\varphi_r(0) \overset{\text{def}}{=} 0$$

$$\varphi_r(n+1) \overset{\text{def}}{=} 1 + \max_{k,m \le n} f(n+k, \Phi_{h(r,k,m)}(n+k)).$$

We can show by induction that φ_r is total. Certainly $\varphi_r(0)\!\downarrow$ by definition, and as argued above, if $\varphi_r(i)\!\downarrow$ for all $0 \le i \le n$, then $\varphi_{h(r,k,m)}(n+k)\!\downarrow$, therefore $\varphi_r(n+1)\!\downarrow$. But for all k, m, for sufficiently large n,

$$\varphi_r(n+1) \quad > \quad f(n+k, \Phi_{h(r,k,m)}(n+k)),$$

thus

$$f(n, \Phi_{h(r,k,m)}(n)) \quad < \quad \varphi_r(n-k+1) \text{ a.e.}$$

In particular, for any i such that φ_i computes g_r, by (J.1) we have

$$f(n, \Phi_{h(r,i+1,m)}(n)) \quad < \quad \varphi_r(n-i) \quad < \quad \Phi_i(n) \text{ a.e.}$$

$\square$

As a final example of an interesting general theorem that holds of all abstract complexity measures, we have the union theorem of McCreight and Meyer [83]. This theorem states that the union of any effective hierarchy of complexity classes is itself a complexity class defined by a single function.

For example, one consequence of the union theorem is that there exists a computable function p such that $DTIME(p(n)) = P$. Again, however, the function p is nothing natural like $2^{(\log n)^2}$ or $n^{\log \log n}$ or anything of the sort. As with t in the gap theorem and g in the speedup theorem, it is constructed by an intricate diagonalization.

Theorem J.3   **(Union Theorem [83])**   *Let $f_0, f_1, \ldots$ be an r.e. list of total recursive functions such that for all i and n, $f_i(n) \le f_{i+1}(n)$. Then there exists a total recursive function f such that*

$$\mathcal{C}_f^\Phi \;=\; \bigcup_i \mathcal{C}_{f_i}^\Phi.$$

In other words, for any recursive function g, there is an index i for g such that $\Phi_i(n) \le f(n)$ a.e. iff there is an index j for g and a number k such that $\Phi_j(n) \le f_k(n)$ a.e.

Proof. We build a total recursive function f by diagonalization satisfying the following two conditions.

(i)   For all k, $f(n) \geq f_k(n)$ a.e.

(ii)   For all i, if $\Phi_i(n) > f_k(n)$ i.o. for every k, then $f(n) < \Phi_i(n)$ i.o.

These two conditions pull f in opposite directions: condition (i) would like f to be large, and (ii) would like f to be small. However, because we only have to satisfy (i) a.e. and (ii) i.o., this gives us some flexibility in the construction. If we can create f satisfying both conditions, we will have achieved our goal, because (i) guarantees that $\mathcal{C}^{\Phi}_{f_j} \subseteq \mathcal{C}^{\Phi}_f$, and (ii) says that if $\Phi_i(n) \leq f(n)$ a.e., then for some j, $\Phi_i(n) \leq f_j(n)$ a.e., thus $\mathcal{C}^{\Phi}_f \subseteq \bigcup_i \mathcal{C}^{\Phi}_{f_i}$.

Now we turn to the construction of f. As in the proof of Theorem J.2, we construct f by diagonalization. We maintain a queue of pairs (i, k), $i \leq k$, which we can view as the conjecture that $\Phi_i(n) \leq f_k(n)$ a.e. When a conjecture is violated, we take some corrective action in terms of the definition of f, retract the conjecture, and replace it with a weaker conjecture.

Stage 0   Define $f(0) := 0$ and initialize the queue to contain the single pair $(0, 0)$.

Stage $n \geq 1$ Find the first conjecture (i, k) on the queue that is violated at n; that is, such that $\Phi_i(n) > f_k(n)$. If such an (i, k) exists, define $f(n) := f_k(n)$, remove (i, k) from the queue, and append $(i, k+1)$ at the back of the queue. If no such (i, k) exists, define $f(n) = f_n(n)$. In either case, append (n, n) at the back of the queue.

For any m, there are only finitely many conjectures (i, k) with $k \leq m$ ever on the queue ($\binom{m+2}{2}$ to be exact). Once a conjecture is deleted from the queue, it never returns. If a conjecture on the queue is violated infinitely often, it is eventually chosen for deletion. If it is violated at some stage, then the only way it would not be chosen for deletion at that stage is if some conjecture ahead of it on the queue is chosen instead, but this can happen only finitely many times. At some stage, all conjectures (i, k) with $k \leq m$ that will ever be deleted from the queue have already been deleted, and thereafter $f(n) \geq f_m(n)$. This establishes (i).

For (ii), if $\Phi_i(n) > f_k(n)$ i.o. for every k, then the conjectures (i, k) for $i \leq k$ all go on the queue eventually and are all deleted eventually. When (i, k) is deleted, $f(n)$ is defined to be $f_k(n) < \Phi_i(n)$, so $f(n) < \Phi_i(n)$. This happens infinitely often, therefore $f(n) < \Phi_i(n)$ i.o. $\square$

The union theorem does not hold without the monotonicity condition on the f_k (Miscellaneous Exercise 126).

More of the theory of abstract complexity measures is explored in Miscellaneous Exercises 116–127.

Lecture 35

The Arithmetic Hierarchy

Let A, B be sets of strings. We say that A is *r.e. in B* if $A = L(M^B)$ for some oracle TM M with oracle B. We say that A is *recursive in B* if $A = L(M^B)$ for some oracle TM M with oracle B such that M^B is total; that is, if membership in A is decidable relative to an oracle for B. We write $A \leq_T B$ if A is recursive in B. The relation $\leq_T$ is called *Turing reducibility*.

We can define a hierarchy of classes above the r.e. sets analogous to the polynomial time hierarchy as follows. Fix the alphabet $\{0, 1\}$ and identify strings in $\{0, 1\}^*$ with the natural numbers. Define

$$\Sigma_1^0 \overset{\text{def}}{=} \{\text{r.e. sets}\},$$

$$\Delta_1^0 \overset{\text{def}}{=} \{\text{recursive sets}\},$$

$$\Sigma_{n+1}^0 \overset{\text{def}}{=} \{L(M^B) \mid B \in \Sigma_n^0\}$$

$$= \{A \mid A \text{ is r.e. in some } B \in \Sigma_n^0\},$$

$$\Delta_{n+1}^0 \overset{\text{def}}{=} \{L(M^B) \mid B \in \Sigma_n^0, \ M^B \text{ total}\}$$

$$= \{A \mid A \text{ is recursive in some } B \in \Sigma_n^0\}$$

$$= \{A \mid A \leq_T B \text{ for some } B \in \Sigma_n^0\},$$

$$\Pi_n^0 \overset{\text{def}}{=} \{\text{complements of sets in } \Sigma_n^0\}.$$

Thus Π_1^0 is the class of co-r.e. sets. The classes Σ_n^0, Π_n^0, and Δ_n^0 comprise what is known as the *arithmetic hierarchy*.

Here is perhaps a more revealing characterization of the arithmetic hierarchy in terms of alternation of quantifiers. This characterization is analogous to the characterization of the polynomial time hierarchy given in Theorem 10.2.

Recall that a set A is r.e. iff there exists a decidable binary predicate R such that

$$A = \{x \mid \exists y \; R(x, y)\}. \tag{35.1}$$

For example, the halting and membership problems can be expressed

$$\mathrm{HP} = \{M \# x \mid \exists t \; M \text{ halts on } x \text{ in } t \text{ steps}\},$$
$$\mathrm{MP} = \{M \# x \mid \exists t \; M \text{ accepts } x \text{ in } t \text{ steps}\}.$$

Note that the predicate "M halts on x" is not decidable, but the predicate "M halts on x in t steps" is, because we can just simulate M on input x with a universal machine for t steps and see if it halts within that time. Alternatively,

$$\mathrm{HP} = \{M \# x \mid \exists v \; v \text{ is a halting computation history of } M \text{ on } x\},$$
$$\mathrm{MP} = \{M \# x \mid \exists v \; v \text{ is an accepting computation history of } M \text{ on } x\}.$$

The class Σ_1^0 is the family of all sets that can be expressed in the form (35.1).

Similarly, it follows from elementary logic that Π_1^0, the family of co-r.e. sets, is the class of all sets A for which there exists a decidable binary predicate R such that

$$A = \{x \mid \forall y \; R(x, y)\}. \tag{35.2}$$

As is well known, a set is recursive iff it is both r.e. and co-r.e. In terms of our new notation,

$$\Delta_1^0 = \Sigma_1^0 \cap \Pi_1^0.$$

These results are special cases of the following theorem, which is analogous to the characterization of PH given in Miscellaneous Exercise 32.

Theorem 35.1 (i) *A set A is in Σ_n^0 iff there exists a decidable $(n+1)$-ary predicate R such that*

$$A = \{x \mid \exists y_1 \; \forall y_2 \; \exists y_3 \; \cdots \; Q y_n \; R(x, y_1, \ldots, y_n)\},$$

where $Q = \exists$ if n is odd, $\forall$ if n is even.

(ii) *A set A is in Π_n^0 iff there exists a decidable $(n+1)$-ary predicate R such that*

$$A = \{x \mid \forall y_1 \; \exists y_2 \; \forall y_3 \; \cdots \; Q y_n \; R(x, y_1, \ldots, y_n)\},$$

where $Q = \forall$ if n is odd, $\exists$ if n is even.

(iii) $\Delta_n^0 = \Sigma_n^0 \cap \Pi_n^0$.

Proof. Miscellaneous Exercise 128.   □

One difference here from our treatment of PH is that we are lacking a characterization of the arithmetic hierarchy in terms of something analogous to alternating TMs. This deficiency is corrected in Lecture 39.

Example 35.2   The set $\mathrm{EMPTY} \overset{\text{def}}{=} \{M \mid L(M) = \varnothing\}$ is in Π_1^0, because

$$\mathrm{EMPTY} \;=\; \{M \mid \forall x \; \forall t \; M \text{ does not accept } x \text{ in } t \text{ steps}\}.$$

The two universal quantifiers $\forall x \; \forall t$ can be combined into one using the computable one-to-one pairing function (33.1) described in Lecture 33. Thus

$$\mathrm{EMPTY} \;=\; \{M \mid \forall z \; M \text{ does not accept } \pi_1^2(z) \text{ in } \pi_2^2(z) \text{ steps}\}.$$

                                                                        □

Example 35.3   The set $\mathrm{TOTAL} \overset{\text{def}}{=} \{M \mid M \text{ is total}\}$ is in Π_2^0, because

$$\mathrm{TOTAL} \;=\; \{M \mid \forall x \; \exists t \; M \text{ halts on } x \text{ in } t \text{ steps}\}.$$

                                                                        □

Example 35.4   The set $\mathrm{FIN} \overset{\text{def}}{=} \{M \mid L(M) \text{ is finite}\}$ is in Σ_2^0, because

$$\begin{aligned}
\mathrm{FIN} \;&=\; \{M \mid \exists n \; \forall x \text{ if } |x| > n \text{ then } x \notin L(M)\} \\
&=\; \{M \mid \exists n \; \forall x \; \forall t \; |x| \le n \text{ or } M \text{ does not accept } x \text{ in } t \text{ steps}\}.
\end{aligned}$$

Again, the two universal quantifiers $\forall x \; \forall t$ can be combined into one using the pairing function (33.1).   □

Example 35.5   A set is *cofinite* if its complement is finite. The set

$$\mathrm{COF} \overset{\text{def}}{=} \{M \mid L(M) \text{ is cofinite}\}$$

is in Σ_3^0, because

$$\begin{aligned}
\mathrm{COF} \;&=\; \{M \mid \exists n \; \forall x \text{ if } |x| > n \text{ then } x \in L(M)\} \\
&=\; \{M \mid \exists n \; \forall x \; \exists t \; |x| \le n \text{ or } M \text{ accepts } x \text{ in } t \text{ steps}\}.
\end{aligned}$$

                                                                        □

Figure 35.1 depicts the inclusions among the lowest few levels of the hierarchy. Each level of the hierarchy is strictly contained in the next; that is, $\Sigma_n^0 \cup \Pi_n^0 \subseteq \Delta_{n+1}^0$, but $\Sigma_n^0 \cup \Pi_n^0 \ne \Delta_{n+1}^0$. We know that there exist r.e. sets that are not co-r.e. (HP, for example) and co-r.e. sets that are not r.e. ($\sim$HP, for example). Thus Σ_1^0 and Π_1^0 are incomparable with respect to set inclusion. One can show in the same way that Σ_n^0 and Π_n^0 are incomparable with respect to set inclusion for any n (Homework 11, Exercise 2).

Reducibility and Completeness

For $A \subseteq \Sigma^*$ and $B \subseteq \Gamma^*$, define $A \leq_m B$ if there exists a total recursive function $\sigma : \Sigma^* \to \Gamma^*$ such that for all $x \in \Sigma^*$,

$$x \in A \quad \Leftrightarrow \quad \sigma(x) \in B.$$

The relation $\leq_m$ is called *many–one reducibility* and is analogous to the reducibility relations $\leq_m^{\log}$ or $\leq_m^p$ we have studied, except without the resource bounds.

The membership problem $\mathrm{MP} \stackrel{\text{def}}{=} \{M\#x \mid M \text{ accepts } x\}$ is not only undecidable but is in a sense a "hardest" r.e. set, because every other r.e. set $\leq_m$-reduces to it: for any Turing machine M, the map $x \mapsto M\#x$ is a trivially computable map reducing $L(M)$ to MP.

We say that a set is *r.e.-hard* if every r.e. set $\leq_m$-reduces to it. In other words, the set B is *r.e.-hard* if for all r.e. sets A, $A \leq_m B$. As just observed, the membership problem MP is r.e.-hard. So is any other problem to which the membership problem $\leq_m$-reduces (for example, the halting problem HP), because the relation $\leq_m$ is transitive.

A set B is said to be *r.e.-complete* if it is both an r.e. set and r.e.-hard. For example, both MP and HP are r.e.-complete.

More generally, if $\mathcal{C}$ is a class of sets, we say that a set B is $\leq_m$-*hard for* $\mathcal{C}$ (or just $\mathcal{C}$-hard) if $A \leq_m B$ for all $A \in \mathcal{C}$. We say that B is $\leq_m$-*complete for* $\mathcal{C}$ (or just $\mathcal{C}$-complete) if B is $\leq_m$-hard for $\mathcal{C}$ and $B \in \mathcal{C}$.

One can prove a theorem corresponding to Lemma 5.3 that says that if $A \leq_m B$ and $B \in \Sigma_n^0$, then $A \in \Sigma_n^0$, and if $A \leq_m B$ and $B \in \Delta_n^0$, then $A \in \Delta_n^0$. Because we know that the arithmetic hierarchy is strict (each level is properly contained in the next), if B is $\leq_m$-complete for Σ_n^0, then $B \notin \Pi_n^0$ (or Δ_n^0 or Σ_{n-1}^0).

It turns out that each of the problems mentioned above is $\leq_m$-complete for the level of the hierarchy in which it naturally falls:

  (i)   HP is $\leq_m$-complete for Σ_1^0,

  (ii) MP is $\leq_m$-complete for Σ_1^0,

 (iii) EMPTY is $\leq_m$-complete for Π_1^0,

 (iv) TOTAL is $\leq_m$-complete for Π_2^0,

  (v) FIN is $\leq_m$-complete for Σ_2^0, and

 (vi) COF is $\leq_m$-complete for Σ_3^0.

Because the hierarchy is strict, none of these problems is contained in any class lower in the hierarchy or reduces to any problem complete for any class lower in the hierarchy. If it did, then the hierarchy would collapse at that level. For example, EMPTY does not reduce to HP and COF does not reduce to FIN.

We have shown (ii) above. We prove (v) here and (vi) in Lecture 36; the others we leave as exercises (Miscellaneous Exercise 130).

We have already argued that FIN $\in \Sigma_2^0$, because finiteness can be expressed with an $\exists \forall$ predicate. To show that FIN is $\leq_m$-hard for Σ_2^0, we need to show that any set in Σ_2^0 reduces to it. We use the characterization of Theorem 35.1. Let

$$A \;=\; \{x \in \Gamma^* \mid \exists y \; \forall z \; R(x, y, z)\}$$

be an arbitrary set in Σ_2^0, where $R(x, y, z)$ is a decidable ternary predicate. Let M be a total machine that decides R. We need to construct a machine N effectively from a given x such that $N \in$ FIN iff $x \in A$. Thus we want N to accept a finite set iff $\exists y \; \forall z \; R(x, y, z)$; equivalently, we want N to accept an infinite set iff $\forall y \; \exists z \; \neg R(x, y, z)$. Let N on input w

(i)   write down all strings y of length at most $|w|$; then

(ii) for each such y, try to find a z such that $\neg R(x, y, z)$ (that is, such that M rejects $x\#y\#z$), and accept if all these trials are successful. The machine N has x and a description of M hard-wired in its finite control.

In step (ii), for each y of length at most $|w|$, N just enumerates strings z in some order and runs M on $x\#y\#z$ until some z is found causing M to reject. Because M is total, N need not worry about timesharing; it can just process the z's in order. If no such z is ever found, N just goes on forever. Surely such an N can be built effectively from M and x.

Now if $x \in A$, then there exists y such that for all z, $R(x, y, z)$ (that is, for all z, M accepts $x\#y\#z$); thus step (ii) fails whenever $|w| \geq |y|$. In this case N accepts a finite set. On the other hand, if $x \notin A$, then for all y there exists a z such that $\neg R(x, y, z)$, and these are all found in step (ii). In this case, N accepts Γ^*.

We have argued that $L(N)$ is a finite set if $x \in A$ and Γ^* if $x \notin A$, therefore the map $x \mapsto N$ constitutes a $\leq_m$-reduction from A to FIN. Because A was an arbitrary element of Σ_2^0, FIN is $\leq_m$-hard for Σ_2^0.

Note that the same reduction shows that TOTAL is $\leq_m$-hard for Π_2^0. This is because in the construction above, N is total iff $x \in \sim A$, and $\sim A \in \Pi_2^0$ because $A \in \Sigma_2^0$.

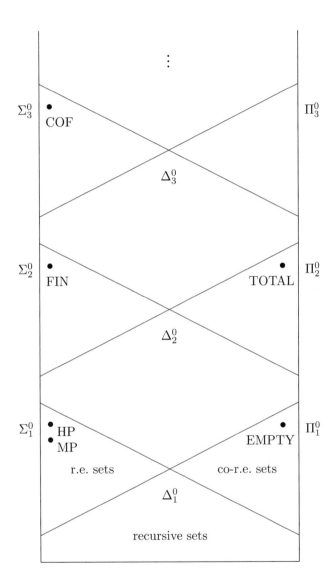

Figure 35.1: The arithmetic hierarchy.

Lecture 36

Complete Problems in the Arithmetic Hierarchy

In this lecture we give some natural problems complete for the third level of the arithmetic hierarchy. Recall our definitions from last time:

$$\Sigma^0_{n+1} \overset{\text{def}}{=} \{A \mid A \text{ is r.e. in some } B \in \Sigma^0_n\}$$
$$= \{L(M^B) \mid B \in \Sigma^0_n\},$$
$$\Delta^0_{n+1} \overset{\text{def}}{=} \{A \mid A \text{ is recursive in some } B \in \Sigma^0_n\}$$
$$= \{L(M^B) \mid B \in \Sigma^0_n,\ M^B \text{ total}\}$$
$$= \{A \mid A \leq_T B \text{ for some } B \in \Sigma^0_n\},$$
$$\Pi^0_n \overset{\text{def}}{=} \{A \mid {\sim}A \in \Sigma^0_n\},$$

and $\Sigma^0_1 = \{\text{r.e. sets}\}$ and $\Delta^0_1 = \{\text{recursive sets}\}$. Define the notation

$M(x){\downarrow}$	M halts on input x,
$M(x){\downarrow}^t$	M halts on input x within t steps,
$M(x){\uparrow}$	M does not halt on input x,
$M(x){\uparrow}^t$	M does not halt on input x for at least t steps.

For example, the *halting problem* is the set $\text{HP} \overset{\text{def}}{=} \{M\#x \mid M(x){\downarrow}\}$.

We can define relativized versions of any of the sets discussed in the last lecture. For example, the finiteness problem relative to oracle A is the set

$$\text{FIN}^A \overset{\text{def}}{=} \{M \mid L(M^A) \text{ is finite}\}.$$

Lemma 36.1 *FIN^{HP} is $\leq_m$-complete for Σ_3^0. More generally, if A is $\leq_m$-complete for Σ_n^0, then FIN^A is $\leq_m$-complete for Σ_{n+2}^0.*

Proof. To show that $\text{FIN}^A \in \Sigma_{n+2}^0$, note that

$$M \in \text{FIN}^A \quad \Leftrightarrow \quad L(M^A) \text{ is finite} \quad \Leftrightarrow \quad \exists y \; \forall z \geq y \; \forall t \; M^A(z){\uparrow}^t \quad (36.1)$$

(without loss of generality, consider machines without reject states, so that halting and accepting are synonymous). The predicate $M^A(z){\uparrow}^t$ is Δ_{n+1}^0 because it is recursive in A, which is Σ_n^0-complete. By Theorem 35.1(iii), it is also in Π_{n+1}^0, and by Theorem 35.1(ii), it can be expressed in the form of $n+1$ alternations of quantifiers beginning with $\forall$ and followed by a recursive predicate. Combining this with the Σ_2 quantifier prefix $\exists y \; \forall z \geq y \; \forall t$ in (36.1), we obtain a Σ_{n+2} quantifier prefix followed by a recursive predicate. This shows that FIN^A is in Σ_{n+2}^0.

To show that FIN^A is Σ_{n+2}^0-hard, let us first recall the proof that FIN is Σ_2^0-hard. We had to reduce an arbitrary set in Σ_2^0 to FIN, which meant we had to construct a reduction

$$\{x \mid \exists y \; \forall z \; R(x,y,z)\} \quad \leq_m \quad \{M \mid L(M) \text{ is finite}\}$$

for an arbitrary recursive predicate $R(x,y,z)$. Thus we needed to give a total recursive function σ such that for all x, $\sigma(x)$ is a description of a machine M such that

$$\exists y \; \forall z \; R(x,y,z) \quad \Leftrightarrow \quad L(M) \text{ is finite}.$$

Given x, we built M that on input w enumerated all y of length at most $|w|$, and for each such y tried to find z such that $\neg R(x,y,z)$. Thus if $\forall y \; \exists z \; \neg R(x,y,z)$, then $L(M) = \Sigma^*$; but on the other hand, if $\exists y \; \forall z \; R(x,y,z)$, then M looped infinitely on all inputs w of length greater than the shortest such y, therefore accepted a finite set.

Now we can do exactly the same construction in the presence of an oracle A. If $R^A(x,y,z)$ is recursive in A, this gives a reduction

$$\{x \mid \exists y \; \forall z \; R^A(x,y,z)\} \quad \leq_m \quad \{M \mid L(M^A) \text{ is finite}\}.$$

We build the oracle machine M^A that on input w enumerates all y of length at most $|w|$, and for each such y tries to find z such that $\neg R^A(x,y,z)$. The machine M^A queries its oracle A as necessary to determine $R^A(x, y, z)$. Again, if $\forall y\ \exists z\ \neg R^A(x, y, z)$, then $L(M^A) = \Sigma^*$; but on the other hand, if $\exists y\ \forall z\ R^A(x, y, z)$, then M^A accepts a finite set.

Now we need to argue that for any A that is Σ_n^0-complete, any Σ_{n+2}^0 set

$$\{x \mid \exists y_1\ \forall y_2\ \exists y_3\ \cdots\ Qy_{n+2}\ S(x, y_1, \ldots, y_{n+2})\}$$

can be written

$$\{x \mid \exists y\ \forall z\ R^A(x, y, z)\}$$

for some R recursive in A. But this follows immediately from the fact that the Σ_n^0 set

$$\{(x, y_1, y_2) \mid \exists y_3\ \cdots\ Qy_{n+2}\ S(x, y_1, \ldots, y_{n+2})\}$$

is recursive in A, because A is Σ_n^0-hard.   □

All the following problems are $\leq_m$-complete for Σ_3^0:

- COF $\overset{\text{def}}{=} \{M \mid L(M)$ is cofinite$\}\ =\ \{M \mid {\sim}L(M)$ is finite$\}$,

- REC $\overset{\text{def}}{=} \{M \mid L(M)$ is recursive$\}$,

- REG $\overset{\text{def}}{=} \{M \mid L(M)$ is regular$\}$,

- CFL $\overset{\text{def}}{=} \{M \mid L(M)$ is context-free$\}$.

These problems can all be shown to be in Σ_3^0 by expressing their defining predicates in the appropriate form. For example,

$$\begin{aligned}
\text{COF} &= \{M \mid \exists y\ \forall z\ |z| \geq |y|\ \rightarrow\ \exists t\ M(z){\downarrow}^t\} \\
&= \{M \mid \exists y\ \forall z\ \exists t\ |z| < |y|\ \vee\ M(z){\downarrow}^t\}.
\end{aligned}$$

We show that COF is $\leq_m$-hard for Σ_3^0 by a reduction from FIN^{HP}, which is Σ_3^0-hard by Lemma 36.1. We wish to construct a reduction

$$\{M \mid L(M^{\text{HP}})\text{ is finite}\}\ \leq_m\ \{M \mid L(M)\text{ is cofinite}\};$$

in other words, we would like to give a total recursive σ such that for all M, $N = \sigma(M)$ is a machine such that

$$L(M^{\text{HP}})\text{ is finite}\ \Leftrightarrow\ L(N)\text{ is cofinite}.$$

Given M, let K^{HP} be an oracle machine with oracle HP that takes the following actions on input y.

(i)   Try to find z greater in length than y accepted by M^{HP}. This is done using a timesharing simulation of M^{HP} on all inputs z such that $|z| > |y|$ in some order, timesharing the computations to make sure that all $M^{\mathrm{HP}}(z)$ are eventually simulated for arbitrarily many steps. If such a z does exist, the simulation will discover it. If not, the simulation will loop forever. As soon as such a z is found, halt the simulation and go on to step (ii).

(ii) Just for fun, verify all "yes" oracle responses from step (i) by simulating H on input w for all strings $H \# w$ on which the oracle was queried and to which it responded "yes". The "no" responses are ignored. All of these simulations terminate, because the oracle responded "yes", therefore H halts on w.

By our construction,

- if $L(M^{\mathrm{HP}})$ is finite, then $L(K^{\mathrm{HP}})$ is finite; and

- if $L(M^{\mathrm{HP}})$ is infinite, then $L(K^{\mathrm{HP}}) = \Sigma^*$.

Now let N be a machine that accepts all strings that are *not* accepting computation histories of K^{HP} on some input. (We encountered computation histories in Lecture 23.) Normally, a string represents a computation history of a given machine if it satisfies the following properties.

1. The string encodes a sequence of configurations of the machine.

2. The first configuration is the start configuration on some input.

3. The last configuration is an accepting configuration.

4. The $i + 1$st configuration follows from the ith according to the transition rules of the machine.

The only difference here is that we need to account for the oracle HP. A string that is purportedly a computation history of K^{HP} is peppered with oracle queries and the corresponding responses of the oracle. Thus we must add the following condition to our definition of computation history.

5. The responses of the oracle as represented in the string are correct.

To check that a string is *not* an accepting computation history of K^{HP}, the machine N has to check that at least one of the conditions 1–5 is violated. Conditions 1–4 present no problem; but N does not have access to the oracle, so how can it check the validity of the oracle responses?

The answer is that N need only check the *negative* oracle responses. The positive ones were checked by K^{HP} itself, and there is a proof of the validity of the oracle response right there in the computation history; that was the purpose of step (ii) above!

Thus, to accept strings that are *not* computation histories, N checks first whether one of the conditions 1–4 is violated. If so, it accepts. If not, it checks that one of the oracle responses as represented in the string is wrong. For the positive responses, it can just check the proof in the computation history itself. For each negative response, say on query $H \# w$, it runs H on input w to see whether it halts. If so, it halts and accepts, because the "no" oracle response on query $H \# w$ as represented in the computation history was incorrect, so condition 5 is violated. It does this in a timesharing fashion for all such oracle queries $H \# w$.

The set $L(K^{\mathrm{HP}})$ is finite iff the set of accepting computation histories of K^{HP} is finite, because there are as many accepted strings as accepting computation histories; and this occurs iff $L(N)$ is cofinite. We have thus built a machine N that accepts a cofinite set iff $L(M^{\mathrm{HP}})$ is finite.

The problems REC, REG, and CFL can be shown Σ_3^0-hard by a similar but only slightly more complicated argument (Miscellaneous Exercise 133).

Lecture 37

Post's Problem

One of the earliest goals of recursive function theory was to understand the structure of the m- and T-degrees of the r.e. sets. The *m-degree* of a set A is the equivalence class of A under many–one reducibility $\leq_m$, and the *T-degree* or *Turing degree* of A is the equivalence class of A under Turing reducibility $\leq_T$. The reason for considering equivalence classes is that equivalent sets contain the same computational information, thus for purposes of computation might as well be identified.

There are at least two distinct r.e. T-degrees, namely the degree of the recursive sets (that is, the degree of $\varnothing$) and the degree of the r.e.-complete sets (that is, the degree of the halting problem). Because every m-degree is contained in a T-degree, there are at least two distinct r.e. m-degrees. Emil Post showed in 1944 [96] that there were more r.e. m-degrees than just these two, and posed the same problem for T-degrees. This became known as *Post's problem*. The problem stood open for 12 years until 1956, when it was solved independently by Friedberg [45] and Muchnik [88].

The Friedberg–Muchnik theorem is a classical result of recursive function theory. We present a proof of this result in Lecture 38. The proof illustrates a technique called a *finite injury priority argument*, which is useful in many other applications as well. For the rest of this lecture, we present a proof of Post's theorem, which solves the problem for m-degrees. For a more comprehensive treatment of this subject, see [104, 114].

For this lecture and the next, we revert to the standard notation of recursive function theory introduced in Lectures 33 and 34, which differs a little from the notation of Lectures 35 and 36. Let $\varphi_0, \varphi_1, \ldots$ be a Gödel numbering of the partial recursive functions. For φ_x a recursive function, write $\varphi_x(y)\downarrow$ if φ_x is defined on y, and define

$$W_x \stackrel{\text{def}}{=} \text{domain of } \varphi_x = \{y \mid \varphi_x(y)\downarrow\}.$$

Every set W_x is an r.e. set, and every r.e. set is W_x for some x. Thus we can consider $W_0, W_1, \ldots$ an indexing of the r.e. sets. Define

$$K \stackrel{\text{def}}{=} \{x \mid \varphi_x(x)\downarrow\} = \{x \mid x \in W_x\}.$$

The set K is easily shown to be r.e.-complete: $W_x \leq_m K$ via the map $\lambda y.\text{comp}(x, \text{const}(y))$, and K itself is W_k, where $k = \text{comp}(u, \text{pair}(i, i))$ and u and i are indices for the universal and identity functions, respectively.

T- and m-degrees

Recall the definitions of the reducibility relations $\leq_m$ and $\leq_T$: for $A, B \subseteq \omega$, $A \leq_m B$ if there exists a total recursive function σ such that for all x,

$$x \in A \iff \sigma(x) \in B,$$

and $A \leq_T B$ if A is recursive in B; that is, if there exists an oracle Turing machine M with oracle B such that M^B is total and $A = L(M^B)$. If $A \leq_m B$ then $A \leq_T B$, because we can build M that on input x computes $\sigma(x)$ and queries the oracle. However, the converse does not hold: $\sim K \leq_T K$ but not $\sim K \leq_m K$.

Define $A \equiv_m B$ if $A \leq_m B$ and $B \leq_m A$, and define $A \equiv_T B$ if $A \leq_T B$ and $B \leq_T A$. The $\equiv_m$- equivalence class and $\equiv_T$-equivalence class of A are called the *m-degree* and *T-degree* of A, respectively. The relation $\equiv_m$ refines the relation $\equiv_T$; in other words, for any A, the m-degree of A is contained setwise in the T-degree of A.

The $\leq_m$-least m-degree consists of all the recursive sets. The $\leq_m$-greatest r.e. m-degree is the m-degree of K, the family of r.e.-complete sets. Post proved in 1944 that there exist other m-degrees besides these two, and conjectured that the same was true for the T-degrees.

Theorem 37.1   (Post 1944 [96])   *There exists a nonrecursive r.e. set that is not r.e.-complete.*

Immune, Simple, and Productive Sets

The proof of Theorem 37.1 involves the concepts of *immune*, *simple*, and *productive* sets.

Definition 37.2   *A set $A \subseteq \omega$ is called* immune *if*

- *A is infinite, and*

- *A contains no infinite r.e. subset.*

Definition 37.3   *A set $B \subseteq \omega$ is called* simple *if*

- *B is r.e., and*

- *$\sim B$ is immune.*

In other words, B is simple if

- *B is r.e.,*

- *$\sim B$ is infinite, and*

- *B intersects every infinite r.e. set.*

Definition 37.4   *A set $C \subseteq \omega$ is* productive *if there exists a total recursive function σ such that whenever $W_x \subseteq C$,*

$$\sigma(x) \in C - W_x.$$

The function σ is called a productive function *for C.*

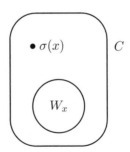

Example 37.5   For example, $\sim K$ is productive with productive function $\lambda x.x$, the identity. To show this, suppose $W_x \subseteq \sim K$. For any x,

$$x \in W_x \iff \varphi_x(x)\downarrow \quad \text{by definition of } W_x$$
$$\iff x \in K \quad \text{by definition of } K,$$

therefore either

- $x \in W_x$ and $x \in K$, or

- $x \notin W_x$ and $x \notin K$.

But the former is impossible by the assumption $W_x \subseteq \sim K$; therefore $x \in \sim K - W_x$. $\square$

Proof of Post's Theorem

The proof of Theorem 37.1 can be broken down into three basic lemmas involving the concepts introduced in the last section.

Lemma 37.6   *There exists a simple set.*

Lemma 37.7   *If B is simple, then $\sim B$ is not productive.*

Lemma 37.8   *If A is r.e.-complete, then $\sim A$ is productive.*

We prove these lemmas below. For now, let us see how to use them to prove Post's theorem.

Proof of Theorem 37.1. Let B be a simple set, which exists by Lemma 37.6. The set B cannot be recursive, because then $\sim B$ would be r.e.; but this contradicts the assumption that B is simple, because $\sim B$ is infinite and B must intersect all infinite r.e. sets. Neither can B be r.e.-complete because of Lemmas 37.7 and 37.8.   □

Proof of Lemma 37.6. We build a simple set B. We have three conditions to fulfill in the construction:

- B must be r.e.,

- $\sim B$ must be infinite, and

- B must intersect every infinite r.e. set.

We describe an enumeration procedure for B. Let M_x be an enumeration machine enumerating the r.e. set W_x. Recall (see [76]) that an *enumeration machine* has a read/write worktape and a write-only output tape but no input tape. It starts with its work and output tapes blank and runs forever, occasionally entering a distinguished *enumeration state*, at which time the string currently written on its output tape is said to be *enumerated*, and the output tape is instantaneously erased and the head returned back to the beginning of the tape.

Our enumeration procedure for B performs a timesharing simulation of all enumeration machines. It keeps a list of machines it is currently simulating and simulates different machines on different blocks of the tape. It starts out by simulating M_0 for one step, then M_0 and M_1 for one step each, then M_0, M_1, and M_2 for one step each, and so on. In each round of simulation, it adds a new machine to the list and allocates a block of its worktape for the simulation of the new machine. If at any time it runs out of space for one of the simulations, it enters a subroutine to move other blocks to create more space.

When and if the simulation of M_x first attempts to enumerate some $y \geq 2x$, our enumeration procedure also enumerates y, then terminates the simulation of M_x and removes it from the list of simulated machines. Define $g(x)$ to be the y that was enumerated by M_x that caused this to happen. It may be that M_x never enumerates any $y \geq 2x$, in which case the simulation of M_x never terminates and $g(x)$ is undefined.

Let B be the set of elements ever enumerated by this procedure. We claim that B is simple. First, B is r.e., because we have just given a procedure to enumerate it. Second, its complement is infinite, because the $2n$-element set $\{0, \dots, 2n-1\}$ can contain at most n elements of B, namely $g(0), \dots, g(n-1)$. No other $g(m)$ can be in this set, because $g(m) \geq 2m$. Finally, B intersects every infinite r.e. set, because any such set is W_x for some x. When M_x enumerates some $y \geq 2x$, which it must eventually because W_x is infinite, then y is enumerated as an element of B.   □

Proof of Lemma 37.7. Any productive set contains an infinite r.e. subset obtained by iterating the productive function starting with $\varnothing$. Let C be a productive set with productive function σ. Let i_0 be an index for the totally undefined function; then

$$W_{i_0} \;=\; \varnothing \;\subseteq\; C.$$

Now suppose we have constructed $W_{i_n} \subseteq C$. Then $\sigma(i_n) \in C - W_{i_n}$. Define

$$W_{i_{n+1}} \;\overset{\text{def}}{=}\; W_{i_n} \cup \{\sigma(i_n)\}.$$

Then $W_{i_{n+1}} \subseteq C$. Moreover, we can get the index i_{n+1} effectively from the index i_n. The set

$$\{\sigma(i_0), \sigma(i_1), \sigma(i_2), \dots\}$$

is therefore an infinite r.e. subset of C.   □

Proof of Lemma 37.8. Suppose A is r.e.-complete. Then $K \leq_{\mathrm{m}} A$, say by the total recursive function σ. Thus for all x,

$$x \in K \;\Leftrightarrow\; \sigma(x) \in A;$$

equivalently,

$$\sim K \;=\; \sigma^{-1}(\sim A), \tag{37.1}$$

where $\sigma^{-1}(\sim A) \overset{\text{def}}{=} \{x \mid \sigma(x) \in \sim A\}$.

Recall from Example 37.5 that $\sim K$ is productive with productive function $\lambda x.x$. We combine this fact with the reduction σ to get a productive function for A.

Suppose $W_i \subseteq \sim A$. Let m be an index for σ and let

$$\tau \quad = \quad \lambda i.\mathsf{comp}(i,m).$$

Then

$$
\begin{aligned}
W_{\tau(i)} \quad &= \quad W_{\mathsf{comp}(i,m)} \\
&= \quad \{x \mid \varphi_{\mathsf{comp}(i,m)}(x)\!\downarrow\} \\
&= \quad \{x \mid \varphi_i(\sigma(x))\!\downarrow\} \\
&= \quad \{x \mid \sigma(x) \in W_i\} \\
&= \quad \sigma^{-1}(W_i) \\
&\subseteq \quad \sigma^{-1}(\sim A) \quad \text{by monotonicity of } \sigma^{-1} \\
&= \quad \sim K \qquad \text{by (37.1).}
\end{aligned}
$$

Because the identity function is a productive function for $\sim K$,

$$\tau(i) \quad \in \quad \sim K - W_{\tau(i)} \quad = \quad \sigma^{-1}(\sim A) - \sigma^{-1}(W_i) \quad = \quad \sigma^{-1}(\sim A - W_i);$$

therefore

$$\sigma(\tau(i)) \quad \in \quad \sim A - W_i.$$

Thus $\sigma \circ \tau$ is a productive function for $\sim A$. $\qquad \square$

See Miscellaneous Exercise 112 for a generalization of Lemma 37.8.

Lecture 38

The Friedberg–Muchnik Theorem

In this lecture we give Friedberg and Muchnik's solution to Post's problem. The proof illustrates a useful technique called a *finite injury priority argument* that is by now quite common in recursive function theory.

Let $\varphi_0, \varphi_1, \ldots$ be a Gödel numbering of the partial recursive functions. Recall from last time our notation

$$W_n \overset{\text{def}}{=} \text{domain of } \varphi_n = \{x \mid \varphi_n(x)\downarrow\}$$
$$K \overset{\text{def}}{=} \{n \mid \varphi_n(n)\downarrow\},$$

where $\downarrow$ means "is defined," and

$$B \leq_{\mathrm{T}} C \overset{\text{def}}{\Longleftrightarrow} B \text{ is recursive in } C.$$

Define

$$B <_{\mathrm{T}} C \overset{\text{def}}{\Longleftrightarrow} B \leq_{\mathrm{T}} C \text{ but } C \not\leq_{\mathrm{T}} B.$$

Theorem 38.1 (Friedberg [45] and Muchnik [88])   *There exists a nonrecursive r.e. set A such that $K \not\leq_{\mathrm{T}} A$. In other words, there exists a set A such that*

$$\varnothing \quad <_{\mathrm{T}} \quad A \quad <_{\mathrm{T}} \quad K.$$

Low Sets

Our proof of Theorem 38.1 follows [114]. The proof involves the concept of a *low set*.

Definition 38.2   *A set A is* low *if it is r.e. and $K^A \leq_T K$.*

In other words, A is low if halting in the presence of the oracle A is no harder to decide than halting without the oracle.

Lemma 38.3   *If A is low, then $A <_T K$.*

Proof. Certainly $A \leq_T K$, because A is r.e. and K is r.e.-complete. Now if $K \leq_T A$, then $K^K \leq_T K^A$. By lowness, $K^A \leq_T K$, and by transitivity of $\leq_T$, $K^K \leq_T K$. But this is impossible, as K^K is complete for Σ_2^0 and K is complete for Σ_1^0.   □

Recall from last time that a set A is *simple* if

- A is r.e.,

- $\sim A$ is infinite, and

- A intersects every infinite r.e. set.

Lemma 38.4   *There exists a low simple set.*

We prove Lemma 38.4 in the next section. For now, let us show how the Friedberg–Muchnik theorem follows.

Proof of Theorem 38.1. Let A be a low simple set, which exists by Lemma 38.4. By Lemma 38.3, $A <_T K$. Because no simple set can be recursive, $\varnothing <_T A$.   □

A Finite Injury Priority Argument

In this section we give a proof of Lemma 38.4. We give a procedure for enumerating a low simple set A as the union of infinitely many finite sets

$$A \;=\; \bigcup_{t \geq 0} A_t,$$

where A_t has t elements and $A_t \subseteq A_{t+1}$, $t \geq 0$.

To ensure that A is low and simple, we need to satisfy several competing conditions. There are some *positive conditions* that want to put elements into A and some *negative conditions* that want to keep elements out of A. At times, in order to satisfy some condition, we may have to break another condition that has already been satisfied. A condition that is broken in this way is said to be *injured*. However, we assign a *priority* to the conditions such that for every condition, there are only finitely many conditions of higher priority, and the condition can be injured only by a higher priority condition and only once by that condition. Thus a condition can only be injured finitely many times, and will eventually be satisfied.

For A to be low and simple, we must ensure that

(i) A is r.e.;

(ii) A is cofinite;

(iii) A intersects every infinite r.e. set;

(iv) $K^A \leq_T K$.

Condition (i) is true automatically, because we are giving a procedure for enumerating A. Condition (ii) is a negative condition, but this presents no difficulty; it is handled in the same way as in the proof of Post's theorem (Theorem 37.1).

Conditions (iii) and (iv) are the interesting ones. For each n, we consider two conditions, one positive and one negative:

P_n: If W_n is infinite, then $A \cap W_n \neq \varnothing$.

N_n: If $\varphi_n^{A_t}(n) \downarrow^t$ for infinitely many t, then $\varphi_n^A(n) \downarrow$.

Here $\downarrow^t$ means "the machine computing this function halts within t steps."

The conditions P_n are the positive conditions; if they are satisfied for all n, then (iii) holds. The conditions N_n are the negative conditions; if they are satisfied for all n, then (iv) holds (we argue this below). We assign priorities

$$P_0 \; > \; N_0 \; > \; P_1 \; > \; N_1 \; > \; P_2 \; > \; N_2 \; > \; \cdots$$

to the conditions. Note that for any condition, there are only finitely many other conditions of higher priority.

Here is an explanation of N_n. Suppose $\varphi_n^{A_t}(n) \downarrow^t$. We would like to avoid putting any new elements into A in later stages that were the subject of oracle queries in this computation. If we can successfully avoid this, then we will have $\varphi_n^A(n) \downarrow$, because A and A_t will agree on the elements queried. Alas, we cannot always prevent this. Thus N_n can be *injured*. But we can ensure that it can only be injured by a higher priority P_k, and only once for each such P_k. Thus the condition N_n can be injured at most finitely many times.

The conjunction of the conditions N_n implies lowness. To see this, observe that if N_n holds, then

$$\overset{\infty}{\exists} t \; \varphi_n^{A_t}(n) \downarrow^t \;\; \Rightarrow \;\; \varphi_n^A(n) \downarrow \;\; \Rightarrow \;\; \overset{\infty}{\forall} t \; \varphi_n^{A_t}(n) \downarrow^t \;\; \Rightarrow \;\; \overset{\infty}{\exists} t \; \varphi_n^{A_t}(n) \downarrow^t,$$

where $\overset{\infty}{\exists}$ means "there exist infinitely many" and $\overset{\infty}{\forall}$ means "for all but finitely many." The first implication is N_n, and the other implications are just basic set theoretic reasoning. Therefore if N_n holds for all n, then

$$
\begin{aligned}
K^A &= \{n \mid \varphi_n^A(n)\downarrow\} &= \{n \mid \overset{\infty}{\exists}t \; \varphi_n^{A_t}(n)\downarrow^t\} \\
&= \{n \mid \overset{\infty}{\forall}t \; \varphi_n^{A_t}(n)\downarrow^t\} &= \{n \mid \forall k \; \exists t \geq k \; \varphi_n^{A_t}(n)\downarrow^t\} \\
&= \{n \mid \exists k \; \forall t \geq k \; \varphi_n^{A_t}(n)\downarrow^t\}. & (38.1)
\end{aligned}
$$

From the form of the quantification in (38.1), we see that

$$
K^A \;\in\; \Sigma_2^0 \cap \Pi_2^0 \;=\; \Delta_2^0 \;=\; \{B \mid B \leq_{\mathrm{T}} K\},
$$

therefore $K^A \leq_{\mathrm{T}} K$, which is exactly condition (iv).

Now we give an event-driven enumeration of A that satisfies all the required conditions. Let $M_0, M_1, M_2, \ldots$ be a list of enumeration machines such that M_m enumerates W_m. Set $A_0 := \varnothing$.

Do a timesharing parallel simulation of $M_0, M_1, \ldots$ as in the proof of Lemma 37.6, maintaining a list of machines currently being simulated. Continue these simulations until one of the M_m enumerates some element x. When that happens, interrupt the simulations and take the following action. Suppose t elements have been put into A so far, so we have constructed A_t.

(a) If $x < 2m$, just resume the simulation.

(b) Otherwise, for all $n < m$, run $\varphi_n^{A_t}(n)$ for t steps. For any that halt, if x was queried of A_t by that computation, just resume the simulation.

(c) Otherwise, put $x \in A$ (that is, set $A_{t+1} := A_t \cup \{x\}$), and cross M_m off the list.

Now we claim that all the desired conditions are satisfied. Certainly (i) is satisfied. As in the proof of Lemma 37.6, (ii) is satisfied, because

$$
|A \cap \{0, 1, \ldots, 2m-1\}| \;\leq\; m
$$

due to action (a).

Now we show that the conditions P_n and N_n are satisfied. For each n, there is a point in time at which every M_m, $m < n$, that will ever be crossed off the list in (c) has already been crossed off. After that point, if ever $\varphi_n^{A_t}(n)\downarrow^t$, then $\varphi_n^{A_s}(n)\downarrow^t$ for every $s \geq t$, because by (b) no changes will ever be made to the oracle that would cause it to do anything different. Therefore $\varphi_n^A(n)\downarrow^t$, so N_n is satisfied.

The condition P_n is also satisfied: if W_n is infinite, then eventually M_n enumerates an element x greater than $2n$ and greater than any oracle query made by any higher priority computation $\varphi_k^{A_t}(k)$ that halts. At that point x becomes an element of A. $\qquad\square$

Lecture 39

The Analytic Hierarchy

The arithmetic hierarchy relates to first-order number theory as the *analytic hierarchy* relates to *second-order number theory*, in which quantification over sets and functions is allowed. We are primarily interested in the first level of this hierarchy, in particular the class Π_1^1 of relations over $\mathbb{N}$ definable with one universal second-order quantifier. A remarkable theorem due to Kleene states that this is exactly the class of relations over $\mathbb{N}$ definable by first-order induction. In the next few lectures we provide a computational characterization of the classes Π_1^1 and Δ_1^1 and sketch a proof of Kleene's theorem.

Definition of Π_1^1

The class Π_1^1 is the class of all relations on $\mathbb{N}$ that can be defined by a universal second-order number-theoretic formula. Here "universal second-order" means using only universal ($\forall$) quantification over functions $f : \mathbb{N} \to \mathbb{N}$. First-order quantification is unrestricted. Using various transformation rules involving pairing and skolemization,[1] we can assume every such formula is of the form

$$\forall f \; \exists y \; \varphi(\overline{x}, y, f), \tag{39.1}$$

---

[1] $\exists x : \mathbb{N} \; \forall f : \mathbb{N} \to \mathbb{N} \; \varphi(f, x) \quad \mapsto \quad \forall g : \mathbb{N} \to (\mathbb{N} \to \mathbb{N}) \; \exists x : \mathbb{N} \; \varphi(g(x), x).$

where φ is quantifier free (Miscellaneous Exercise 141). This formula defines the n-ary relation

$$\{\bar{a} \in \mathbb{N}^n \mid \forall f \ \exists y \ \varphi(\bar{a}, y, f)\}.$$

Inductive Definability and the Programming Language IND

Traditionally, the *first-order inductive relations* on a structure are defined in terms of least fixpoints of monotone maps defined by first-order formulas. For example, the reflexive transitive closure R^* of a binary relation R on a set is the least fixpoint of the monotone map

$$X \mapsto \{(a, c) \mid a = c \lor (\exists b \ (a, b) \in R \land (b, c) \in X)\} \tag{39.2}$$

(see Lecture A). The theory of first-order inductive definability is quite well established; see for example [87].

Our approach to the subject is more computational. We introduce a programming language IND and use it to define the inductive and hyperelementary relations and the recursive ordinals. This turns out to be equivalent to the more traditional approach, as we argue below. However, keep in mind that the relations "computed" by IND programs are highly noncomputable. IND programs were defined by Harel and Kozen [53] (see also [54]).

An IND program consists of a finite sequence of labeled statements. Each statement is of one of three forms:

- Assignment: $\ell : x := \exists$ $\ell : y := \forall$

- Conditional test: $\ell :$ if $R(t_1, \ldots, t_n)$ then goto ℓ'

- Halt statement: $\ell :$ accept $\ell :$ reject.

The semantics of programs is very much like alternating Turing machines, except that the branching is infinite. The execution of an assignment statement causes countably many subprocesses to be spawned, each assigning a different element of $\mathbb{N}$ to the variable. If the statement is $x := \exists$, the branching is existential; if it is $y := \forall$, the branching is universal. The conditional jump tests the atomic formula $R(t_1, \ldots, t_n)$, and if true, jumps to the indicated label. The accept and reject commands halt and pass a Boolean value back up to the parent.

Computation proceeds as in alternating Turing machines. The input is an initial assignment to the program variables. Execution causes a countably branching computation tree to be generated downward, and Boolean accept (1) or reject (0) values are passed back up the tree, a Boolean $\land$ being computed at each existential node and a Boolean $\lor$ being computed at each universal node. The program is said to *accept* the input if the root of the computation tree ever becomes labeled with the Boolean value 1 on that input; it is said to *reject* the input if the root ever becomes labeled with the Boolean value 0 on that input; and it is said to *halt* on an input if it either accepts or rejects that input. An IND program that halts on all inputs is said to be *total*.

These notions are completely analogous to alternating Turing machines, so we forgo the formalities in favor of some revealing examples.

First, we show how to simulate a few other useful programming constructs with those listed above. An unconditional jump

    goto ℓ

is simulated by the statement

    if $x = x$ then goto ℓ

More complicated forms of conditional branching can be effected by manipulation of control flow. For example, the statement

    if $R(\bar{t})$ then reject else ℓ

is simulated by the program segment

        if $R(\bar{t})$ then goto ℓ'
        goto ℓ
    ℓ':   reject

A simple assignment is effected by guessing and verifying:

    $x := y + 1$

is simulated by

    $x := \exists$
    if $x \neq y + 1$ then reject

The process spawns infinitely many subprocesses, all but one of which immediately reject!

Any first-order relation is definable by a loop-free program. For example, the set of natural numbers x such that

    $\exists y \, \forall z \, \exists w \; x \leq y \wedge x + z \leq w$

is defined by the program

    $y := \exists$
    $z := \forall$
    $w := \exists$
    if $x > y$ then reject
    if $x + z \leq w$ then accept
    reject

The converse is true too: any loop-free program defines a first-order relation.

However, IND can also define inductively definable relations that are not first-order. For example, the reflexive transitive closure R^* of a relation R is definable by the following program, which takes its input in the variables x, z and accepts if $(x, z) \in R^*$.

```
ℓ:   if x = z then accept
     y := ∃
     if ¬R(x, y) then reject
     x := y
     goto ℓ
```

Compare this program to (39.2).

Here is another example. Recall from Lecture 8 that a *two-person perfect information game* consists of a binary predicate MOVE. The two players alternate. If the current board configuration is x and it is player I's turn, player I chooses y such that MOVE(x, y); then player II chooses z such that MOVE(y, z); and so on. A player wins by *checkmate*; that is, by forcing the opponent into a position from which there is no legal next move. Thus a checkmate position is an element y such that $\forall z \ \neg$MOVE(y, z).

We would like to know for a given board position x whether the player whose turn it is has a forced win from x. As in Lecture 8, we might define this as the least solution WIN of the recursive equation

$$\text{WIN}(x) \quad \Leftrightarrow \quad \exists y \ (\text{MOVE}(x, y) \land \forall z \ \text{MOVE}(y, z) \rightarrow \text{WIN}(z)).$$

(The base case involving an immediate win by checkmate is included: if y is a checkmate position, then the subformula $\forall z \ \text{MOVE}(y, z) \rightarrow \text{WIN}(z)$ is vacuously true.) The least solution to this recursive equation is the least fixpoint of the monotone map τ defined by

$$\tau(R) \quad \overset{\text{def}}{\Longleftrightarrow} \quad \{x \mid \exists y \ \text{MOVE}(x, y) \land \forall z \ \text{MOVE}(y, z) \rightarrow R(z)\}.$$

We can express WIN(x) with an IND program as follows.

```
ℓ:   y := ∃
     if ¬MOVE(x, y) then reject
     x := ∀
     if ¬R(y, x) then accept
     goto ℓ
```

Our last example involves well-founded relations. As is well known (see Miscellaneous Exercise 24), induction and well-foundedness go hand in hand. Here is an IND program that tests whether a strict partial order $<$ is well-founded:

$$\ell: \quad \begin{aligned} x &:= \forall \\ y &:= \forall \\ &\text{if } \neg(y < x) \text{ then accept} \\ x &:= y \\ &\text{goto } \ell \end{aligned}$$

Any property that is expressed as a least fixpoint of a monotone map defined by a positive first-order formula can be computed by an IND program. Here is what we mean by this. Let $\varphi(\overline{x}, R)$ be a first-order formula with free individual variables $\overline{x} = x_1, \dots, x_n$ and free n-ary relation variable R. Assume further that all occurrences of R in φ are *positive*; that is, occur under an even number of negation symbols $\neg$. For any n-ary relation B, define

$$\tau(B) \stackrel{\text{def}}{=} \{\overline{a} \mid \varphi(\overline{a}, B)\}.$$

That is, we think of φ as a set operator τ mapping a set of n-tuples B to another set of n-tuples $\{\overline{a} \mid \varphi(\overline{a}, B)\}$. One can show that the positivity assumption implies that the set operator τ is monotone, therefore by Theorem A.9 has a least fixpoint F_φ, which is an n-ary relation. The traditional treatment defines a first-order inductive relation as a projection of such a fixpoint; that is, a relation of the form

$$\{(a_1, \dots, a_m) \mid F_\varphi(a_1, \dots, a_m, b_{m+1}, \dots, b_n)\},$$

where $b_{m+1}, \dots, b_n$ are fixed elements of the structure. Given the formula φ and the elements $b_{m+1}, \dots, b_n$, one can construct an IND program that assigns $b_{m+1}, \dots, b_n$ to the variables $x_{m+1}, \dots, x_n$, then checks whether the values of $x_1, \dots, x_n$ satisfy F_φ by decomposing the formula top-down, executing existential assignments at existential quantifiers, executing universal assignments at universal quantifiers, using control flow for the propositional connectives, using conditional tests for the atomic formulas, and looping back to the top of the program at occurrences of the inductive variable R. The examples above involving reflexive transitive closure, games, and well-foundedness illustrate this process.

Conversely, any relation computed by an IND program is inductive in the traditional sense, essentially because the definition of acceptance for IND programs involves the least fixpoint of an inductively defined set of labelings of the computation tree.

Inductive and Hyperelementary Relations

Many of the sample IND programs of the previous section make sense when interpreted over any structure, not just $\mathbb{N}$. We define the *inductive relations* of any structure $\mathfrak{A}$ to be those relations computable by IND programs over

$\mathfrak{A}$. We define the *hyperelementary relations* of $\mathfrak{A}$ to be those relations computable by total IND programs over $\mathfrak{A}$, that is, programs that halt on all inputs. An *elementary relation* of $\mathfrak{A}$ is just a first-order relation. All elementary relations are hyperelementary, because they are computed by loop-free programs, which always halt. Over $\mathbb{N}$, the hyperelementary and elementary relations are called hyperarithmetic and arithmetic, respectively. An example of a hyperarithmetic set that is not arithmetic is first-order number theory $\text{Th}(\mathbb{N})$ (Miscellaneous Exercise 142).

One can show that a relation over $\mathfrak{A}$ is hyperelementary iff it is both inductive and coinductive: if there is an IND program that accepts R and another IND program that accepts $\sim R$, then one can construct a total IND program that runs the two other programs in parallel, as with the corresponding result for Turing machines.

Lecture 40

Kleene's Theorem

In this lecture we restrict our attention to the structure of arithmetic $\mathbb{N}$. Over this structure, the hyperelementary relations are sometimes called the *hyperarithmetic relations*.

Recursive Trees, Recursive Ordinals, and ω_1^{ck}

An ordinal is *countable* if there exists a bijection between it and ω. The ordinals $\omega \cdot 2$ and ω^2, although greater than ω, are still countable. The smallest uncountable ordinal is called ω_1.

Traditionally, a *recursive ordinal* is defined as one for which there exists a *computable* bijection between it and ω under some suitable encoding of ordinals and notion of computability (see Rogers [104]). The smallest nonrecursive ordinal is called ω_1^{ck}. It is a countable ordinal, but it looks uncountable to any computable function.

We define recursive ordinals in terms of inductive labelings of *recursive ω-trees*. An *ω-tree* is a nonempty prefix-closed subset of ω^*. In other words, it is a set T of finite-length strings of natural numbers such that

- $\epsilon \in T$, and

- if $xy \in T$ then $x \in T$.

A *path* in T is a maximal subset of T linearly ordered by the prefix relation. The tree T is *well-founded* if there are no infinite paths. A *leaf* is an element of T that is not a prefix of any other element of T. An ω-tree T is *recursive* if the set T, suitably encoded, is a recursive set.

Given a well-founded tree T, we define a labeling $o : T \to \text{Ord}$ inductively as follows:

$$o(x) \overset{\text{def}}{=} \sup_{\substack{n \in \omega \\ xn \in T}} (o(xn) + 1).$$

Thus $o(x) = 0$ if x is a leaf, and if x is not a leaf, then $o(x)$ is determined by first determining $o(xn)$ for all $xn \in T$, then taking the supremum of all the successors of these ordinals.

For example, consider the tree consisting of ε and all sequences of the form $(n, \underbrace{0, 0, \ldots, 0}_{m})$ for $n \geq 0$ and $m \leq n$. The leaves are labeled 0 by o, the next elements above the leaves are labeled 1, and so on. The root ε is labeled ω.

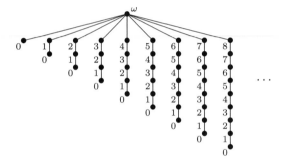

For a well-founded tree T, let $o(T)$ be the ordinal assigned to the root of T. Every $o(T)$ is a countable ordinal, and $\sup_T o(T) = \omega_1$.

Now define an ordinal to be *recursive* if it is $o(T)$ for some recursive tree T. The supremum of the recursive ordinals is ω_1^{ck}.

An alternative definition of recursive ordinals is the set of all running times of IND programs. The running time of an IND program on some input is the time it takes to label the root of the computation tree with 1 or 0. This is the closure ordinal of the inductive definition of labelings of the computation tree in the formal definition of acceptance. It is very similar to the definition of the labelings o of recursive trees. The ordinal ω_1^{ck} is the supremum of all running times of IND programs.

Kleene's Theorem

Theorem 40.1   (Kleene [74])   *Over $\mathbb{N}$, the inductive relations and the Π_1^1 relations coincide, and the hyperelementary and Δ_1^1 relations coincide.*

Proof sketch. First we show that every inductive relation is Π^1_1. This direction holds in any structure $\mathfrak{A}$, not just $\mathbb{N}$. Let $\varphi(\overline{x}, R)$ be a positive first-order formula with fixpoint $F_\varphi \subseteq A^n$, where A is the carrier of $\mathfrak{A}$. We can describe F_φ as the intersection of all relations closed under φ:

$$F_\varphi(\overline{x}) \quad \Leftrightarrow \quad \forall R \; (\forall \overline{y} \; \varphi(\overline{y}, R) \rightarrow R(\overline{y})) \rightarrow R(\overline{x}).$$

This is a Π^1_1 formula.

Conversely, consider any Π^1_1 formula over $\mathbb{N}$. As previously noted, using various rules for manipulating formulas, we can assume without loss of generality that the formula is of the form

$$\forall f \; \exists x \; \varphi(x, f), \tag{40.1}$$

where φ is quantifier free (Miscellaneous Exercise 141).

Regarding a function $f : \omega \rightarrow \omega$ as the infinite string of its values $f(0), f(1), f(2), \ldots$, the functions f are in one-to-one correspondence with paths in the complete tree ω^*. Moreover, for any x, the truth of $\varphi(x, f)$ is determined by any finite prefix of this path that includes all arguments to f corresponding to terms appearing in $\varphi(x, f)$. Let $f \upharpoonright n$ denote the finite prefix of f of length n. We can think of $f \upharpoonright n$ either as a string of natural numbers of length n or as a partial function that agrees with f on domain $\{0, 1, \ldots, n-1\}$.

Let $\varphi'(n, x, f)$ be a formula that has the same truth value as $\varphi(x, f)$ if $f \upharpoonright n$ has enough information to determine whether $\varphi(x, f)$, and is 0 otherwise. We can obtain φ' easily from φ. For example, if $\varphi(x, f)$ is $x = f(f(x))$, which is equivalent to the loop-free IND program

```
if x = f(f(x)) then accept else reject,
```

take $\varphi'(n, x, f)$ to be a first-order formula equivalent to the loop-free IND program

```
if x < n {          //is f⌈n(x) defined?
    y := f(x);      //if so, let y be its value
    if y < n {      //is f⌈n(f⌈n(x)) defined?
        z := f(y);  //if so, let z be its value
        if x = z accept; //test whether x = f(f(x))
    }
}
reject;
```

Now instead of (40.1), we can write

$$\forall f \; \exists n \; \exists x \; \varphi'(n, x, f \upharpoonright n). \tag{40.2}$$

Note that if $\exists x \; \varphi'(n, x, f)$, then $\exists x \; \varphi'(m, x, f)$ for all $m \geq n$. This says that (40.2) is essentially a well-foundedness condition: if we label the vertices $f \upharpoonright n$ of the infinite tree with the truth value of $\exists x \; \varphi'(n, x, f)$, (40.2) says that along every path in the tree we eventually encounter the value 1. And as observed in the Lecture 39, well-foundedness is inductive.

We have shown that the inductive and Π_1^1 relations over $\mathbb{N}$ coincide. Because the hyperarithmetic relations are those that are both inductive and coinductive and the Δ_1^1 relations are those that are both Π_1^1 and Σ_1^1, the hyperarithmetic and Δ_1^1 relations coincide as well. □

Inductive Is Existential over Hyperelementary

Over $\mathbb{N}$, it is apparent from the characterization of Π_1^1 as those sets accepted by IND programs and Δ_1^1 as those sets accepted by total IND programs that there is a strong analogy between the inductive and the r.e. sets and between the hyperelementary and the recursive sets.

It may seem odd that the class analogous to Σ_1^0 at the analytic level should be Π_1^1 and not Σ_1^1. This is explained by the following result, which corresponds to the characterization of the r.e. sets given by (35.1).

Theorem 40.2 *A set $A \subseteq \mathbb{N}$ is inductive iff there is a hyperelementary relation R such that*

$$
\begin{aligned}
A &= \{x \mid \exists \alpha < \omega_1^{\mathrm{ck}} \; R(x, \alpha)\} \\
&= \{x \mid \exists y \; y \text{ encodes a recursive ordinal and } R(x, y)\}. \quad (40.3)
\end{aligned}
$$

Proof sketch. If R is hyperelementary, then we can build an IND program for (40.3) consisting of the statement $y := \exists$ followed by a program that in parallel checks that the Turing machine with index y accepts a well-founded recursive tree and that $R(x, y)$.

Conversely, if A is inductive, say accepted by an IND program p, then we can describe A by an existential formula that says, "there exists a recursive ordinal α such that p halts and accepts x in α steps." More concretely, one would say, "there exists a well-founded recursive tree T such that on input x, p halts and accepts in $o(T)$ steps." The quantification is then over indices of Turing machines. To show that the predicate "p halts and accepts x in $o(T)$ steps" is hyperelementary, one would construct an IND program that runs p together with a program q that halts in $o(T)$ steps (q just enumerates the tree T using existential branching and rejects) and takes whichever action happens first. □

Lecture 41

Fair Termination and Harel's Theorem

You may have gotten the impression from Lectures 39 and 40 that ω_1^{ck} and Π_1^1 have little to do with computer science. Here is an example of a real application in which they arise: proving *fair termination of concurrent programs*.

Termination proofs typically rely on induction to show that progress toward termination is made with each step. For ordinary sequential programs, induction on the natural numbers ω is usually sufficient.

For example, consider the following program for computing the greatest common divisor (gcd) of two given positive integers x, y. The gcd is the value of the variable x upon termination.

```
while (y ≠ 0) {
    z := x mod y;
    x := y;
    y := z;
}
```

This program eventually terminates for any nonnegative integers x, y because each iteration of the loop causes the value of y to remain nonnegative but strictly decrease, thus progress is made toward termination. To prove this formally, ordinary induction on ω suffices.

For *concurrent programs*, the story is a little more complicated. There may be several processes operating simultaneously. These processes may be competing for resources, such as execution time on a shared processor or access to a shared variable. In modeling the behavior of concurrent programs, nondeterminism is often involved, because we may not know exactly how such contention will be resolved in each instance, although we may know the range of possibilities. Thus the computation can be modeled by a branching tree, each path of which is a possible computation path of the system.

Unfortunately, under the usual semantics of nondeterminism, some computation paths may fail to terminate for uninteresting reasons. For example, consider the following nondeterministic program.

$$
\begin{aligned}
&x := 0; \\
&y := 0; \\
&\texttt{while } (x < 10 \lor y < 10) \; \{ \\
&\quad x := x + 1 \parallel y := y + 1; \\
&\}
\end{aligned}
\tag{41.1}
$$

Here $\parallel$ means, "do either p or q." According to the usual semantics of nondeterminism, there are nonhalting computations of this program; for example, the one in which the left branch in the body of the loop is always chosen. However, any scheduler that always chooses the left branch would be considered unfair, because the right branch was never allowed to execute, even though it was infinitely often enabled.

Thus we might want to assume that the agent that resolves the nondeterminism at choice points does so in a fair way, although we may not know (or care) exactly how this is accomplished. We abstract away from the exact nature of the scheduler and just assume some formal fairness property. For example, we might want to assume that if a statement of the form $p \parallel q$ is infinitely often enabled, then each of p and q is infinitely often chosen for execution. Under this assumption, all infinite computation paths of the example above are unfair; in other words, all fair paths terminate. Such assumptions allow us to study correctness properties such as termination independent of the implementation details of the scheduler.

Fair Termination

The *fair termination problem* is the problem of determining whether a given concurrent program terminates under the assumption of a fair scheduler. Intuitively, the concurrent program (41.1) terminates fairly, because any sequence of actions causing nontermination would never be chosen by a fair scheduler.

Formally, we define a *fairness condition* to be a pair (ρ, σ) of properties that are true or false of a state of the computation. We think of ρ as expressing a request for a resource. Thus ρ is true of a state s (written

$s \models \rho$) if some resource associated with ρ, whatever it may be, is being requested. We think of σ as expressing the condition that the request has been satisfied.

An infinite computation path is *unfair* with respect to the fairness condition (ρ, σ) if ρ is true infinitely often along the path but σ is true only finitely often. This models the idea that a request is made infinitely often along the path, but from some point on is never satisfied. The path is *unfair* with respect to a set of fairness conditions if it is unfair with respect to any one of them. (Notice the similarity to the Rabin acceptance condition for automata on infinite strings; see Lecture 26.) A path is *fair* if it is not unfair. By our definition, all finite paths are fair.

A (nondeterministic) computation is *fairly terminating* if there are no infinite fair paths; equivalently, if all infinite paths are unfair. Intuitively, we do not care if the computation tree contains infinite paths as long as they are unfair, because a fair scheduler would never allow them to occur anyway.

In our example (41.1) above, let ℓ be the statement in the body of the while loop. Maintain in the state an extra bit telling whether the left or the right branch was taken the last time ℓ was executed. Let ρ be the property, "about to execute ℓ", and let σ_0 (respectively, σ_1) be the property, "the left (respectively, right) branch was taken the last time ℓ was executed". Consider the fairness conditions (ρ, σ_0) and (ρ, σ_1). The program terminates fairly with respect to these conditions, because any computation path that satisfies both σ_0 and σ_1 sufficiently many times eventually satisfies the condition for exiting the while loop.

Proof Rules for Fair Termination

There is a rather large literature on fairness and fair termination (see [44] and references therein). Much of that work was devoted to deriving proof rules for establishing correctness and termination in various logical formalisms and under various fairness assumptions. A central notion is the idea of *helpful directions* that move a computation toward termination. This notion ultimately reduces to well-foundedness, but is not simply a decrease of an integer parameter. It was observed that transfinite induction on ordinals higher than ω was necessary in general.

Harel's Theorem

The situation was significantly clarified in 1986 by Harel [52]. He showed that fair termination of finitely branching recursive trees is equivalent to the well-foundedness of countably branching recursive trees. Because deciding well-foundedness of countably branching recursive trees is Π^1_1-complete

(Homework 12, Exercise 1(b)), and the supremum of the ordinals of well-founded countably branching recursive trees is ω_1^{ck}, a corollary of Harel's theorem is that the fair termination problem is Π_1^1-complete, and that the ordinals involved in fair termination proofs can be as high as ω_1^{ck} in general.

We prove Harel's theorem in a special case, which nevertheless conveys the main idea. Define a *binary tree* to be a nonempty prefix-closed subset of $\{0,1\}^*$. Similarly, an *ω-tree* is a nonempty prefix-closed subset of ω^*. A *path* in either type of tree is a maximal subset linearly ordered by the prefix relation.

For binary trees T, consider the fairness condition $(\text{true}, \text{last}(0))$, where for $x \in \{0,1\}^*$, $x \vDash \text{last}(0)$ if $x = y0$ for some string y; that is, if 0 is the last letter of x. The unfair paths of T with respect to this fairness condition are the sets of all finite prefixes of infinite strings of the form $x1^\omega$, where $x \in \{0,1\}^*$. Thinking of 0 as "go left" and 1 as "go right", an unfair path is one that goes right at all but finitely many points. The tree T is fairly terminating if there is no infinite path that goes left infinitely often.

Now we describe an effective map from binary trees to ω-trees such that a binary tree is fairly terminating iff the corresponding ω-tree is well-founded.

For $x \in \{0,1\}^*$, define $\tau(x) = x_1 x_2 \cdots x_n$, where x can be uniquely parsed as $1^{x_1} 0 1^{x_2} 0 \cdots 0 1^{x_n}$. For example,

$$\tau(110100111110100111) = 2105103.$$

We can define τ inductively:

$$\tau(\varepsilon) \stackrel{\text{def}}{=} 0$$

$$\tau(x0) \stackrel{\text{def}}{=} \tau(x) \cdot 0$$

$$\tau(x1) \stackrel{\text{def}}{=} \text{lastinc}(\tau(x)),$$

where $\text{lastinc}(xn) \stackrel{\text{def}}{=} x \cdot (n+1)$. It is easily shown that τ is one-to-one and onto except for the empty string in ω^*.

Now if $T \subseteq \{0,1\}^*$ is a binary tree, let

$$\tau(T) \stackrel{\text{def}}{=} \{\tau(x) \mid x \in T\} \cup \{\varepsilon\}.$$

Then $\tau(T)$ is nonempty and prefix-closed, therefore is an ω-tree.

For example, consider the following binary tree $T \subseteq \{0,1\}^*$.

The maximal elements of T are 00, 010, 011, 101, and 11. The prefixes of these strings not already listed are 01, 10, 0, 1, and ε. Applying τ to the maximal strings gives 000, 010, 02, 11, and 2, and applying τ to the remaining prefixes gives 01, 10, 00, 1, and 0. We must also include ε. This gives the following ω-tree $\tau(T) \subseteq \omega^*$.

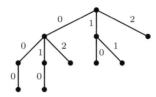

An intuitive way to view the construction of $\tau(T)$ from T is to reorient the edges of T so that those labeled 0 in T go down to the leftmost child in $\tau(T)$ and those labeled 1 in T go right to the next sibling in $\tau(T)$. The rightmost spine of T corresponds to the children of the root in $\tau(T)$.

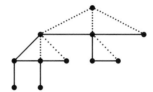

The ω-trees of the form $\tau(T)$ have the property that if $x \cdot (n+1) \in \tau(T)$, then $x \cdot n \in \tau(T)$. Let us say that an ω-tree is *full* if it has this property. Then every $\tau(T)$ is full, and every nontrivial full ω-tree is $\tau(T)$ for some binary tree T.

We now argue that T is fairly terminating iff $\tau(T)$ is well-founded. We wish to show that T has an infinite path with infinitely many 0's iff $\tau(T)$ has an infinite path. If T has an infinite path with infinitely many 0's, then that path is the set of finite prefixes of an infinite string of the form $1^{x_0}01^{x_1}01^{x_2}0\cdots$. Then all finite prefixes of the infinite string $x_0x_1x_2\cdots$ are members of $\tau(T)$, and this is an infinite path. Conversely, if $\tau(T)$ has an infinite path, then it must be the set of finite prefixes of an infinite string of the form $x_0x_1x_2\cdots$, thus $1^{x_0}01^{x_1}01^{x_2}\cdots01^{x_n} \in T$ for all n. The set of prefixes of these strings is contained in T and constitutes an infinite unfair path.

We have shown

Theorem 41.1   (Harel [52])   *The map τ constitutes a recursive one-to-one correspondence between binary trees and nontrivial full ω-trees such that the binary tree T is fairly terminating with respect to the fairness condition* $(\mathsf{true}, \mathsf{last}(0))$ *if and only if $\tau(T)$ is well-founded.*

Corollary 41.2   *Fair termination is Π^1_1-complete.*

Proof. Miscellaneous Exercise 143.   □

Exercises

Homework 1

1. (a) Give a one-tape deterministic $O(n \log n)$-time-bounded Turing machine accepting a nonregular set.

    (b) Show that any set accepted by a one-tape deterministic TM in time $o(n \log n)$ is regular.

2. A *k-counter automaton with linearly bounded counters* is a one-tape TM with two-way read-only input head and a fixed finite number of integer counters that can hold an integer between 0 and n, the length of the input string. In each step, the machine may test each of its counters for zero. Based on this information, its current state, and the input symbol it is currently scanning, it may add one or subtract one from each of the counters, move its read head left or right, and enter a new state.

    (a) Give a formal definition of these machines, including a definition of acceptance.

    (b) Show that a set is in $LOGSPACE$ ($NLOGSPACE$) iff it is accepted by a deterministic (nondeterministic) k-counter machine with linearly bounded counters for some k.

3. Show that if $P = NP$ then $NEXPTIME = EXPTIME$. (*Hint.* Pad the input with extra #'s).

Homework 2

1. Show that the reducibility relation $\leq_m^{\log}$ is transitive: if $A \leq_m^{\log} B$ and $B \leq_m^{\log} C$, then $A \leq_m^{\log} C$. (*Warning.* This is nontrivial! You don't have enough space to write down an intermediate result in its entirety.)

2. A Boolean formula is in *2-conjunctive normal form* (2CNF) if it is a conjunction of disjuncts of the form $\ell \vee \ell'$, where ℓ and ℓ' are *literals* (Boolean variables or negations of variables). The problem of deciding satisfiability of Boolean formulas in 2CNF is denoted 2SAT. Show that 2SAT is complete for co-*NLOGSPACE* under $\leq_m^{\log}$.

3. Show that the value of a given Boolean formula under a given truth assignment can be computed in deterministic logspace.

4. Consider the nonregular set

$$B \;=\; \{ \$ b_k(0) \$ b_k(1) \$ b_k(2) \$ \cdots \$ b_k(2^k - 1) \$ \mid k \geq 0 \} \;\subseteq\; \{0, 1, \$\}^*,$$

where $b_k(i)$ denotes the k-bit binary representation of i. Show that this set is in $DSPACE(\log \log n)$. Note that it is not enough to give a deterministic TM for B in which every *accepting* computation takes $O(\log \log n)$ space. According to the official definition, in order to show that B is in the complexity class $DSPACE(\log \log n)$, we must give a deterministic TM for B in which *every* computation, either accepting, rejecting, or looping, takes $O(\log \log n)$ space.

Homework 3

1. (a) The set of strings of balanced parentheses is the set generated by the context-free grammar

    $$S \quad \rightarrow \quad (S) \mid SS \mid \varepsilon$$

    Prove that this set is in $LOGSPACE$.

    (b) How about strings of balanced parentheses of two types

    $$S \quad \rightarrow \quad (S) \mid [S] \mid SS \mid \epsilon \ ?$$

2. Suppose that the game of generalized geography is altered so that vertices can be reused. That is, Players I and II alternate choosing edges along a directed path (not necessarily simple) starting from a given vertex s and try to force each other into a position from which there is no next move. What is the complexity of determining whether Player I has a winning strategy?

3. An *alternating finite automaton* (AFA) is a 5-tuple

    $$M \quad = \quad (Q, \Sigma, \delta, F, \alpha),$$

    where Q is a finite set of *states*, Σ is a finite *input alphabet*, $F : Q \rightarrow \{0, 1\}$ is the characteristic function of a set of *final* or *accept states*, that is,

    $$F(q) \quad = \quad \begin{cases} 1, & \text{if } q \text{ is a final state} \\ 0, & \text{otherwise,} \end{cases}$$

    δ is the *transition function*

    $$\delta : (Q \times \Sigma) \quad \rightarrow \quad ((Q \rightarrow \{0, 1\}) \rightarrow \{0, 1\}),$$

    and α is the *acceptance condition*

    $$\alpha : (Q \rightarrow \{0, 1\}) \quad \rightarrow \quad \{0, 1\}.$$

    Intuitively, F gives a labeling of 0 or 1 at the leaves of the computation tree, and for all $q \in Q$ and $a \in \Sigma$, the Boolean function

    $$\delta(q, a) : (Q \rightarrow \{0, 1\}) \quad \rightarrow \quad \{0, 1\}$$

takes a labeling on states and computes a new label for state q; this is used to pass Boolean labels 0 or 1 back up the computation tree. Formally, the transition function δ uniquely determines a map

$$\widehat{\delta} : (Q \times \Sigma^*) \quad \rightarrow \quad ((Q \rightarrow \{0,1\}) \rightarrow \{0,1\}),$$

defined inductively as follows. For $q \in Q$, $a \in \Sigma$, and $x \in \Sigma^*$,

$$\widehat{\delta}(q, \epsilon)(u) \;=\; u(q)$$
$$\widehat{\delta}(q, ax)(u) \;=\; \delta(q, a)(\lambda p.(\widehat{\delta}(p, x)(u))).$$

The machine is said to *accept* $x \in \Sigma^*$ if

$$\alpha(\lambda p.(\widehat{\delta}(p, x)(F))) \;=\; 1.$$

Prove that a set $A \subseteq \Sigma^*$ is accepted by a k-state alternating finite automaton if and only if its *reverse*

$$A^R \;=\; \{a_n \cdots a_1 \mid a_1 \cdots a_n \in A\}$$

is accepted by a 2^k-state deterministic finite automaton (DFA).

Homework 4

1. Prove the following generalization of Savitch's theorem. For $S(n) \geq \log n$,

$$STA(S(n), *, A(n)) \quad \subseteq \quad DSPACE(A(n)S(n) + S(n)^2).$$

(*Hint.* Let type $: Q \to \{\wedge, \vee\}$ be the map in the specification of the alternating machine telling whether a state in Q is universal or existential. Extend type to configurations in the obvious way. Consider the predicate $R(\alpha, \beta, k) =$ "There is a computation path of length at most k leading from configuration α to configuration β in which all configurations γ, except possibly β, satisfy type$(\gamma) =$ type(α)." Use a Savitch-like argument.)

2. Give a formal definition of a hierarchy over *PSPACE* with levels Σ_k^{PSPACE} and Π_k^{PSPACE} analogous to *PH*. Show that this hierarchy collapses to *PSPACE*. (*Hint.* Use Exercise 1.)

3. Let H_k be the complete set for Σ_k^P defined in Lecture 9:

$$H_k \quad = \quad \{M\$x\$^d \mid M \text{ is a } \Sigma_k \text{ machine accepting } x \text{ in time at most } d\}.$$

Let $\#(y)$ denote the number represented by y in binary. Show that the set

$$H_\omega \quad = \quad \{y\$z \mid z \in H_{\#(y)}\}$$

is $\leq_m^{\log}$-complete for *PSPACE*.

4. Define a class of sets G_k similar to H_k for space instead of time:

$$G_k \quad = \quad \{M\$x\$^d \mid M \text{ is a } \Sigma_k \text{ machine accepting } x \text{ in space at most } d\}.$$

Show that G_k is $\leq_m^{\log}$-complete for Σ_k^{PSPACE}, and that the set

$$G_\omega \quad = \quad \{y\$z \mid z \in G_{\#(y)}\}$$

is complete for exponential time. How do you reconcile this with Exercise 2?

Homework 5

1. Show that if $NP=$ co-NP, then PH collapses to NP. More generally, show that if $\Sigma_k^{\mathrm{p}} = \Pi_k^{\mathrm{p}}$ then PH collapses to Σ_k^{p}.

2. Suppose there exists a sequence of polynomial-size circuits $B_0, B_1, B_2, \ldots$ for the Boolean satisfiability problem. That is, B_n has n inputs, one output, and $n^{O(1)}$ gates, and given (an encoding over $\{0,1\}^*$ of) a Boolean formula x with $|x| = n$, x is satisfiable iff $B_n(x) = 1$.

   (a) Show that if the sequence $B_0, B_1, \ldots$ is polynomial-time uniform (that is, B_n can be produced from 0^n in polynomial time), then $P = NP$.

   (b) Show that even if the sequence is not polynomial-time uniform, then PH collapses to Σ_3^{p}.

3. In Lecture 5 and Homework 1, Exercise 2, we considered k-counter automata whose counter values could not exceed n, the length of the input. Without this restriction, it is known that two-counter automata are as powerful as arbitrary Turing machines (see [61, 76]). Show that the membership problem for nondeterministic (unbounded) one-counter automata is complete for $NLOGSPACE$. (*Warning.* The difficult part is to detect looping. Unlike the bounded counter case, there are infinitely many possible configurations on inputs of length n.)

Homework 6

1. In Lecture 3 we showed the existence of an unbounded space-constructible function $S(n) \leq O(\log \log n)$. In this exercise we show that the function $\lceil \log \log n \rceil$ itself is not space constructible.

   (a) Prove that for any space-constructible function $S(n) \leq o(\log n)$, there exists a value k such that $S(n) = k$ for infinitely many n. In other words,

   $$\liminf_{\substack{n \geq 0 \, m \geq n}} S(m) \quad < \quad \infty.$$

   (*Hint.* If $S(n)$ is space-constructible, there must be a machine that on any input of length n lays off exactly $S(n)$ space on its worktape without using more than $S(n)$ space and halts. Count configurations of state, worktape contents, and worktape head positions (*not* read head positions). Consider what happens as the machine scans across a very long input of the form 0^n.)

   (b) Conclude from (a) that the function $\lceil \log \log n \rceil$ is not space-constructible.

2. Show that the problem of deciding whether $L(M) = \Sigma^*$ for a given non-deterministic finite automaton (NFA) M is *PSPACE*-complete. (*Hint.* Use computation histories

   $$\#\alpha_0 \# \alpha_1 \# \cdots \# \alpha_N \#,$$

   where each $\alpha_i \in \Delta^*$ is an encoding of a configuration of some TM N running in *PSPACE*, and α_{i+1} follows from α_i in one step according to the transition rules of N.)

3. What is the complexity of Problem 2 when M is deterministic? Give proof.

Homework 7

1. Determine the complexity of the first-order theory of the structure $(\omega, \leq)$, where ω is the set of natural numbers and $\leq$ is the usual linear order on ω.

2. Consider the following Ehrenfeucht–Fraïssé game G_n between two players, Sonja (also known in the literature as the *Spoiler*) and David (also known as the *Duplicator*). Each player has n pebbles, one of each of n distinct colors. The players alternate placing the pebbles on elements of two linear orders $\mathcal{A}$ and $\mathcal{B}$. In each round, Sonja plays one of her remaining pebbles on some element of either $\mathcal{A}$ or $\mathcal{B}$. David must then play his pebble of the same color on some element of the other structure. At the end of n rounds, David is declared the winner if the pebbles occur in the same order in $\mathcal{A}$ as in $\mathcal{B}$ (pebble colors are significant when determining this). Otherwise, Sonja is the winner.

   (a) Show that if $\mathcal{A}$ is the set of rational numbers and $\mathcal{B}$ is the set of integers, then Sonja has a forced win in G_3.

   (b) Show that if $\mathcal{A}$ is the set of rationals and $\mathcal{B}$ is the set of reals, then David has a forced win in G_n for any n.

   (c) Show that David has a forced win in G_n if and only if $\mathcal{A}$ and $\mathcal{B}$ agree on all first-order sentences of quantifier depth n. (The *quantifier depth* of a formula is the maximum number of quantifiers in whose scope some symbol appears. For example, the formula

   $$\exists x\ ((\forall y\ y \leq x) \wedge (\exists z\ z \leq x))$$

   has quantifier depth two.)

Homework 8

1. Show that the set of finite subsets of ω, represented as a set of strings in $\{0, 1\}^\omega$, is accepted by a nondeterministic Büchi automaton, but by no deterministic Büchi automaton. (Recall that in Büchi acceptance, $F \subseteq Q$, and a run σ is *accepting* if $IO(\sigma) \cap F \neq \varnothing$.)

2. (a) Show that integer addition (that is, the predicate "$x = y+z$") is not definable in S1S. (*Hint.* Use a pumping technique from automata theory. See [61, Section 4.1] or [76, Lectures 11, 12].)

   (b) On the other hand, for a finite set $A \subseteq \omega$, define

   $$n(A) \;=\; \sum_{x \in A} 2^x.$$

   Let $\varphi(A, B, C)$ be the predicate

   "A, B, C are finite sets, and $n(A) = n(B) + n(C)$".

   Show how to express this in S1S.

3. (a) For $n \geq 1$, let $B_n \subseteq \omega$ be the set

   $$B_n \;=\; \{x \mid \text{if } x = mn + k,\, 0 \leq k < n,\, \text{then } \lfloor \tfrac{m}{2^k} \rfloor \text{ is odd}\}.$$

   In other words, for any $m \geq 0$, the mnth, $mn + 1$st, $\ldots$, $mn + n - 1$st bits in the $\{0, 1\}^\omega$ representation of B_n represent $m \bmod 2^n$ in binary, lowest-order bit first. For example,

   $$B_7 \;=\; \underbrace{0000000}_{n}\,\underbrace{1000000}_{n}\,\underbrace{0100000}_{n}\,\underbrace{1100000}_{n}\,\underbrace{0010000}_{n}\ldots\,.$$

   Let $\varphi_n(x, y)$ be the predicate "$x \equiv y \pmod{n}$", and let $\psi_n(B)$ be the predicate "$B = B_n$". Construct S1S formulas for φ_1 and ψ_1, and show inductively how to get short S1S formulas for φ_{n2^n} and ψ_{n2^n}, given formulas for φ_n and ψ_n.

   (b) Explain informally how you might use (a) to show that S1S is nonelementary.

Homework 9

1. Given a sentence φ of the first-order language of number theory (addition and multiplication allowed) and a number $n \geq 2$ in binary, what is the complexity of determining whether φ holds in the ring $\mathbb{Z}_n$ of integers modulo n? Give proof.

2. Show that for any nondeterministic Muller automaton M with input alphabet $\{0, 1\}$, there is a short formula $\varphi_M(X)$ of S1S with one free set variable X such that

   $$L(M) \;=\; \{A \subseteq \omega \mid \varphi_M(A)\}.$$

   (Recall that in Muller acceptance, $\mathcal{F} \subseteq 2^Q$, and a run σ is accepting if $IO(\sigma) \in \mathcal{F}$.)

3. Let Σ be a finite alphabet with at least two letters. A set $A \subseteq \Sigma^*$ is *sparse* if there is a constant $c > 0$ such that

   $$|A \cap \Sigma^n| \;\leq\; n^c \text{ a.e.}$$

   In other words, for all but finitely many n, the number of elements of A of length n is bounded by a polynomial. Show that P^{sparse}, the class of sets computable by deterministic polynomial-time oracle machines with sparse oracles, is exactly the class of sets for which there exist polynomial-size circuits $B_0, B_1, \ldots$, not necessarily uniform.

Homework 10

1. Using the s_n^m functions and the universal function U, construct total recursive functions **pair** and **const** such that

$$\varphi_{\mathsf{pair}(i,j)} \;=\; \langle \varphi_i, \varphi_j \rangle \qquad\qquad \varphi_{\mathsf{const}(i)} \;=\; \kappa_i.$$

The construction should be similar to the construction of **comp** given in Lecture 33.

2. In the recursion theorem (Theorem 33.1), we proved that for any total recursive function σ, there exists an index i such that $\varphi_{\sigma(i)} = \varphi_i$. Show that such an i can be obtained effectively from an index for σ. That is, show that there is a total recursive function fix such that for all j such that φ_j is total,

$$\varphi_{\varphi_j(\mathsf{fix}(j))} \;=\; \varphi_{\mathsf{fix}(j)}.$$

3. Show that every total recursive function σ has infinitely many fixpoints; moreover, an infinite list of such fixpoints can be enumerated effectively.

Homework 11

1. The *jump operation* ($'$) is defined as follows:

$$A' \;=\; K^A \;=\; \{x \mid \varphi_x^A(x)\!\downarrow\}.$$

This is the halting problem relativized to A. Define

$$
\begin{aligned}
A^{(0)} &= A \\
A^{(n+1)} &= (A^{(n)})'.
\end{aligned}
$$

Show that $\varnothing^{(n)}$ is $\leq_m$-complete for Σ_n^0, $n \geq 1$.

2. (a) Show that if A is $\leq_m$-complete for Σ_n^0, $n \geq 1$, then A' as defined in the previous exercise is not in Σ_n^0.

   (b) Conclude from (a) that Σ_n^0 and Π_n^0 are incomparable with respect to set inclusion for all $n \geq 1$.

3. Consider the following three relativized versions of the recursion theorem.

   (a) For any total recursive function $\sigma : \omega \to \omega$, there is an n such that $\varphi_n^A = \varphi_{\sigma(n)}^A$.

   (b) For any total function $\sigma^A : \omega \to \omega$ recursive in A, there is an n such that $\varphi_n = \varphi_{\sigma^A(n)}$.

   (c) For any total function $\sigma^A : \omega \to \omega$ recursive in A, there is an n such that $\varphi_n^A = \varphi_{\sigma^A(n)}^A$.

   Two are true and one is false. Which is which? Give two proofs and a counterexample.

4. Recall that a directed graph is *strongly connected* if for any pair of vertices (u, v) there exists a directed path from u to v. Show that the following problem is $\leq_m$-complete for Π_2^0: given a recursive binary relation $E \subseteq \omega^2$, is the infinite graph (ω, E) strongly connected?

Homework 12

1. (a) Show that for every IND program over the natural numbers, there is an equivalent IND program with simple assignments $y := e(\overline{y})$ but without existential assignments $y := \exists$. (*Hint.* Convert countable existential branching to finite existential branching first, using the construct $\ell_i \vee \ell_j$; then get rid of finite existential branching.) Why can we not eliminate $y := \forall$ in the same way?

   (b) Recall that a binary relation is *well-founded* if there are no infinite descending chains. Using (a), show that the following problem is Π_1^1-complete. Given a recursive binary relation $R \subseteq \omega^2$, is it well-founded?

2. Gödel defined the *μ-recursive functions* to be the primitive recursive functions (see Miscellaneous Exercise 91) with the addition of an extra programming construct, namely *unbounded minimization*: if $f : \omega^2 \to \omega$ is a μ-recursive function, then so is

$$g(x) \overset{\text{def}}{=} \mu y.(f(x,y) = 0),$$

   where the expression on the right-hand side denotes the least y such that $f(x,z)\downarrow$ for all $z \leq y$ and $f(x,y) = 0$, if such a y exists, and is undefined otherwise. Prove *axiomatically* (that is, using the constructs based on the s_n^m and universal function properties as described in Lecture 33) that if f is a partial recursive function, then so is g, and an index for g can be obtained from an index for f effectively. You may use the conditional test $\varphi_{\mathsf{cond}(i,j)}$ of Miscellaneous Exercise 111 without proof.

3. Show that there exists a function $T(n)$ such that $DTIME(T(n)) = DSPACE(T(n))$.

Miscellaneous Exercises

The annotation ^{H} indicates that there is a hint for this exercise in the Hints section beginning on p. 361, and ^{S} indicates that there is a solution in the Solutions section beginning on p. 367. The number of stars indicates the approximate level of difficulty.

1. Prove Theorem 3.2.

2. Show that the set $\{0^n 1^n \mid n \geq 0\}$ requires $\Omega(\log n)$ space.

3. Prove the following simulation results.

   (a) For constants $k > 1$ and $\varepsilon > 0$, any k-tape TM running in time $T(n)$ can be simulated by a k-tape TM running in time $\varepsilon T(n) + O(n)$.

   (b) For constant $\varepsilon > 0$, any 1-tape TM running in time $T(n)$ can be simulated by a 1-tape TM running in time $\varepsilon T(n) + O(n^2)$.

   (c) For $T(n) \geq n$ and constant $k > 1$, any k-tape TM running in time $T(n)$ can be simulated by a 1-tape TM running in time $T(n)^2$.

   *(d) For $T(n) \geq n$ and constant $k > 1$, any k-tape TM running in time $T(n)$ can be simulated by a 2-tape TM running in time $T(n) \log T(n)$.

4. Prove that $NSPACE(n^s) \subsetneq NSPACE(n^t)$ for any fixed $t > s \geq 1$ using the padding technique of Lecture 3.

**5. Prove the following generalization of Miscellaneous Exercise 4 using the padding technique of Lecture 3. Let $S_1, S_2 : \mathbb{R} \to \mathbb{R}$ be real-valued functions of a real variable satisfying appropriate constructibility conditions such that

(a) S_1 and S_2 are *monotonically increasing*; that is, if $m < n$, then $S_1(m) < S_1(n)$ and $S_2(m) < S_2(n)$; and

(b) there exists $\varepsilon > 0$ such that $S_1(n)^{1+\varepsilon} \leq O(S_2(n))$.

Then $NSPACE(S_1(n)) \subsetneq NSPACE(S_2(n))$.

S6. Sometimes it is useful to refine complexity analysis to distinguish the time or space usage of a TM on different inputs of the same length. For $G : \Sigma^* \to \mathbb{N}$, define

$$DTIME(G(x)) \stackrel{\text{def}}{=} \{L(M) \mid M \text{ is a deterministic TM that takes no more than } G(x) \text{ steps on input } x\}$$

$$DSPACE(G(x)) \stackrel{\text{def}}{=} \{L(M) \mid M \text{ is a deterministic TM that uses no more than } G(x) \text{ work-tape cells on input } x\}.$$

For $S : \mathbb{N} \to \mathbb{N}$, $DTIME(S(n))$ and $DSPACE(S(n))$ are defined as usual (see Lecture 2). Prove that if $T : \mathbb{N} \to \mathbb{N}$ is monotone and

$$DSPACE(n) \subseteq DTIME(T(n)),$$

then for any $G : \Sigma^* \to \mathbb{N}$ such that $G(x) \geq |x|$, constructible or not,

$$DSPACE(G(x)) \subseteq DTIME(G(x)T(G(x))).$$

7. Show that any TM that writes at most $o(\log n)$ symbols on inputs of length n accepts a regular set.

8. (a) Show that if M writes no more than $t(n)$ symbols and runs for no more than $T(n)$ steps, then $T(n)$ is $O(t(n)^2)$.

H(b) Show that this bound for $T(n)$ in terms of $t(n)$ is the best possible for one-tape TMs.

9. A *k-headed finite automaton* (*k*-FA) is a one-tape TM with k read-only input heads that can move left or right but cannot move off the input string.

   (a) Give a formal definition of these machines, including a definition of acceptance.

   (b) Show that a set is in *LOGSPACE* (*NLOGSPACE*) iff it is accepted by a deterministic (nondeterministic) *k*-FA for some k.

10. A Boolean formula is in *3-conjunctive normal form* (3CNF) if it is a conjunction of clauses of the form $\ell_1 \vee \ell_2 \vee \ell_3$, where the ℓ_i are *literals* (Boolean variables or negations of variables). Show that 3CNF-satisfiability is *NP*-complete under $\leq_{\mathrm{m}}^{\log}$.

11. Give a set that is complete for $NSPACE(n)$ with respect to linear-time many–one reductions. Conclude that your set is in $DSPACE(n)$ iff $NSPACE(n) = DSPACE(n)$.

12. Show that if $NSPACE(n) = DSPACE(n)$, then $NSPACE(S(n)) = DSPACE(S(n))$ for all $S(n) \geq n$.

13. Show that $NSPACE(\log n) \cap \{a\}^* = DSPACE(\log n) \cap \{a\}^*$ iff $NSPACE(n) = DSPACE(n)$.

^{H}14. Prove that $DSPACE(n) \neq P$ and $DSPACE(n) \neq NP$.

^{H}15. Show that the following two problems are $\leq_{\mathrm{m}}^{\log}$-complete for *NLOGSPACE*.

   (a) Given a nondeterministic finite automaton, does it accept any string?

   (b) Given a nondeterministic finite automaton, does it accept infinitely many strings?

16. Let $\leq_{\mathrm{T}}^{\mathrm{P}}$ denote the polynomial-time-bounded Turing reducibility relation. Show that

   (a) $\leq_{\mathrm{T}}^{\mathrm{P}}$ is transitive, and that

   (b) $\leq_{\mathrm{m}}^{\mathrm{P}}$ refines $\leq_{\mathrm{T}}^{\mathrm{P}}$.

**S17. An *auxiliary pushdown automaton* (APDA) is a TM equipped with a single stack in addition to its worktape. In each step it can check for an empty stack, and if it is nonempty, read the top element. It also reads the symbols currently being scanned on its input and worktape. Based on this information and its current state, it can push or pop a symbol from a finite stack alphabet, write a symbol on its worktape, move its input and worktape heads one cell in either direction, and enter a new state. It may not read an element of the stack without popping the elements above it off. Show that deterministic or nondeterministic APDAs with $S(n)$ workspace accept exactly the sets in $DTIME(2^{O(S(n))})$.

18. A map $h : \Sigma^* \to \Gamma^*$ is a *homomorphism* if $h(xy) = h(x)h(y)$ for all strings $x, y \in \Sigma^*$. It follows that $h(\varepsilon) = \varepsilon$. A homomorphism is *nonerasing* if $h(a) \neq \varepsilon$ for any $a \in \Sigma$. A family of sets $\mathcal{C}$ is *closed under nonerasing homomorphisms* if for any nonerasing homomorphism h, $\{h(x) \mid x \in A\} \in \mathcal{C}$ whenever $A \in \mathcal{C}$.

   (a) Show that NP is closed under nonerasing homomorphisms.

   (b) Show that P is closed under nonerasing homomorphisms iff $P = NP$.

19. Recall from Supplementary Lecture A that a *complete partial order* is a set U with a partial order $\leq$ defined on it such that every subset $A \subseteq U$ has a supremum (least upper bound) $\sup A$. Show that every subset $A \subseteq U$ also has a unique infimum (greatest lower bound) $\inf A$ satisfying the following properties.

   (a) For all $y \in A$, $\inf A \leq y$ ($\inf A$ is a lower bound for A).

   (b) If for all $y \in A$, $x \leq y$, then $x \leq \inf A$ ($\inf A$ is the greatest lower bound).

**S20. Prove that a set operator is finitary iff it is chain-continuous.

21. (a) Prove that every chain-continuous operator on any complete partial order is monotone.

   ^S(b) Give an example of a monotone set operator that is not chain-continuous.

22. Give an example of a complete partial order U, a monotone operator τ on U, and a set $A \subseteq U$ of prefixpoints such that $\sup A$ is not a prefixpoint; thus $\sup A < \inf PF_\tau(\sup A)$.

^{S}23. Prove that if τ is chain-continuous, then its closure ordinal is at most ω, but not for monotone operators in general.

24. Induction and well-foundedness go hand in hand. A binary relation $<$ on a set X is *well-founded* if there is no infinite descending chain $x_0, x_1, \ldots$ in X such that $x_{i+1} < x_i$ for all $i \in \omega$. For example, the natural order $<$ on ω is well-founded, as is the strict subset order $\subset$ on any finite set. The strict subset order on 2^ω is not well-founded, because $\omega \supset \omega - \{0\} \supset \omega - \{0,1\} \supset \cdots$.

The *induction principle* for a binary relation $<$ states that for any set $A \subseteq X$,

$$(\forall x \ ((\forall y \ y < x \to y \in A) \to x \in A)) \quad \to \quad \forall x \ x \in A.$$

Show that this induction principle is valid iff $<$ is well-founded. Feel free to use the axiom of choice (see Lecture A).

25. (a) Show that for an alternating machine M running in $S(n)$ space, the root of the computation tree is labeled 0 or 1 within time $c^{S(n)}$ for some c (depending only on M and not on n), or not at all.

(b) Using (a), prove that $ASPACE(S(n)) = \bigcup_c DTIME(c^{S(n)})$.

26. Show that an alternating Turing machine M without negations accepts x iff there is a *finite accepting subtree* of the computation tree on input x; that is, a finite subtree T of the computation tree containing the start configuration such that every $\vee$-configuration has at least one successor in T and every $\wedge$-configuration has all its successors in T.

^{H}27. Show that if A is $\leq_m^{\log}$-hard for linear space ($DSPACE(n)$), then A is also $\leq_m^{\log}$-hard for $PSPACE$.

^{S}28. Let $S(n) \geq \log n$ and $T(n) \geq n$. Show that any set accepted by a nondeterministic $S(n)$-space and $T(n)$-time bounded TM can be accepted by a deterministic TM in space $S(n) \log T(n)$. In other words,

$$STA(S(n), T(n), \Sigma 1) \quad \subseteq \quad STA(S(n) \log T(n), *, 0).$$

Do not forget to worry about constructibility.

[S]29. Show that the problem of deciding whether $\bigcap_{i=1}^{n} L(M_i) = \varnothing$ for a given set of deterministic finite automata M_i, $1 \leq i \leq n$ is *PSPACE*-complete.

[S]30. The security of cryptosystems is based on the existence of one-way functions. For the purposes of this problem, let us define a *one-way function* to be a deterministic-polynomial-time-computable length-preserving map $f : \Sigma^* \to \Sigma^*$ that is not invertible in deterministic polynomial time. Here *invertible* means: given y, either produce some x such that $f(x) = y$, or say that no such x exists. Show that one-way functions exist if and only if $P \neq NP$.

[H]31. Of the following three problems, one is $\leq_{\mathrm{m}}^{\log}$-complete for co-*NLOGSPACE* and the other two are $\leq_{\mathrm{m}}^{\log}$-complete for *PSPACE*. Which is which? Give proofs.

   (a) Given a regular expression α, is $L(\alpha) = \varnothing$?

   (b) Given a regular expression α, is $L(\alpha) = \Sigma^*$?

   (c) Given two regular expressions α and β, is $L(\alpha) = L(\beta)$?

32. [S](a) Prove that a set A is in *NP* iff there is a deterministic polynomial-time computable binary predicate R and constant c such that

   $$A = \{x \mid \exists y \, |y| \leq |x|^c \wedge R(x, y)\}.$$

   (b) More generally, prove Theorem 10.2: a set A is in Σ_k^{p} iff there is a deterministic polynomial-time computable $(k+1)$-ary predicate R and constant c such that

   $$A = \{x \mid \exists^{|x|^c} y_1 \, \forall^{|x|^c} y_2 \, \exists^{|x|^c} y_3 \cdots Q^{|x|^c} y_k \, R(x, y_1, \dots, y_k)\}.$$

   The bounded quantifiers $\exists^t$ and $\forall^t$ are defined at the end of Lecture 10.

33. Let $S(n) \geq \log n$. Prove that

   $$\bigcup_k STA(S(n), *, \Sigma k) \cup STA(S(n), *, \Pi k) = NSPACE(S(n)).$$

34. Define $\Delta_{k+1}^{\mathrm{p}} = P^{\Sigma_k^{\mathrm{p}}}$, the family of all sets accepted by deterministic polynomial-time-bounded oracle machines with an oracle in Σ_k^{p}.

(a) Show that

$$\Sigma_k^p \cup \Pi_k^p \;\subseteq\; \Delta_{k+1}^p \;\subseteq\; \Sigma_{k+1}^p \cap \Pi_{k+1}^p.$$

(b) Show that the set of Boolean formulas with *exactly one* satisfying assignment is in Δ_2^p.

(c) Give a $\leq_m^{\log}$-complete problem for Δ_k^p, and prove that it is complete.

35. Prove that PH has a $\leq_m^{\log}$-complete set if and only if it collapses.

^{H}36. Prove that the circuit value problem for constant-depth circuits is $\leq_m^{\log}$-complete for $NLOGSPACE$.

37. A *Boolean decision diagram* (BDD) is a directed acyclic graph with a single source and two sinks, one labeled 0 and the other labeled 1, such that all non-sink nodes have exactly two exiting edges, one labeled x and the other labeled $\bar{x}$ for some Boolean variable x. The *value* of a BDD on a truth assignment σ is the label of the sink node of the unique σ-enabled path from the source to a sink, where an edge with literal ℓ is σ-*enabled* if $\sigma(\ell) = 1$. For BDDs, what is the complexity of

(a) determining the value for a given σ?

(b) satisfiability?

38. Every decision problem has a family of Boolean circuits of size at most $O(n2^n)$ by writing the nth circuit in disjunctive normal form. Prove that every decision problem has a family of Boolean circuits of size at most

^{H}(a) $O(2^n)$,

*(b) $O(2^n/n)$.

**39. ([18]) A *rectilinear maze* is a connected subset of the infinite checkerboard. That is, it is a connected undirected graph whose vertices are ordered pairs of integers and whose edges are all of the form $((x, y), (x, y + 1))$ or $((x, y), (x + 1, y))$. Show that the MAZE problem for rectilinear mazes is solvable in deterministic logspace.

^{H}40. Give an NC algorithm for finding a topological sort in a given directed acyclic graph (V, E). That is, find a total ordering $<$ of V that extends E in the sense that if uEv, then $u < v$.

[S]41. (a) Show that any deterministic logspace transducer can be simulated by a family of NC circuits with multiple output wires.

   (b) We showed in Theorem 6.1 that the circuit value problem (CVP) is $\leq_{\mathrm{m}}^{\log}$-complete for P. Conclude from this and part (a) that CVP $\in$ NC if and only if $P = NC$.

[H]42. Consider the following Standard ML program to compute the greatest common divisor (gcd) of two numbers m, n, not both of which are 0.

```
fun Euclid(m:int, n:int) : int * int * int =
  if n = 0 then (1,0,m)
  else let
    val q = m div n
    val r = m mod n
    val (s,t,g) = Euclid(n,r)
  in
    (t,s-t*q,g)
  end
```

Here `div` computes the quotient and `mod` the remainder when dividing m by n using ordinary integer division. Prove that the output of the program is a triple (s, t, g), where g is the gcd of m and n and s, t are integers such that $sm + tn = g$.

[H]43. Let a and n be positive integers. Prove that the following are equivalent.

   (i) a has an order modulo n; that is, there exists m such that $a^m \equiv 1 \pmod{n}$.

   (ii) a is relatively prime to n; that is, $\gcd(a, n) = 1$.

   (iii) a is invertible modulo n; that is, there exists b, $1 \leq b \leq n - 1$, such that $ab \equiv 1 \pmod{n}$.

44. Prove the following amplification lemma for IP and PCP analogous to Lemma 14.1 for BPP and RP. If L has an IP (respectively, PCP) protocol that uses $r(n)$ random bits and makes $q(n)$ queries, then for any $\varepsilon > 0$, L has an IP (respectively, PCP) protocol (P, V) that uses $kr(n)$ random bits and makes $kq(n)$ queries and has error probability bounded by ε, where k is $O(-\log \varepsilon)$. In the case of PCP, this means

   (i)   if $x \in L$ then $\Pr((P, V) \text{ accepts } x) = 1$,

(ii)   if $x \notin L$ then for any P', $\mathrm{Pr}((P', V) \text{ accepts } x) \leq \varepsilon$,

and in the case of IP,

(i)   if $x \in L$ then $\mathrm{Pr}((P, V) \text{ accepts } x) \geq 1 - \varepsilon$,

(ii)   if $x \notin L$ then for any P', $\mathrm{Pr}((P', V) \text{ accepts } x) \leq \varepsilon$.

45. In this exercise we complete the proof of Lemma 18.2. Let B be a Boolean formula in 3CNF with m clauses over n variables $x_1, \ldots, x_n$ such that each clause contains exactly three literals with distinct variables. Let S_i and S be as in the proof of Lemma 18.2. As argued in that proof, $\mathcal{E}(S) = 7m/8$. Let $a_1, \ldots, a_n \in \{0, 1\}$ be the truth assignment to $x_1, \ldots, x_n$ obtained from the greedy algorithm. For a random truth assignment $r_1, \ldots, r_n$, let E_k be the event

$$E_k \overset{\text{def}}{=} \bigwedge_{i=1}^{k} r_i = a_i.$$

Then $\mathcal{E}(S \mid E_{k-1} \wedge r_k = a_k) \geq \mathcal{E}(S \mid E_{k-1} \wedge r_k = \bar{a}_k)$, because that was how we chose a_k in the greedy algorithm.

(a) Show how to calculate $\mathcal{E}(S \mid E_k)$ efficiently.

[H](b) Prove that

$$\begin{aligned} \mathcal{E}(S \mid E_{k-1}) \quad = \quad & \mathcal{E}(S \mid E_{k-1} \wedge r_k = a_k) \cdot \mathrm{Pr}(r_k = a_k \mid E_{k-1}) \\ & + \mathcal{E}(S \mid E_{k-1} \wedge r_k = \bar{a}_k) \cdot \mathrm{Pr}(r_k = \bar{a}_k \mid E_{k-1}). \end{aligned}$$

(c) From (b), conclude that $\mathcal{E}(S \mid E_{k-1}) \leq \mathcal{E}(S \mid E_k)$.

(d) Using (c), show that the greedy assignment $a_1, \ldots, a_n$ satisfies at least $7m/8$ clauses.

[H]46. Show that if $P = NP$, then

(a) MAX-3SAT

(b) MAX-CLIQUE

can be solved exactly in polynomial time (see Lecture 18).

47. Complete the proof of Theorem 18.3 by proving the following claims about the construction in that proof.

(a) Show that if $a \neq b$, then (y, a) and (y, b) are not consistent, thus $((y, a), (y, b))$ cannot be an edge of the graph.

[H](b) Show that if $x \in L$, then the maximum clique of G is of size n^c, whereas if $x \notin L$, the maximum clique of G is of size strictly less than αn^c.

48. Let $\mathbb{F} = \mathbb{Z}_p$ for some prime p. Give a $PCP(n^3, 1)$ protocol for determining for given oracles $f : \mathbb{F}^n \to \mathbb{F}$ and $f : \mathbb{F}^{n^3} \to \mathbb{F}$ whether f is close to a function of the form $r \mapsto r \bullet a$ and h is close to a function of the form $t \mapsto t \bullet (a \otimes a \otimes a)$ for some $a \in \mathbb{F}^n$ (see Lecture 20).

49. Come up with a definition of *trivial* for first-order theories in terms of constraints on the signature. Prove that for your definition of trivial,

[HS](a) Every nontrivial theory is *PSPACE*-hard.

(b) Every trivial theory is decidable in polynomial time.

[HS]50. An Ehrenfeucht–Fraïssé game is said to be *finite* if from each board position, there are only finitely many legal next moves. Show that *every* first-order theory is characterized by a finite Ehrenfeucht–Fraïssé game. Why does this not show that every first-order theory is decidable?

51. Prove that the set of sets of infinite strings accepted by deterministic Büchi automata is closed under union and intersection.

52. What is the complexity of the emptiness problem for nondeterministic Büchi automata? Give proof.

53. How hard is it to determine whether a given

(a) deterministic

*[H](b) nondeterministic

Büchi automaton accepts all strings? Give proof.

54. Show that every set accepted by a Büchi automaton is a finite union of sets of the form AB^ω, where A and B are regular sets. Here B^ω denotes the set of infinite words of the form $w_0 w_1 w_2 \cdots$, where $w_i \in B - \{\varepsilon\}$ for all $i \geq 0$.

55. Prove that equicardinality of sets, that is, the predicate

$$\varphi(A, B) \quad \overset{\text{def}}{\Longleftrightarrow} \quad A \text{ and } B \text{ have the same number of elements,}$$

cannot be expressed in S1S.

56. Consider the following two acceptance conditions for automata on infinite words.

   (a) *Streett condition*: As with the Rabin acceptance condition, there is a finite set of pairs (G_i, R_i), $1 \leq i \leq k$, where G_i and R_i are subsets of Q. A run σ is *accepting* if for all i, $\mathrm{IO}(\sigma) \cap G_i \neq \varnothing$ implies $\mathrm{IO}(\sigma) \cap R_i \neq \varnothing$.

   (b) *Parity condition*: Assume that the states of the automaton are numbered $\{0, 1, \ldots, n-1\}$. A run σ is *accepting* if the least-numbered state that occurs infinitely often is even.

   Show that for nondeterministic automata, these two acceptance conditions are equivalent to the other acceptance conditions discussed in Lectures 25 and 26.

57. ^{S}(a) Prove that $(1 - 1/z)^z \leq e^{-1}$ for all $z > 0$ and that $(1 - 1/z)^z$ approaches e^{-1} in the limit as $z \to \infty$, where $e = 2.718\ldots$ is the base of natural logarithms.

   (b) Prove that $(1 + 1/z)^z \leq e$ for all $z > 0$ and that $(1 + 1/z)^z$ approaches e in the limit as $z \to \infty$.

58. Construct a recursive oracle A such that $NP^A \neq \text{co-}NP^A$.

59. (Cai and Ogihara [25]) In this exercise, we give an alternative proof of Mahaney's theorem (Theorem 29.2). Suppose that S is a sparse NP-hard set. For a Boolean formula φ of length n and truth assignment $t \in \{0, 1\}^n$ to the variables of φ, let $\varphi(t)$ denote the truth value obtained by evaluating φ at t. Let $\leq_{\text{lex}}$ denote lexicographic order on truth assignments of the same length. Define the set

$$E \quad \overset{\text{def}}{=} \quad \{(\varphi, s) \mid \exists t \; s \leq_{\text{lex}} t \wedge \varphi(t) = 1\}.$$

Observe that if $(\varphi, t) \in E$ and $s \leq_{\text{lex}} t$, then $(\varphi, s) \in E$. We give a polynomial-time decision procedure for E.

(a) Show that E is *NP*-complete. Thus the existence of a polynomial-time decision procedure for E implies $P = NP$.

(b) Let σ be a polynomial-time many–one reduction from E to S with time bound n^c. Show that there exists a polynomial n^d such that $|S \cap \Sigma^{\leq n^c}| \leq n^d$.

Let φ be a Boolean formula of length n. Let $A_0 = \{0,1\}^n$, the set of all truth assignments to the variables of φ. We construct a sequence of subsets $A_0 \supseteq A_1 \supseteq A_2 \supseteq \cdots$ of decreasing size, maintaining the following invariant: if φ is satisfiable, then A_i contains a satisfying assignment.

Here is how we get A_{i+1} from A_i. Suppose $A_i = \{t_0, \ldots, t_{m-1}\} \subseteq \{0,1\}^n$, where $t_0 \leq_{\text{lex}} t_1 \leq_{\text{lex}} \cdots \leq_{\text{lex}} t_{m-1}$, and $m \geq n^d + 1$ (the d is the same d as in part (b)). Let

$$J \stackrel{\text{def}}{=} \{\lfloor km/(n^d + 1)\rfloor \mid 0 \leq k \leq n^d\} \subseteq \{0,1,2,\ldots,m-1\}.$$

This is a set of $n^d + 1$ evenly spaced indices of elements of A_i.

(c) Show that $\lfloor km/(n^d + 1)\rfloor < \lfloor (k+1)m/(n^d + 1)\rfloor$ for all k, thus J contains $n^d + 1$ distinct elements.

Compute $\sigma(\varphi, t_j)$ for each $j \in J$. If there exist $i, j \in J$ with $i < j$ and $\sigma(\varphi, t_i) = \sigma(\varphi, t_j)$, remove the interval $\{t_k \mid i \leq k < j\}$ from A_i. If on the other hand all $\sigma(\varphi, t_j)$ are distinct for $j \in J$, remove the last interval $\{t_k \mid k \geq n^d m/(n^d + 1)\}$ from A_i.

(d) Argue that in either case, the invariant is maintained.

(e) In either case, we remove at least $|A_i|/(n^d + 1)$ elements from A_i. Using Miscellaneous Exercise 57(a), show that after at most $(n^d + 1)(n \ln 2 - \ln(n^d + 1)) = O(n^{d+1})$ steps, we get down to a small enough set that we can evaluate φ on all the remaining truth assignments directly.

(f) The sets A_i produced in this construction are of exponential size in general. However, they can be represented efficiently so that the above procedure can be carried out in polynomial time. Describe a representation for the sets A_i that allows $|A_i|$ and the maps $j \mapsto t_j$ to be calculated efficiently.

60. (a) Recall that a square matrix is *nonsingular* if all its columns are linearly independent. Show there are exactly $\prod_{i=0}^{n-1}(q^n - q^i)$ nonsingular $n \times n$ matrices over $\mathbb{GF}_q$.

*H(b) Show how to generate a random $n \times n$ nonsingular matrix over $\mathbb{GF}_q$ efficiently, all nonsingular matrices equally likely.

61. If E is a subspace of an n-dimensional vector space V, let $\dim E$ and $\operatorname{codim} E$ denote the dimension and codimension (n minus the dimension) of E, respectively. Show that $\operatorname{codim} E \cap F \leq \operatorname{codim} E + \operatorname{codim} F$.

62. Prove that the lower bound $3/4$ established in Lemma F.1 is tight for all n.

63. (Papadimitriou and Zachos [93]) The class $\oplus P$ ("parity P") is defined to be the class of sets A for which there exists a polynomial-time nondeterministic TM M such that $x \in A$ iff the number of accepting computation paths of M on input x is odd. Equivalently, we can think of the machine M as a kind of alternating machine that at branching nodes computes the mod-2 sum of the labels of its children. Another characterization of $\oplus P$ is given in Lecture G. Prove that $\oplus P^{\oplus P} = \oplus P$.

64. Show that the following two characterizations of $\#P$ are equivalent.

   (a) A function $f : \{0,1\}^* \to \mathbb{N}$ is in $\#P$ iff there exist $A \in P$ and $k \geq 0$ such that for all x, $f(x) = |W(n^k, A, x)|$.

   (b) A function $f : \{0,1\}^* \to \mathbb{N}$ is in $\#P$ iff there is a polynomial-time nondeterministic TM M such that the number of accepting computation paths of M on input x is $f(x)$.

65. In the proof of Toda's theorem (Theorem G.1), we need to know that the polynomial $h^m(z)$ satisfies the property

$$z \text{ is odd} \quad \Rightarrow \quad h^m(z) \equiv -1 \ (\operatorname{mod} 2^{2^m})$$
$$z \text{ is even} \quad \Rightarrow \quad h^m(z) \equiv 0 \ (\operatorname{mod} 2^{2^m}),$$

where

$$h(z) \overset{\text{def}}{=} 4z^3 + 3z^4$$
$$h^0(z) \overset{\text{def}}{=} z$$
$$h^{m+1}(z) \overset{\text{def}}{=} h(h^m(z)).$$

Show that this is so.

66. Prove the following clauses of Lemma G.2. If $\mathcal{C}$ is closed downward under $\leq_{\mathrm{T}}^{\mathrm{P}}$, then

   ^{S}(ii) $\Pi^{\log} \cdot \oplus \cdot \mathcal{C} \subseteq \oplus \cdot \mathcal{C}$;

   $^{*\mathrm{HS}}$(iii) $\oplus \cdot BP \cdot \mathcal{C} \subseteq BP \cdot \oplus \cdot \mathcal{C}$;

   ^{H}(iv) $BP \cdot BP \cdot \mathcal{C} \subseteq BP \cdot \mathcal{C}$;

   ^{H}(v) $\oplus \cdot \oplus \cdot \mathcal{C} \subseteq \oplus \cdot \mathcal{C}$;

   ^{H}(vi) $BP \cdot \mathcal{C}$ and $\oplus \cdot \mathcal{C}$ are closed downward under $\leq_{\mathrm{T}}^{\mathrm{P}}$.

67. Let $\#P$ denote the class of all polynomial-time counting functions. A formal definition of this class is given in Lecture G. Show that $\#P$ is closed under the following pointwise operations.

   (a) Addition: if $f, g \in \#P$, then so is $f + g = \lambda x.f(x) + g(x)$.

   (b) Multiplication: if $f, g \in \#P$, then so is $f \cdot g = \lambda x.f(x)g(x)$.

   *(c) If f is in $\#P$, then so is $\lambda x.f(x)^{|x|^d}$ for any constant d.

   $^{*\mathrm{S}}$(d) Let $g : \{0,1\}^* \to \mathbb{N}[z]$ be a function that on input $x \in \{0,1\}^*$ gives a polynomial with indeterminate z and positive integer coefficients representing a polynomial function $\mathbb{N} \to \mathbb{N}$. Suppose g is polynomial-time computable in the sense that $g(x)$ is of degree at most polynomial in $|x|$, the coefficients of $g(x)$ can be represented in binary with at most polynomially many bits, and the coefficient of z^i can be produced from x and i in polynomial time. Show that if f is in $\#P$, then so is $\lambda x.g(x)(f(x))$.

68. (Papadimitriou [92]) Some computations produce an output $f(x)$ on input x that is in the worst case much larger than $|x|$. For such problems, we would like to have algorithms that are polynomial in $|x| + |f(x)|$ to decide for a given x, y whether $f(x) = y$. The class of all such problems is called *output polynomial time*.

   ^{H}(a) Consider the function that for a given regular expression produces the minimum-state equivalent deterministic finite automaton. Prove that this function is in output polynomial time iff $P = PSPACE$.

(b) Prove that it is impossible to produce in output polynomial time an equivalent deterministic automaton that has at most polynomially more states than the minimum-state automaton unless $P = PSPACE$.

69. (a) Prove that

$$(z \wedge u) \vee (\overline{z} \wedge v) \ \equiv\ (z \rightarrow u) \wedge (\overline{z} \rightarrow v).$$

   (b) Let $\psi(x)$ be any Boolean formula with a single positive occurrence of the Boolean variable x along with possibly other variables. Here *positive* means in the scope of an even number of negations. Show that

$$\psi(x_0 \vee x_1) \ =\ \psi(x_0) \vee \psi(x_1)$$
$$\psi(x_0 \wedge x_1) \ =\ \psi(x_0) \wedge \psi(x_1),$$

   where x_0 and x_1 do not occur in $\psi(x)$.

   ^{HS}(c) Show that any Σ_k oracle machine can be efficiently simulated by another Σ_k oracle machine that makes all its oracle queries at the end of the computation.

^H70. Let $\mathrm{excl}(z, u, v)$ denote the Boolean formula on the left-hand side of the equation in Exercise 69(a) (or the equivalent formula on the right-hand side). Let $\varphi(x)$ be any Boolean formula with Boolean variable x and possibly other variables. Prove that

$$\varphi(\mathrm{excl}(z, u, v)) \ \equiv\ \mathrm{excl}(z, \varphi(u), \varphi(v)).$$

71. A *minterm* of a Boolean formula φ is a $\leq$-maximal term M such that $M \leq \varphi$. If φ is a CNF formula, show that M is a minterm of φ iff M contains at least one literal from each clause of φ and no proper subterm of M has this property.

72. Let φ be a Boolean formula. Show that the disjunction of all minterms of φ is a DNF formula equivalent to φ.

73. Prove that if the G_i are mutually exclusive events that cover F, then

$$\Pr(E \mid F) \ =\ \sum_i \Pr(E \mid G_i \wedge F) \cdot \Pr(G_i \mid F)$$
$$\leq\ \max_i \Pr(E \mid G_i \wedge F).$$

74. Prove that $\Pr(E \wedge F \mid G) = \Pr(E \mid F \wedge G) \cdot \Pr(F \mid G)$.

75. Prove that if $E_0 \wedge F_0$ and $E_1 \wedge F_1$ are independent and F_0 and F_1 are independent, then

$$\Pr(E_0 \wedge E_1 \mid F_0 \wedge F_1) \;\; = \;\; \Pr(E_0 \mid F_0) \cdot \Pr(E_1 \mid F_1).$$

76. (a) Prove that if G is independent of both E and $E \wedge F$, then
$$\Pr(E \mid F \wedge G) \;\; = \;\; \Pr(E \mid F).$$

    (b) Give a counterexample showing that the following statement is false. If G is independent of both E and F, then
$$\Pr(E \mid F \wedge G) \;\; = \;\; \Pr(E \mid F).$$

77. (a) Prove that
$$\Pr(E \mid F) \leq \Pr(E) \;\; \Leftrightarrow \;\; \Pr(F \mid E) \leq \Pr(F).$$

    (b) More generally, prove that
$$\Pr(E \mid F \wedge G) \leq \Pr(E \mid G) \;\; \Leftrightarrow \;\; \Pr(F \mid E \wedge G) \leq \Pr(F \mid G).$$

78. Prove that if $M \leq \psi \leq \varphi$ and M is a minterm of φ, then M is also a minterm of ψ.

79. Prove Lemma H.2(iii) and (iv): if ρ is a partial valuation and W a set of variables such that $\rho(W) = W$, then

    (a) $\rho(\varphi)^{-W} = \rho(\varphi^{-W})$,
    (b) $\rho(\varphi) = 1$ iff $\rho(\varphi^{-W}) = 1$.

^{S}80. For definitions and notation, see Supplementary Lecture H. Let φ be a CNF formula over variables X and $\rho : X \rightarrow X \cup \{0, 1\}$ a partial valuation. Let Y be the set of variables that are not assigned by ρ, and let N be a term over Y. If N is a minterm of $\rho(\varphi)$, then there exists a minterm M of φ such that $N = \rho(M)$.

^{S}81. For definitions and notation, see Supplementary Lecture H. Consider random partial valuations ρ that assign 0 or 1 to each variable independently with probability $\frac{1}{2}(1 - p)$ each or leave it unassigned with probability p. Give CNF formulas φ and ψ and $s \geq 0$ such that

$$\Pr(\exists M \in \mathrm{m}(\rho(\varphi)) \,|M| \geq s)$$
$$< \quad \Pr(\exists M \in \mathrm{m}(\rho(\varphi)) \,|M| \geq s \mid \rho(\psi) = 1).$$

Thus conditioning on a formula becoming equivalent to 1 does not always make it less likely that there exists a large minterm.

82. (a) Let X_i be independent real-valued random variables, $1 \leq i \leq n$. Show that the expected value of their product is the product of their expected values.

   (b) Show that (a) is false without the independence assumption.

83. Prove the *Markov bound* (I.1): for a nonnegative random variable X with mean $\mu = \mathcal{E}X$,

$$\Pr(X \geq k) \quad \leq \quad \mu/k.$$

84. Prove the *Chebyshev bound* (I.2): for a random variable X with mean $\mu = \mathcal{E}X$ and standard deviation $\sigma = \sqrt{\mathcal{E}((X - \mu)^2)}$, for any $\delta \geq 1$,

$$\Pr(|X - \mu| \geq \delta\sigma) \quad \leq \quad \delta^{-2}.$$

85. Prove that the standard deviation of n Bernoulli trials with success probability p is $\sqrt{np(1 - p)}$.

86. Using (a) the Chebyshev bound (I.2), and (b) the Chernoff bound (I.7), estimate the probability that the number of successes in n Bernoulli trials with success probability p is less than half the mean.

87. Show that the three alternative forms of the Chernoff bound (I.3), (I.4) and (I.6) are equivalent.

88. Prove the Chernoff bounds (I.7)–(I.9) for the lower tail.

^{H}89. Prove the Chernoff bound (I.10): for the sum of Poisson trials with mean μ, for $0 \leq \delta \leq 1$,

$$\Pr\left(X < (1 - \delta)\mu\right) \;<\; e^{-\delta^2 \mu / 2}.$$

^{S}90. Consider a simple programming language with variables $x, y, \dots$ ranging over $\mathbb{N}$ containing the following constructs.

(i)   *Simple assignment:* $x := 0$ $x := y + 1$ $x := y$

(ii)   *Sequential composition:* $p \,;\, q$

(iii)   *Conditional test:*   if $x < y$ then p else q

(iv)   *For loop:*   for y do p

(v)   *While loop:*   while $x < y$ do p

In (iii) and (v), the relation $<$ can be replaced by any one of $>$, $\geq$, $\leq$, $=$, or $\neq$.

Programs built inductively from these constructs are called *while programs*. Programs built without the *while* construct (v) are called *for programs*.

The semantics of the *for* loop is as follows. Upon entry to the loop, the variable y is evaluated, and the body of the loop p is executed that many times. No assignment to y in the loop changes the number of times the loop is executed, nor does execution of the loop change the value of y in any way, except by explicit assignment.

Show that every *while* program over $\mathbb{N}$ is equivalent to one containing at most one *while* loop. The program may contain as many *for* loops as you like, and you are allowed to declare extra local variables, which are not counted in the definition of equivalence.

^{S}91. Gödel defined the *primitive recursive functions* to be the smallest class $\mathcal{P}$ of number-theoretic functions $\mathbb{N}^k \to \mathbb{N}$ containing the constant zero function $\mathsf{zero}(\ ) = 0$, the successor function $\mathsf{s}(x) = x + 1$, and the projections $\pi_k^n : \mathbb{N}^n \to \mathbb{N}$ given by $\pi_k^n(x_1, \dots, x_n) = x_k$, $1 \leq k \leq n$, and closed under the following operations.

(a) Composition:

If $f : \mathbb{N}^k \to \mathbb{N}$ and $g_1, \dots, g_k : \mathbb{N}^n \to \mathbb{N}$ are in $\mathcal{P}$, then so is the function $f \circ (g_1, \dots, g_k) : \mathbb{N}^n \to \mathbb{N}$ that on input $\overline{x} = x_1, \dots, x_n$ gives

$$(f \circ (g_1, \dots, g_k))(\overline{x}) \;\overset{\text{def}}{=}\; f(g_1(\overline{x}), \dots, g_k(\overline{x})).$$

(b) Primitive recursion:

If $h_i : \mathbb{N}^{n-1} \to \mathbb{N}$ and $g_i : \mathbb{N}^{n+k} \to \mathbb{N}$ are in $\mathcal{P}$, $1 \le i \le k$, then so are the functions $f_i : \mathbb{N}^n \to \mathbb{N}$, $1 \le i \le k$, defined by mutual induction as follows:

$$f_i(0, \overline{x}) \stackrel{\text{def}}{=} h_i(\overline{x}),$$

$$f_i(x+1, \overline{x}) \stackrel{\text{def}}{=} g_i(x, \overline{x}, f_1(x, \overline{x}), \dots, f_k(x, \overline{x})),$$

where $\overline{x} = x_2, \dots, x_n$.

Define the class $\mathcal{C}$ to be the smallest class of number-theoretic functions containing the constant zero function, the successor function, and the projection functions, and closed under the following operations.

(a) Composition:

If $f : \mathbb{N}^m \to \mathbb{N}^n$ and $g : \mathbb{N}^n \to \mathbb{N}^k$ are in $\mathcal{C}$, then so is the function $g \circ f : \mathbb{N}^m \to \mathbb{N}^k$ defined by

$$(g \circ f)(\overline{x}) \stackrel{\text{def}}{=} g(f(\overline{x})).$$

Note the difference from the composition rule in Gödel's definition above.

(b) Tupling:

If $f_1, \dots, f_n : \mathbb{N}^m \to \mathbb{N}$ are in $\mathcal{C}$, then so is the function $(f_1, \dots, f_n) : \mathbb{N}^m \to \mathbb{N}^n$ defined by

$$(f_1, \dots, f_n)(\overline{x}) \stackrel{\text{def}}{=} (f_1(\overline{x}), \dots, f_n(\overline{x})).$$

(c) Iterated composition:

If $g : \mathbb{N}^n \to \mathbb{N}^n \in \mathcal{C}$, then the function $f : \mathbb{N}^{n+1} \to \mathbb{N}^n$ defined by

$$f(x, \overline{y}) \stackrel{\text{def}}{=} g^x(\overline{y})$$

is in $\mathcal{C}$, where g^n is g composed with itself n times:

$$g^0(\overline{y}) \stackrel{\text{def}}{=} \overline{y}$$

$$g^{n+1}(\overline{y}) \stackrel{\text{def}}{=} g(g^n(\overline{y})).$$

Show that for functions with a single output, $\mathcal{P}$ and $\mathcal{C}$ coincide.

92. (Meyer and Ritchie [85]) Show that *for* programs over $\mathbb{N}$ as defined in Miscellaneous Exercise 90 compute exactly the primitive recursive functions. Use either of the equivalent definitions of Miscellaneous Exercise 91.

HS93. (Meyer and Ritchie [85]) Show that a function is primitive recursive iff it is computed by a *while* program with a primitive recursive time bound.

H94. *Ackermann's function* is defined inductively as

$$A(0, n) \overset{\text{def}}{=} n + 1$$
$$A(m + 1, 0) \overset{\text{def}}{=} A(m, 1)$$
$$A(m + 1, n + 1) \overset{\text{def}}{=} A(m, A(m + 1, n)).$$

Thus

$$
\begin{aligned}
A(0, n) &= n + 1 \\
A(1, n) &= n + 2 \\
A(2, n) &= 2n + 3 \\
A(3, n) &= 2^{n+3} - 3 \\
A(4, n) &= \underbrace{2^{2^{2^{\cdot^{\cdot^{\cdot^2}}}}}}_{n+3} - 3.
\end{aligned}
$$

Prove that $\lambda x. A(x, 2)$ grows asymptotically faster than any primitive recursive function. Use any of the three equivalent characterizations of primitive recursive functions given in Miscellaneous Exercises 91 and 92.

95. Write a program in any programming language except C or UNIX shell that prints itself out. Your program must be nonnull, may not do any file I/O, and must be syntactically correct. Programs in C and UNIX shell were given in Lecture 34.

H96. Prove that the Post correspondence problem (PCP) is undecidable (see Lecture 25).

97. Construct a recursive set A for which any Turing machine uses more than polynomial time almost everywhere. In other words, for any M such that $L(M) = A$ and any constant k, M runs for at least $|x|^k$ steps for all but finitely many inputs x. Can you make $A \in EXPTIME$?

HS98. Show that the set of all *minimal indices*

$$\{x \mid \forall y < x \ \varphi_y \neq \varphi_x\}$$

is immune.

H99. Show that every infinite r.e. set has an infinite recursive subset.

100. (a) Show that there does not exist an r.e. set A of Turing machine indices such that

  - if $i \in A$ then M_i is total; and
  - every recursive set has an index in A.

  *H(b) Show that there exists an r.e. set A of Turing machine indices such that

  - if $i \in A$ then $L(M_i)$ is recursive; and
  - every recursive set has an index in A.

101. (a) Using the recursion theorem, one can easily find a program that computes its own index: construct σ total such that $\varphi_{\sigma(n)}(x) = n$, then take a fixpoint. Show that there exist two different programs that compute each other's indices. That is, show that there exist $m, n \in \omega$, $m \neq n$, such that $\varphi_n(x) = m$ and $\varphi_m(x) = n$.

  (b) Write two different programs in your favorite programming language that print each other out.

102. (a) Let φ and ψ be Gödel numberings. Show that there exist indices $m, n \in \omega$ such that $\varphi_n(x) = m$ and $\psi_m(x) = n$.

  (b) Write a program in each of your two favorite programming languages that print each other out.

*103. Let A be a recursive set that cannot be computed in polynomial time. Show that there exists an infinite recursive subset $B \subseteq A$ on which any TM computing A takes more than polynomial time a.e. on B.

104. (a) Show that if $P = NP$, then given nondeterministic polynomial-time machine with explicit time bound n^k, one can find an equivalent deterministic polynomial time machine *effectively* (that is, by a total recursive function).

  **(b) Show that this is still true even if the time bound n^k is not known.

105. The following problems refer to the proof of the speedup theorem (Theorem 32.2).

  (a) Show that it is undecidable for a given machine M_i whether M_i falls in case A or case B.

(b) Show that the values $m(i)$ cannot be obtained effectively.

**(c) Show that it is impossible to obtain a machine for A running in time $f^{n-k}(2)$ effectively from k.

106. Let P be a nontrivial property of the linear-time sets that is true for all regular sets. Show that P is undecidable.

*HS107. This is a general isomorphism theorem that has both the Rogers isomorphism theorem (Theorem 34.3, Miscellaneous Exercise 108) and the Myhill isomorphism theorem (Miscellaneous Exercise 109) as special cases. Let $\circ$ denote relational composition and $^{-1}$ the reverse operator on binary relations on ω. That is, for $R, S \subseteq \omega \times \omega$, define

$$R \circ S \overset{\text{def}}{=} \{(u, w) \mid \exists v \ (u, v) \in R, \ (v, w) \in S\}$$
$$R^{-1} \overset{\text{def}}{=} \{(u, v) \mid (v, u) \in R\}.$$

For a function $f : \omega \to \omega$, define

$$\text{graph}\, f \overset{\text{def}}{=} \{(x, f(x)) \mid x \in \omega\}.$$

Let R be a binary relation on ω such that $R \circ R^{-1} \circ R \subseteq R$. Let $f, g : \omega \to \omega$ be one-to-one total recursive functions such that $\text{graph}\, f \subseteq R$ and $\text{graph}\, g \subseteq R^{-1}$. Show that there exists a one-to-one and onto total recursive function $h : \omega \to \omega$ such that $\text{graph}\, h \subseteq R$.

H108. Here we use Miscellaneous Exercise 107 to give an alternative proof of the Rogers isomorphism theorem (Theorem 34.3). Assuming there exist $\sigma, \tau : \omega \to \omega$ such that for all i, $\varphi_i = \psi_{\sigma(i)}$ and $\psi_i = \varphi_{\tau(i)}$, show that there exists a one-to-one and onto total recursive function $\rho : \omega \to \omega$ such that for all i, $\varphi_i = \psi_{\rho(i)}$.

H109. The *Cantor–Schröder–Bernstein theorem* of set theory says that if there is a one-to-one function $A \to B$ and a one-to-one function $B \to A$, then A and B are of the same cardinality. Here is an effective version due to Myhill [90] (see [104]) known as the *Myhill isomorphism theorem*. Show that any two sets that are reducible to each other via one-to-one reductions are recursively isomorphic. In other words, let $A, B \subseteq \omega$ and let $f, g : \omega \to \omega$ be one-to-one total recursive functions such that for all $x \in \omega$, $x \in A \Leftrightarrow f(x) \in B$ and $x \in B \Leftrightarrow g(x) \in A$. Show that there exists a one-to-one and onto total recursive function $h : \omega \to \omega$ such that for all $x \in \omega$, $x \in A \Leftrightarrow h(x) \in B$.

110.*^{S}(a) Give a recursive set A such that both A and $\sim A$ are infinite, but neither A nor $\sim A$ contains an infinite polynomial-time computable subset.

*(b) Let $\mathcal{L}$ be a family of recursive sets such that there exists an r.e. list of TMs that always halt and accept all and only sets in $\mathcal{L}$. Give a recursive set A such that both A and $\sim A$ are infinite, but no infinite subset of A or $\sim A$ is in $\mathcal{L}$.

111. Use the constructs based on the s_n^m and universal function properties as described in Lecture 33 to construct a lazy conditional test

$$
\varphi_{\text{cond}(i,j)}(x,y) \;\;=\;\; \begin{cases} \varphi_i(y), & \text{if } x = 0, \\ \varphi_j(y), & \text{if } x \neq 0. \end{cases}
$$

The conditional test should be *lazy* in the sense that it should not attempt to evaluate $\varphi_i(y)$ if $x \neq 0$ nor $\varphi_j(y)$ if $x = 0$.

112. Prove the following generalization of Lemma 37.8. If $A \leq_{\mathrm{m}} B$ and A is productive, then so is B.

^{H}113. Give a set A such that both A and $\sim A$ are productive.

114. Two disjoint sets A and B are *recursively separable* if there exists a recursive set C such that $A \subseteq C$ and $B \subseteq \sim C$. They are *recursively inseparable* if no such C exists.

(a) Show that any pair of disjoint co-r.e. sets are recursively separable.

*^{H}(b) Construct a pair of recursively inseparable r.e. sets.

^{H}115. Define $x \equiv y$ if $\varphi_x = \varphi_y$. Show that any pair of distinct $\equiv$-equivalence classes are recursively inseparable. (See Miscellaneous Exercise 114.)

^{H}116. Let Φ be an abstract complexity measure (see Lecture J). Prove that Φ_i is a partial recursive function, and that it is possible to obtain an index for Φ_i effectively from i. That is, there exists a total recursive function σ such that $\varphi_{\sigma(i)} = \Phi_i$.

117. Let Φ be an abstract complexity measure. Prove that for any total recursive function g, there exists a $0,1$-valued total recursive function f such that for all indices i for f, $\Phi_i(n) > g(n)$ a.e.

118. ^{H}(a) Let Φ be an abstract complexity measure. Prove that there does not exist a total recursive function f such that for all i, $\Phi_i(n) \leq f(n, \varphi_i(n))$ a.e.; that is, the complexities of recursive functions are not uniformly recursively bounded by their values.

   (b) On the other hand, show that the values of recursive functions are uniformly recursively bounded by their complexities; that is, there exists a total recursive function g such that for all i, $\varphi_i(n) \leq g(n, \Phi_i(n))$ a.e.

119. (a) Show that the domain of φ_i is recursive iff there exists a total recursive function f such that $\Phi_i(n) \leq f(n)$ whenever $\varphi_i(n){\downarrow}$.

   (b) Conclude from (a) that any TM accepting a nonrecursive r.e. set must use more space than any total TM on infinitely many accepted inputs.

120. Prove the gap theorem for abstract complexity measures (Theorem J.1).

^{H}121. (Blum [17]) Prove the following *slowdown theorem*. For all total recursive functions f, g, there exists an index i for f such that $\Phi_i(n) > g(n)$ for all n.

122. Show that any two abstract complexity measures are uniformly recursively bounded by each other. Formally, let Φ and Ψ be abstract complexity measures. Give a total recursive function f such that for all i, $\Psi_i(n) \leq f(n, \Phi_i(n))$ a.e. and $\Phi_i(n) \leq f(n, \Psi_i(n))$ a.e.

123. (a) (Combining Lemma) Let Φ be an abstract complexity measure. Let c be a total recursive operator of two variables such that for all i, j, if $\varphi_i(n){\downarrow}$ and $\varphi_j(n){\downarrow}$, then $\varphi_{c(i,j)}(n){\downarrow}$. Show that there exists a total recursive function h such that for all i, j,

$$\Phi_{c(i,j)}(n) \leq h(n, \Phi_i(n), \Phi_j(n)) \text{ a.e.}$$

(b) In particular, let c be a total recursive operator such that for all i, if $\varphi_i(n)\downarrow$, then $\varphi_{c(i)}(n)\downarrow$. Show that there exists a total recursive function h such that for all i,

$$\Phi_{c(i)}(n) \leq h(n, \Phi_i(n)) \text{ a.e.}$$

[H]124. Let Φ be an abstract complexity measure. Show that given some complexity class $\mathcal{C}_f^\Phi$ in terms of an index for f, one can uniformly and effectively find a strictly larger complexity class. That is, there exists a total recursive function σ such that $\mathcal{C}_{\varphi_{\sigma(i)}}^\Phi$ strictly contains $\mathcal{C}_{\varphi_i}^\Phi$.

125. An abstract complexity measure Φ is *completely honest* if there exists a total recursive function σ such that for all i, $\varphi_{\sigma(i)} = \Phi_i$ and $\Phi_{\sigma(i)}(n) \leq \Phi_i(n)$ a.e. In other words, we can effectively find an index for the complexity of a given partial recursive function φ_i, and it is no more difficult to compute the complexity than to compute the function itself.

(a) Argue that space complexity of Turing machines is completely honest.

[H](b) Show that not all abstract complexity measures are completely honest.

[*S]126. The functions f_k in the statement of the union theorem (Theorem J.3) are postulated to satisfy a monotonicity condition, which states that for all i and n, $f_i(n) \leq f_{i+1}(n)$. Show that the theorem can fail without this condition.

127. A set of total recursive functions is *recursively enumerable* (r.e.) if there exists an r.e. set of indices representing all and only functions in the set. For example, the complexity class P is r.e., because we can represent it by an r.e. list of TMs with polynomial-time clocks.

(a) Let Φ be an abstract complexity measure. Show that any Φ-complexity class containing all functions that are almost everywhere constant is r.e.

(b) Suppose the complexity measure Φ is *invariant under finite modifications*; that is, if $f(n) = g(n)$ a.e., then $f \in \mathcal{C}_t^\Phi$ iff $g \in \mathcal{C}_t^\Phi$. Show that all Φ-complexity classes are r.e.

[H](c) Show that there exists a measure Φ and a total recursive function f such that $\mathcal{C}_f^{\Phi}$ is not r.e.

128. Prove Theorem 35.1.

129. Prove that $\Sigma_n^0 \cup \Pi_n^0 \neq \Delta_{n+1}^0$ (see Lecture 35).

130. Prove the following completeness results (see Lecture 35 for definitions).

(a) HP is $\leq_m$-complete for Σ_1^0.

(b) EMPTY is $\leq_m$-complete for Π_1^0.

(c) TOTAL is $\leq_m$-complete for Π_2^0.

131. Let M_i and φ_i denote Turing machines and partial recursive functions, respectively. Show that the sets

(a) ALL $\overset{\text{def}}{=}$ $\{i \mid L(M_i) = \Sigma^*\}$

[S](b) EQUAL $\overset{\text{def}}{=}$ $\{(i,j) \mid \varphi_i = \varphi_j\}$

are $\leq_m$-complete for Π_2^0.

[H]132. Let $\mathcal{L}$ be any family of recursive sets containing all the regular sets such that there exists an r.e. list of TMs that always halt and accept all and only sets in $\mathcal{L}$. Show that $\{i \mid L(M_i) \in \mathcal{L}\}$ is $\leq_m$-complete for Σ_3^0.

[H]133. Prove that the following decision problems are $\leq_m$-complete for Σ_3^0.

(a) Given a Turing machine M, is $L(M)$ a regular set?

(b) Given a Turing machine M, is $L(M)$ a context-free language?

(c) Given a Turing machine M, is $L(M)$ a recursive set?

134. [S](a) Let PA denote Peano arithmetic (or your favorite proof system for number theory). Does there exist a polynomial-time machine that cannot be proved in PA to run in polynomial time?

^{*S}(b) Is it always possible, given a machine that runs in polynomial time, to effectively compute a bound of the form n^k?

^{*H}135. Prove that there exists a total computable function $f : \mathbb{N} \to \mathbb{N}$ that is not provably total in Peano arithmetic.

^{*S}136. Formalize and prove the following statement. "In any formal deductive system for number theory, there is a decidable problem for which no algorithm can be proved totally correct."

^{H}137. Write $A \leq_1 B$ if $A \leq_m B$ via a reduction σ that is one-to-one. Show that $K = \{x \mid M_x(x)\downarrow\}$ is Σ^0_1-complete with respect to $\leq_1$.

^{H}138. (a) In Miscellaneous Exercise 130, we showed that

$$\text{TOTAL} \quad \overset{\text{def}}{=} \quad \{M \mid \forall x \; \exists k \; M(x)\downarrow^k\}$$

is $\leq_m$-complete for Π^0_2. How about the set

$$\text{WAYTOTAL} \quad \overset{\text{def}}{=} \quad \{M \mid \exists k \; \forall x \; M(x)\downarrow^k\}?$$

*(b) Is it always possible, given $M \in \text{WAYTOTAL}$, to effectively compute a bound k?

^{**HS}139. One of the following problems is Σ^0_2-complete and the other is Π^1_1-complete. Which is which? Provide conclusive evidence that your choice is correct.

(a) Given a nondeterministic Turing machine M and state q of M, does there exist a computation path of M on input ϵ along which M enters state q only finitely often?

(b) Given a nondeterministic Turing machine M and state q of M, does M on input ϵ enter state q only finitely often along *every* computation path?

140. (a) Prove that the following problem is in *LOGSPACE*. Given a finite binary relation, is it transitive?

^{S}(b) Prove that the following problem is $\leq_{\text{m}}$-complete for Π_1^0. Given a recursive binary relation $R \subseteq \omega^2$, say by a total Turing machine accepting the set of pairs in R, is R transitive?

141. Argue that any Π_1^1 formula over $\mathbb{N}$ can be put into the form (39.1) using pairing and the skolemization rule

$$\forall x : \mathbb{N} \; \exists f : \mathbb{N} \to \mathbb{N} \; \varphi(f, x) \quad \mapsto \quad \exists g : \mathbb{N} \to (\mathbb{N} \to \mathbb{N}) \; \forall x : \mathbb{N} \; \varphi(g(x), x).$$

142. Show that first-order number theory is hyperelementary over $\mathbb{N}$ but not elementary (see Lecture 39). Here *first-order number theory* refers to the set

$$\text{Th}(\mathbb{N}) \quad \overset{\text{def}}{=} \quad \{\varphi \mid \varphi \text{ is a sentence of the language of first-order number theory and } \mathbb{N} \vDash \varphi\}.$$

143. Prove Corollary 41.2.

**144. ([77]) As argued in Lectures 39 and 40, IND programs accept exactly the Π_1^1 sets over $\mathbb{N}$. Show that IND programs with dictionaries accept exactly the Π_1^1 sets over any countable structure. A *dictionary* is an abstract data structure that allows data items to be associated with keys and retrieved by key lookup.

Hints and Solutions

Homework 1 Solutions

1. (a) One possible nonregular set accepted in time $O(n \log n)$ is

    $$\{a^{2^n} \mid n \geq 0\}.$$

    Repeatedly scan the input, checking that the number of a's remaining is even and erasing every second one, until one a remains.

   (b) For this proof, crossing sequences are sequences of states only. Assume M moves all the way to the right before accepting. Let q be the number of states of M. By an argument similar to the proof of Lemma 1.4, if $L(M)$ is nonregular, then there can be no fixed finite bound on the length of crossing sequences generated by M on accepted inputs. Then for each $k > 0$, there is a string in $L(M)$ for which a crossing sequence of length $\geq k$ is generated. Let x_k be the shortest such string and let $n = |x_k|$. As in the proof of Theorem 1.3, M must generate at least $n/2$ distinct crossing sequences on input x_k, otherwise we could remove a substring of x_k and get a smaller string generating the longest crossing sequence, contradicting the minimality of x_k.

    Let S_0 be the set of the $n/2$ shortest possible crossing sequences in lexicographic order. The set S_0 must contain all crossing sequences of length up to $m - 1$, where m is the least number such that

    $$\sum_{i=1}^{m} q^i \geq \frac{n}{2}.$$

    The running time on x_k is bounded below by the sum of the lengths of the crossing sequences in S, because it takes one step to generate each element of each crossing sequence; so

    $$T(n) \geq \sum_{c \in S} |c| \geq \sum_{c \in S_0} |c| \geq \sum_{i=1}^{m-1} i q^i.$$

    The bound

    $$T(n) \geq \Omega(n \log n)$$

    follows from these inequalities and some arithmetic.

2. (a) A *(nondeterministic) k-headed finite automaton (k-FA)* is a 7-tuple

    $$M = (Q, \Sigma, \vdash, \dashv, \delta, s, f),$$

    where

    - Q is a finite set (the *states*),

- Σ is a finite set (the *input alphabet*),
- $\vdash$, $\dashv$ are symbols not in Σ (the *left* and *right endmarkers*, respectively),
- $\delta \subseteq (Q \times (\Sigma \cup \{\vdash, \dashv\})^k) \times (Q \times \{-1, 0, +1\}^k)$ (the *transition relation*),
- $s \in Q$ (the *start state*), and
- $f \in Q$ (the *accept state*),

such that if

$$((p, a_1, \ldots, a_k), (q, d_1, \ldots, d_k)) \in \delta, \tag{1}$$

then if $a_i = \vdash$ then $d_i \neq -1$ and if $a_i = \dashv$ then $d_i \neq +1$. The machine M is *deterministic* if δ is single-valued.

Informally, (1) means that if the machine is in state p scanning symbol a_i under its ith head, $1 \leq i \leq k$, then it can move its ith head in direction d_i, $1 \leq i \leq k$, and enter state q. The purpose of the condition involving $\vdash$ and $\dashv$ is to keep the heads from moving off the input.

Let $x \in \Sigma^*$, say $x = x_1 x_2 \cdots x_n$ where $x_i \in \Sigma$, $1 \leq i \leq n$. Let $x_0 = \vdash$ and $x_{n+1} = \dashv$. A *configuration* of M on input x is an element of

$$Q \times \{0, \ldots, n+1\}^k.$$

Informally, a configuration specifies a current state and the positions of the k heads. If α and β are configurations, we write

$$\alpha \overset{1}{\to} \beta$$

and say β *follows from* α *in one step* if

$$\alpha = (p, i_1, i_2, \ldots, i_k)$$
$$\beta = (q, i_1 + d_1, i_2 + d_2, \ldots, i_k + d_k)$$

and

$$((p, x_{i_1}, x_{i_2}, \ldots, x_{i_k}), (q, d_1, d_2, \ldots, d_k)) \in \delta.$$

The reflexive transitive closure of the relation $\overset{1}{\to}$ is denoted $\overset{*}{\to}$. The *start configuration* and *accept configuration* of M on input x are the configurations

$$(s, 0, 0, \ldots, 0) \qquad (f, n+1, n+1, \ldots, n+1),$$

respectively. The machine is said to *accept input* x if

$$(s, 0, 0, \ldots, 0) \overset{*}{\to} (f, n+1, n+1, \ldots, n+1).$$

(b) ($\Leftarrow$) Given a k-FA M, we construct an $O(\log n)$ space-bounded Turing machine N that simulates M as follows. The worktape of N is partitioned into k tracks, each of which holds an $O(\log n)$-bit binary number between $-(n+1)$ and $n+1$, inclusive. These numbers are used to record the positions of the k simulated heads of M relative to the position of N's read head. The state of M will be kept in N's finite control. To simulate a move of M, N needs to know what symbol each head of M is currently scanning. Starting with its read head all the way to the left, N repeatedly moves its read head one cell to the right and decrements each of the k counters. Whenever any counter contains 0, that indicates that the simulated head of M corresponding to that counter is scanning the same input tape cell that N is currently scanning. N reads the symbol on its input tape and remembers it in its finite control. When N's head reaches the right side of the tape, it has seen all the symbols under the k simulated heads of M. It adjusts the counters and changes the simulated state of M according to the transition relation of M, which it has stored in its finite control. It then moves its head all the way back to the left, incrementing the k counters as it goes, and simulates another step.

($\Rightarrow$) Let N be an $O(\log n)$ space-bounded TM with worktape alphabet $\{0,1\}$. We show how to simulate N with a machine M with a two-way read-only input head and finitely many counters, each of which can hold a nonnegative integer. In each step, M can add one or subtract one from any counter and test for 0. Such a machine can easily be simulated by a k-FA for some k using the positions of the heads for the counters, provided the counter values never get bigger than n.

We first show how to implement the following counter operations.

- Duplicate the value of a counter.
- Double the value of a counter.
- Halve the value of a counter.
- Check whether the value of a counter is even.
- Add or subtract the value of one counter from another.

To duplicate the value of counter c, zero out two scratch counters d and e and then repeatedly decrement c while incrementing d and e. To double c, zero out a scratch counter d and repeatedly decrement c and increment d twice. Halving is similar. To test for evenness, halve the counter and see if there is one left over. To add, increment one counter while decrementing the other. To subtract, decrement both.

For any configuration of N, the contents of N's worktape can be viewed as a $c \log n$-bit binary number. We break this number up into c blocks of $\log n$ bits each, and record these $\log n$-bit numbers in c counters of M. Another counter of M is used to store the position of N's worktape head: the block on the worktape being scanned by N is stored in M's finite control, and the position i within the block is represented by 2^i in the counter. The position of N's input head is represented directly by the position of M's input head. A finite number of other counters of M are used for temporary storage. The state of N is represented in the state of M.

In order to simulate a move of N, M must know the symbols currently being scanned by N on its two tapes. It can read the symbol on the input tape directly. For the worktape, it must be able to determine and change the ith bit of a number contained in some counter c, where 2^i is the number contained in some other counter d. To do this, first duplicate c and d so we do not lose their contents. Repeatedly halve c and d until d contains 1, then check whether c is even. This determines the ith bit of c. We can modify the bit by adding or subtracting the original value of d from the original value of c.

3. Assume that $P = NP$. Let $L \in NEXPTIME$, say accepted by some 2^{n^c}-time-bounded nondeterministic Turing machine M. Let

$$\widehat{L} \;=\; \{x\#^k \mid x \in L \text{ and } k = 2^{|x|^c}\}.$$

Then $\widehat{L} \in NP$, because on input $x\#^k$ we can just count the $\#$'s to make sure there are $2^{|x|^c}$, then erase them and run M. By the assumption $P = NP$, we have $\widehat{L} \in P$, say accepted by deterministic polynomial-time machine N. Then $L \in EXPTIME$, because on input x we can append $2^{|x|^c}$ $\#$'s and run N.

Homework 2 Solutions

1. If the function f is computable by a logspace transducer, then $|f(x)|$ is polynomial in $|x|$, because the transducer can run for at most polynomial time before repeating a configuration. Suppose f and g are computable by logspace transducers M and N, respectively. To compute $g(f(x))$, we simulate the computation of N on input $f(x)$. The string $f(x)$ is not computed in advance, but provided to N symbol by symbol on a demand basis. Whenever N wishes to read the ith symbol of its input, the number i is provided to a subroutine that simulates M on input x from scratch, counting and throwing away all output symbols until the ith, which it returns to the calling procedure. We need only enough worktape space to hold the worktapes of M and N and a counter that can count up to $|f(x)|$. For more details, see [63, Lemmas 13.1 and 13.2].

2. We first show that 2SAT is co-$NLOGSPACE$-hard by reducing the MAZE problem, known to be $NLOGSPACE$-hard, to the complementary problem, 2CNF unsatisfiability. Given an instance $G = (V, E, s, t)$ of MAZE, take V as a set of Boolean variables and consider the 2CNF formula

$$\varphi_G \;=\; s \wedge \Big(\bigwedge_{(u,v)\in E} (u \to v) \Big) \wedge \neg t.$$

If there is a path from s to t in G, then the clauses in φ_G corresponding to the edges in this path imply $s \to t$, thus φ_G implies $s \wedge (s \to t) \wedge \neg t$, which is unsatisfiable. Conversely, if there is no path from s to t, assign all vertices reachable from s the value 1 and assign all other variables 0. This assignment satisfies φ_G, because s is assigned 1, t is assigned 0, and there is no clause $u \to v$ with u assigned 1 and v assigned 0. Hence 2SAT is hard for co-$NLOGSPACE$.

We now show that 2SAT is in co-$NLOGSPACE$, or equivalently, that 2CNF unsatisfiability is in $NLOGSPACE$. Given a 2CNF formula $\mathcal{B}$, the clauses in $\mathcal{B}$ contain at most two literals, and we can assume exactly two without loss of generality by replacing any clause of the form (u) with $(u \vee u)$. Now we think of every two-literal clause $(u \vee v)$ as a pair of implications

$$(\neg u \to v) \quad \text{and} \quad (\neg v \to u). \tag{2}$$

Construct a directed graph $G = (V, E)$ with a vertex for every literal and directed edges corresponding to the implications (2). It is not difficult to show (see, for example, [75, p. 119]) that $\mathcal{B}$ is unsatisfiable if and only if there exists a cycle of G containing two complementary literals u, $\neg u$. The latter condition can be tested in $NLOGSPACE$ by guessing u, then guessing and tracing a cycle to verify that it is a cycle and contains u and $\neg u$. This requires only logspace to remember u, where we are in the cycle, and the starting point of the cycle (finite fingers!).

3. Evaluate the formula recursively as follows. Start at the root. To evaluate a subformula $\varphi \wedge \psi$, first evaluate φ; if the value is 1 then the value of the entire expression is the value of ψ, otherwise it is 0 and ψ need not be evaluated. Dually, to evaluate $\varphi \vee \psi$, first evaluate φ; if the value is 0 then the value of the entire expression is the value of ψ, otherwise it is 1 and ψ need not be evaluated. To evaluate $\neg\varphi$, we evaluate φ and then negate the result. We need only finitely many fingers to walk the tree, so this can be implemented in logspace.

4. Let

$$
\begin{aligned}
B &= \{\#b_k(0)\#b_k(1)\#b_k(2)\#\cdots\#b_k(2^k-1)\# \mid k \geq 0\} \\
B_j &= \{\#u_0\#u_1\#\cdots\#u_{m2^j-1}\# \mid m \geq 0,\ b_j(i) \equiv u_i \ (\mathrm{mod}\ 2^j), \\
&\qquad 0 \leq i \leq m2^j - 1\} \\
F_k &= \#0^k(\#((0+1)^k - 0^k - 1^k))^*\#1^k\#,
\end{aligned}
$$

where $b_j(i)$ denotes the j-bit binary representation of i mod 2^j. Strings in B_j consist of sequences of strings $u \in (0+1)^*$ of length at least j separated by $\#$ such that the low-order j bits of the successive strings u represent successive integers mod 2^j from 0 to $m2^j - 1$ for some m. Strings in F_k consist of successive strings of length k separated by $\#$ such that the first string is 0^k, the last is 1^k, and none of the intermediate strings are either 0^k or 1^k.

Note that

$$
B_0 \supseteq B_1 \supseteq B_2 \supseteq \cdots \supseteq B_k,
$$

so

$$
B = B_k \cap F_k = \bigcap_{j=0}^{k} B_j \cap F_k.
$$

To check whether a given string x is in the set B, we check that x is in $B_0, B_1, \ldots, B_k, F_k$ in that order. We do it this way so that we do not use too much space even if the input is not in B.

The set B_j can be recognized in space $\log j$. We just check for successive substrings u and v whether the low-order j bits represent successive integers mod 2^j by comparing the corresponding low-order j bits of u and v. We need $\log j$ space to count the distance of a bit from the closest # to its right. We also need to check that the first substring u is $0 \bmod 2^j$ and the last is $-1 \bmod 2^j$.

For $j = 0, 1, 2, \ldots$, we lay off $\log j$ tape cells and test membership in B_j. All strings in B_j are of length at least $j2^j$, so if the test succeeds, then we have used only $\log j \leq \log \log n$ space. If we ever encounter a j for which the test fails, we reject immediately; but even in this case, because $x \in B_{j-1}$, we have used only

$$\log j \;\leq\; 1 + \log(j-1) \;\leq\; 1 + \log \log n$$

space. If all tests $x \in B_j$ are successful, then we have laid off $\log k$ space, which is sufficient to check membership in F_k.

Homework 3 Solutions

1. (a) Let $\#L(x)$ (respectively, $\#R(x)$) be the number of left (respectively, right) parentheses in the string x. One can show by induction that a string x is balanced iff

      (i) $\#L(x) = \#R(x)$, and
      (ii) for every prefix y of x, $\#L(y) \geq \#R(y)$,

      hence we can just scan left to right, counting $\#L(y) - \#R(y)$.

    (b) A string x of two types of parentheses is balanced iff it satisfies the conditions (i) and (ii) of part (a) irrespective of parenthesis type, and each matching pair has the same type. One can find matching pairs by counting. In the string $x\,[y]\,z$, the brackets shown match iff y satisfies (i) and (ii) irrespective of parenthesis type.

2. This version of the game is complete for $ALOGSPACE = P$. Because vertices can be reused, a board position consists only of the vertex currently being visited and a bit to tell whose move it is. We do not have to remember which vertices have already been played as with the previous version. It requires only logspace to maintain the current board, so an alternating logspace machine can determine whether the first player has a forced win, thus the problem is in $ALOGSPACE = P$.

    To show that the problem is hard for P, we reduce the circuit value problem to it. Given an instance of CVP, first transform it into an instance in which there are no negations and the alternation is strict. To get rid of negations, produce the dual of the circuit:

Original	Dual
$c_i := c_j \wedge c_k$	$c_i' := c_j' \vee c_k'$
$c_i := c_j \vee c_k$	$c_i' := c_j' \wedge c_k'$
$c_i := 0$	$c_i' := 1$
$c_i := 1$	$c_i' := 0$
$c_i := \neg c_j$	$c_i' := \neg c_j'$

    Then replace all $c_i := \neg c_j$ with $c_i := c_j'$ and $c_i' := \neg c_j'$ with $c_i' := c_j$. To make the alternation strict, simultaneously replace each statement $c_i := c_j \wedge c_k$ with the two statements $c_i := d \vee d$ and $d := c_j \wedge c_k$, where d is a new variable, and each statement $c_i := c_j \vee c_k$ with the three statements $c_i := d \vee e$, $d := c_j \wedge c_j$, $e := c_k \wedge c_k$ where d and e are new variables. This at most triples the number of variables.

    Now we make this into a geography game. Note that a player wins by trapping the opponent in a *cul-de-sac* (a vertex with outdegree 0). We produce a graph with vertices $\{c_i \mid 0 \le i \le n\} \cup \{\bot\}$, directed edges (c_i, c_j) and (c_i, c_k) for each statement $c_i := c_j \wedge c_k$ or $c_i := c_j \vee c_k$, and edges $(c_i, \bot)$ for each statement $c_i := 1$. Thus the cul-de-sacs are $\bot$ and the c_i such that the statement $c_i := 0$ appears in the circuit. The starting position is c_n, the variable with the highest index. The first player has a forced win iff the value of the circuit is 1.

3. See [63] or [76] for background on finite automata. In deterministic as well as alternating finite automata, it is technically convenient to consider F to be the *characteristic function* of the set of final states rather than the set of final states itself. That is, $F : Q \to \{0, 1\}$ such that

$$F(q) \;=\; \begin{cases} 1, & \text{if } q \text{ is a final state} \\ 0, & \text{otherwise.} \end{cases}$$

To construct a DFA from an AFA, let

$$A \;=\; (Q_A, \Sigma, \delta_A, F_A, \alpha_A)$$

be the given AFA, $|Q_A| = k$. Let Q_D be the set of all functions $Q_A \to \{0, 1\}$. Define the DFA

$$D \;=\; (Q_D, \Sigma, \delta_D, F_D, s_D),$$

where

$$\delta_D(u, a)(q) \;=\; \delta_A(q, a)(u) \tag{3}$$
$$F_D \;=\; \alpha_A \tag{4}$$
$$s_D \;=\; F_A. \tag{5}$$

To construct an AFA from a DFA, let

$$D \;=\; (Q_D, \Sigma, \delta_D, F_D, s_D)$$

be the given DFA, $|Q_D| = k$. Let Q_A be any set of size $\lceil \log k \rceil$ and identify each element of Q_D with a distinct function $Q_A \to \{0, 1\}$. Define the AFA

$$A \;=\; (Q_A, \Sigma, \delta_A, F_A, \alpha_A),$$

where δ_A, F_A, and α_A are defined such that (3)–(5) hold. For $u \notin Q_D$, define $\delta_A(q, a)(u)$ arbitrarily.

In both reductions, one can show by induction on $|x|$ that for any $q \in Q_A$, $u \in Q_D$, and $x \in \Sigma^*$,

$$\widehat{\delta}_D(u,x)(q) = \widehat{\delta}_A(q, \text{rev } x)(u),$$

where $\widehat{\delta}_D : Q_D \times \Sigma^* \to Q_D$ is the multistep version of δ_D derived by induction on the length of the input string:

$$\widehat{\delta}_D(u, \varepsilon) \overset{\text{def}}{=} u$$
$$\widehat{\delta}_D(u, xa) \overset{\text{def}}{=} \delta_D(\widehat{\delta}_D(u, x), a).$$

Thus

$$
\begin{aligned}
x \in L(D) \quad &\Leftrightarrow \quad F_D(\widehat{\delta}_D(s_D, x)) = 1 \\
&\Leftrightarrow \quad \alpha_A(\widehat{\delta}_D(F_A, x)) = 1 \\
&\Leftrightarrow \quad \alpha_A(\lambda q.(\widehat{\delta}_D(F_A, x)(q))) = 1 \\
&\Leftrightarrow \quad \alpha_A(\lambda q.(\widehat{\delta}_A(q, \text{rev } x)(F_A))) = 1 \\
&\Leftrightarrow \quad \text{rev } x \in L(A).
\end{aligned}
$$

Homework 4 Solutions

1. First assume $A(n)$ and $S(n)$ are space constructible. Let M be an $A(n)$-alternation-bounded, $S(n)$-space-bounded machine. Let C_n be the set of configurations of M on inputs of length n. There is a fixed constant c depending only on M such that $|C_n| \leq c^{S(n)}$.

   Let type $: C_n \to \{\wedge, \vee\}$ tell whether a configuration is universal or existential. Accept configurations are universal configurations without successors and reject configurations are existential configurations without successors. For $\alpha, \beta \in C_n$, write

   $$R(\alpha, \beta, k)$$

   if there is a computation path from α to β of length at most k such that all configurations γ along the path except β satisfy $\text{type}(\gamma) = \text{type}(\alpha)$, and $\text{type}(\beta) \neq \text{type}(\alpha)$. For $\alpha, \beta \in C_n$, the predicate

   $$R(\alpha, \beta, c^{S(n)})$$

   is decidable in nondeterministic space $S(n)$, therefore it is decidable in deterministic space $S(n)^2$ by Savitch's theorem.

   Then, for input x of length n with initial existential configuration α_0, M accepts iff

   $$\exists \alpha_1 \ R(\alpha_0, \alpha_1, c^{S(n)}) \wedge$$
   $$\forall \alpha_2 \ R(\alpha_1, \alpha_2, c^{S(n)}) \to$$
   $$\exists \alpha_3 \ R(\alpha_2, \alpha_3, c^{S(n)}) \wedge$$
   $$\forall \alpha_4 \ R(\alpha_3, \alpha_4, c^{S(n)}) \to$$
   $$\cdots$$
   $$Q\alpha_{A(n)} \ R(\alpha_{A(n)-1}, \alpha_{A(n)}, c^{S(n)}).$$

   This can be checked by a Boolean-valued recursive procedure $S(\alpha)$ that works as follows. If α is existential, it cycles through all β lexicographically, checking for the existence of a β such that $R(\alpha, \beta, c^{S(n)})$ and $S(\beta)$. It checks the former using the Savitch algorithm, and if that succeeds, it checks the latter by a recursive call. Similarly, if α is universal, it cycles through all β, checking that if $R(\alpha, \beta, c^{S(n)})$ then $S(\beta)$.

   Each recursive instantiation of the procedure needs $S(n)$ space to save the current configuration α across recursive calls, and the depth of the recursion is $A(n)$, so $A(n)S(n)$ space is needed in all for this purpose. In addition, the Savitch procedure to compute R requires $S(n)^2$ space, which can be reused. This gives a total space bound of $A(n)S(n) + S(n)^2$.

When $A(n)$ and $S(n)$ are not space constructible, we try iteratively all values of A and S such that $AS + S^2 = 1, 2, \ldots$.

2. Define a hierarchy over $PSPACE$ by setting

$$
\begin{aligned}
\Sigma_k^{PSPACE} &= STA(n^{O(1)}, *, \Sigma k) \\
\Pi_k^{PSPACE} &= STA(n^{O(1)}, *, \Pi k).
\end{aligned}
$$

Then by the previous exercise,

$$
\begin{aligned}
\Sigma_k^{PSPACE} &= STA(n^{O(1)}, *, \Sigma k) \\
&= \bigcup_{c>0} STA(n^c, *, \Sigma k) \\
&\subseteq \bigcup_{c>0} DSPACE(kn^c + n^{2c}) \\
&\subseteq \Sigma_0^{PSPACE}.
\end{aligned}
$$

3. The set $H_\omega = \{y\$z \mid z \in G_{\#(y)}\}$ is in $PSPACE$, because a universal alternating machine on input $y\$M\$x\$^d$ can simulate M on x, checking off one $\$$ for every step simulated and decrementing the binary number y for every alternation. Each step of M requires at most polynomial time in d and the length of M to simulate, and there are at most d steps. If either the time bound or the bound on the number of alternations is exceeded, that process of the simulating machine rejects.

To show that H_ω is hard for $PSPACE$, it suffices to reduce an arbitrary set in $APTIME$ to H_ω. Let M be any alternating machine running in time n^c. Then the map

$$
x \;\mapsto\; y\$M\$x\$^{|x|^c},
$$

where $\#(y) = |x|^c$, constitutes a $\leq_m^{\log}$ reduction from $L(M)$ to H_ω.

4. The set

$$
G_k = \{M\$x\$^d \mid M \text{ accepts } x \text{ in at most } d \text{ space and } k \\ \text{alternations, beginning with } \vee\}
$$

is in $STA(n^2, *, \Sigma k)$. On input $M\$x\$^d$, simulate M on input x. Because each tape symbol requires at most $|M|$ space to represent on the tape of the simulating machine (assuming the tape symbols are represented explicitly in the description of the machine), the simulation requires no more than $d \cdot |M|$ space to represent the tape of M. The set G_k is also hard for Σ_k^{PSPACE}: let M be any Σ_k machine running in space n^c. Then the map

$$x \;\; \mapsto \;\; M\$x\$^{|x|^c}$$

constitutes a $\leq_m^{\log}$ reduction from $L(M)$ to G_k.

Now let

$$
\begin{aligned}
G_\omega \;\; &= \;\; \{y\$z \mid z \in G_{\#(y)}\} \\
&= \;\; \{y\$M\$x\$^d \mid M \text{ accepts } x \text{ in at most } d \text{ space and } \#(y) \\
&\qquad\qquad \text{alternations, beginning with } \vee\}.
\end{aligned}
$$

The set G_ω is in $APSPACE = EXPTIME$ via a simulation similar to the one in the previous problem. To show that it is hard for $APSPACE$, let M be an n^c space bounded ATM and let e be a constant depending only on M such that the number of distinct configurations of M on inputs of length n is bounded by e^{n^c}. Then the map

$$x \;\; \mapsto \;\; y\$M\$x\$^{|x|^c},$$

where $\#(y) = e^{|x|^c}$, constitutes a $\leq_m^{\log}$ reduction from $L(M)$ to G_ω.

Homework 5 Solutions

1. Assume that $\Pi_k^p \subseteq \Sigma_k^p$. By Theorem 10.2, any set $A \in \Sigma_{k+1}^p$ can be written

$$A = \{x \mid \exists y \, |y| \le |x|^c \wedge R(x, y)\}$$

for some constant c, where R is a Π_k^p-predicate. By the assumption, R is also a Σ_k^p-predicate, hence so is $\exists y \, |y| \le |x|^c \wedge R(x, y)$, and $A \in \Sigma_k^p$. Thus $\Sigma_{k+1}^p \subseteq \Sigma_k^p$. The collapse follows by induction on k.

2. (a) We show that under the given assumptions, SAT $\in P$. On input x of length n, generate the circuit B_n and evaluate $B_n(x)$. Generating the circuit can be done in P by the assumption of polynomial-time uniformity, and evaluation is in P because it is just an instance of CVP, the circuit value problem, which by Theorem 6.1 is P-complete.

   (b) Using Exercise 1, it suffices to show that $\Pi_3^p \subseteq \Sigma_3^p$. Let $A \in \Pi_3^p$, and let M be an n^k-time bounded Π_2^p oracle machine such that $L(M^{\text{SAT}}) = A$. We construct a Σ_3^p machine accepting A as follows. Let $m = n^k$. On input x of length n, we can construct B_m by a Σ_3^p computation: guess the circuit, and then verify that the circuit is correct by verifying that for all encodings y of Boolean formulas such that $|y| = m$,

$$B_m(y) = 1 \quad \Leftrightarrow \quad y \in \text{SAT}. \tag{6}$$

   The circuit is guessed using $\vee$-branching, then the set of all formulas y of length m are generated using $\wedge$-branching, and finally the condition (6) can be checked in Δ_2^p (recall $\Delta_2^p = P^{NP} \subseteq \Sigma_2^p \cap \Pi_2^p$). Thus the whole computation is in Σ_3^p.

   Now to verify $x \in A$ in Σ_3^p, we first guess the circuit B_m (any correct circuit will do) as described above using $\vee$-branching, then check that both

   (i) B_m is correct, and

   (ii) M accepts x, using B_m to answer the oracle queries.

   These two facts can be checked simultaneously in Π_2^p.

3. First we reduce the MAZE problem to the membership problem for nondeterministic one-counter automata. This shows that the membership problem is hard for $NLOGSPACE$. Given an instance of the MAZE

problem consisting of a graph with two distinguished vertices s and t, transform it into an instance in which the names of the vertices are integers written in unary notation (that is, n is represented as 0^n). This can be done in logspace. Given such an encoding, a nondeterministic one-counter automaton can guess a path from s to t, using its counter to hold the name of the vertex it is current visiting.

Now we show that the membership problem is in $NLOGSPACE$. This is the harder part of the problem; it is made hard by the fact that the counter is unbounded. However, we show that if there is an accepting computation path, then there is one of length at most $O(n^3)$, where n is the size of the input. This will allow us to simulate the one-counter automaton with a logspace TM, maintaining the counter in binary on the worktape and counting the number of simulated steps, and halting if the simulation has not finished within the allotted time.

Let M be a nondeterministic one-counter automaton. Assume without loss of generality that if M wants to accept, it empties its counter before doing so. Let q be the number of states of the finite control of M. There are $m = q(n+2)$ possible configurations of state and input head position on inputs of length n.

Suppose there is an accepting computation path α of M on input x, $|x| = n$. Let $c(t)$ be the value of the counter at time t and let $q(t)$ be the pair (state, input head position) at time t on α. A *matched interval* is a pair of times (s, t) such that $s < t$, $c(s) = c(t)$, and $c(u) > c(t)$ for all u in the range $s < u < t$. Let N be the maximum value ever contained in the counter, and let t_N be a time such that $c(t_N) = N$. For $1 \leq i < N$, let s_i be the latest time before t_N that $c(s_i) = i$, and let t_i be the earliest time after t_N that $c(t_i) = i$. Then

$$s_1 < s_2 < \cdots < s_{N-1} < t_N < t_{N-1} < \cdots < t_2 < t_1$$

and (s_i, t_i) is a matched interval, $1 \leq i < N$. If $N > m^2 + 1$, there must exist $1 \leq i < j < N$ such that $q(s_i) = q(s_j)$ and $q(t_i) = q(t_j)$. Then there is a shorter accepting computation obtained by deleting the portion of α between s_i and s_j and between t_j and t_i.

We have shown that a minimal accepting computation on x has no counter value in excess of $m^2 + 1$. There are at most $m(m^2 + 2)$ configurations of state, input head position, and counter value for computations that do not exceed this bound. If the length of such a computation exceeds $m(m^2 + 2)$, then there must be a repeated configuration, and a portion of the computation can be deleted to give a shorter accepting computation. Thus a computation of minimal length is of length at most $m(m^2 + 2) = O(n^3)$.

Homework 6 Solutions

1. (a) If $S(n)$ is space-constructible, there is a machine M that on any input of length n lays off exactly $S(n)$ space on its worktape without using more than $S(n)$ space and halts. If M has q states and d worktape symbols, and if $S(n) \leq o(\log n)$, then the number of configurations of state, worktape contents, and worktape head positions is at most $qS(n)d^{S(n)}$, which is less than $n/2$ for sufficiently large n. If M ever scans all the way to the midpoint of a very long input string 0^n, it must be in a loop by the time it hits the midpoint. That is, after the last time it sees the left endmarker and before it gets to the midpoint, it must be in the same configuration c while scanning two different input tape locations i and j, $i < j < n/2$, without seeing the left endmarker in between. Because it sees nothing but 0's on the input tape after i, it will go through the same sequence of configurations starting from j as it did from i, continuing all the way across until it sees the right endmarker, which may cause it to change behavior. If we insert a string of 0's of length a multiple of $p_1 = j - i \leq n/2$, the machine will not be able to tell the difference; it will hit the right endmarker in the same configuration. The same is true if it scans back again from right to left; by the time it hits the midpoint, it is in a loop of period $p_2 \leq n/2$, and so on. We can thus insert a string of 0's of length $m!$ for any $m \geq n$, which is a multiple of all the possible periods p_i of these loops, and the machine will not behave any differently; in particular, it will not lay off any more or less space on its worktape. This says that $S(n + m!) = S(n)$ for all $m \geq n$.

   (b) $\liminf \lceil \log \log n \rceil = \infty$.

2. To show that the problem is in *PSPACE*, we guess a string that is *not* accepted by M and verify that it is not accepted. We start with a pebble on the start state of M, then guess an input string symbol by symbol, moving pebbles on the states of M to mark all states reachable from the start state under the string guessed so far. We do not have to remember the guessed string, just the states that M could currently be in. We accept if we ever get to a situation with no pebble on an accept state. Because the pebble configuration can be represented in polynomial space, this is a nondeterministic *PSPACE* computation. It can be made deterministic by Savitch's theorem.

   (*Note.* It is not true that if an NFA accepts all strings of length less than or equal to the number of states, then it accepts all strings, even over a single letter alphabet. For example, consider an automaton with a start state going to n disjoint loops with pairwise relatively prime lengths $p_1, p_2, \ldots, p_n$. Make all states accept states except for the one in each loop farthest from the start state. Then there are $1 + p_1 + p_2 + \cdots + p_n$ states, but the shortest string not accepted is of length $p_1 p_2 \cdots p_n$.)

To show that the problem is hard for *PSPACE*, let N be an arbitrary deterministic n^k-space bounded TM. Assume without loss of generality that N has a unique accept configuration. Given an input x, we build a nondeterministic finite automaton M with $O(n^k)$ states accepting all strings that are *not* accepting computation histories of N on input x. Then $L(F) = \Sigma^*$ iff N does not accept x. (Thus we are really reducing $L(N)$ to $\{M \mid L(M) \neq \Sigma^*\}$.)

An accepting computation history of N on input x of length n is a string of the form

$$\#\alpha_0 \#\alpha_1 \#\alpha_2 \# \quad \cdots \quad \#\alpha_{m-1}\#\alpha_m\#, \tag{7}$$

where each α_i is a string of length n^k over some finite alphabet Δ encoding a configuration of N on input x, such that

(i) α_0 is the start configuration of N on x,

(ii) α_m is the accept configuration of N on x, and

(iii) each α_{i+1} follows from α_i according to the transition rules of N.

If a string is not an accepting computation history, then either it is not of the form (7), or one of the three conditions (i), (ii), (iii) is violated. The NFA M guesses nondeterministically which of these to check. Checking that the input string is not of the form (7) requires checking that the input string is not in the regular set $(\#\Delta^{n^k})^*\#$, plus some other simple format checks (exactly one state of N per configuration, each configuration begins and ends with endmarkers, etc.). This requires $O(n^k)$ states of M. Checking (i) or (ii) involves just checking whether the input begins or ends with a certain fixed string of length n^k. These strings are encoded in the finite control of M. Finally, to check (iii), recall from the proof of the Cook–Levin theorem that there is a finite set of local conditions involving the $j - 1$st, jth, and $j + 1$st symbols α_i and the jth symbol of α_{i+1} such that (iii) holds iff these local conditions are satisfied for all j, $1 \leq j \leq n^k$. The local conditions depend only on the description of N. To check that (iii) is violated, M scans across the input, and at some point guesses nondeterministically where the violation occurs. It remembers the next three symbols in its finite control, skips over the next n^k symbols, and accepts if the next symbol is not correct.

3. The problem is complete for $NLOGSPACE$. To show that the problem is in $NLOGSPACE$, by exchanging accept and reject states it suffices to ask whether the set accepted is nonempty. This is true iff there exists a path from the start state to some final state. If there are k final states, this is essentially k instances of MAZE.

To show that the problem is hard for $NLOGSPACE$, we reduce MAZE to it. Given an instance $G = (V, E, s, t)$ of MAZE, we can assume without loss of generality that every vertex v has at least one outgoing edge. If not, add an edge from v to s; this does not affect the reachability of t from s. Let m be the maximum outdegree of any vertex and let $\Sigma = \{0, 1, \ldots, m-1\}$. Build a DFA M with states V, input alphabet Σ, start state s, unique final state t, and transitions obtained by labeling the edges with elements of Σ in such a way that for each $a \in \Sigma$ and $v \in V$, there is exactly one edge out of v with label a. (Edges may have more than one label.) Then M is deterministic, and $L(M)$ is nonempty iff there is a path in G from s to t.

Homework 7 Solutions

1. The problem is *PSPACE*-complete. Because *every* nontrivial first-order theory is *PSPACE*-hard (Miscellaneous Exercise 49), the interesting part is showing that it is in *PSPACE*.

   For k-tuples $a_1, \ldots, a_k$ and $b_1, \ldots, b_k$ of natural numbers, let $a_0 = b_0 = 0$ and define

   $$a_1, \ldots, a_k \quad \equiv_k^m \quad b_1, \ldots, b_k$$

   if there is a permutation $\pi : \{0, 1, \ldots, k\} \to \{0, 1, \ldots, k\}$ such that

   $$a_{\pi(0)} \leq a_{\pi(1)} \leq \cdots \leq a_{\pi(k)}$$
   $$b_{\pi(0)} \leq b_{\pi(1)} \leq \cdots \leq b_{\pi(k)}$$

   (that is, if the a's and b's occur in the same order) and for all $0 \leq i \leq k - 1$,

   $$\min\{2^m,\ a_{\pi(i+1)} - a_{\pi(i)}\} \quad = \quad \min\{2^m,\ b_{\pi(i+1)} - b_{\pi(i)}\}.$$

   In other words, for any adjacent pair of a's and the corresponding adjacent pair of b's, either their respective distances are less than 2^m and are equal, or both are at least 2^m.

   Lemma *If*

   $$a_1, \ldots, a_k \quad \equiv_k^m \quad b_1, \ldots, b_k$$

   then for all a_{k+1} there exists b_{k+1} such that

   $$a_1, \ldots, a_k, a_{k+1} \quad \equiv_{k+1}^{m-1} \quad b_1, \ldots, b_k, b_{k+1}.$$

   Proof. Let π be the permutation giving the order of the a's and the b's. Let a_{k+1} be arbitrary, and suppose i is the largest number such that $a_{\pi(i)} \leq a_{k+1}$. Thus either $i < k$ and a_{k+1} lies between $a_{\pi(i)}$ and $a_{\pi(i+1)}$, or $i = k$ and a_{k+1} is the maximum of $a_0, \ldots, a_{k+1}$. Define

   $$b_{k+1} = \begin{cases} b_{\pi(i+1)} - a_{\pi(i+1)} + a_{k+1}, & \text{if } i < k \text{ and } a_{k+1} \text{ is closer} \\ & \quad \text{to } a_{\pi(i+1)} \text{ than to } a_{\pi(i)}, \\ b_{\pi(i)} + a_{k+1} - a_{\pi(i)}, & \text{otherwise.} \end{cases}$$

Define a new permutation $\rho : \{0, 1, \ldots, k+1\} \rightarrow \{0, 1, \ldots, k+1\}$ by

$$
\rho(j) = \begin{cases}
\pi(j), & j < i, \\
k+1, & j = i, \\
\pi(j-1), & j > i.
\end{cases}
$$

Then

$$
a_{\rho(0)} \leq a_{\rho(1)} \leq \cdots \leq a_{\rho(k+1)}
$$
$$
b_{\rho(0)} \leq b_{\rho(1)} \leq \cdots \leq b_{\rho(k+1)}
$$

and for all $0 \leq i \leq k$,

$$
\min\{2^{m-1}, \ a_{\rho(i+1)} - a_{\rho(i)}\} = \min\{2^{m-1}, \ b_{\rho(i+1)} - b_{\rho(i)}\},
$$

therefore

$$
a_1, \ldots, a_k, a_{k+1} \equiv^{m-1}_{k+1} b_1, \ldots, b_k, b_{k+1}.
$$

$\square$

Now $a_1, \ldots, a_k \equiv^0_k b_1, \ldots, b_k$ implies that $a_i \leq a_j$ iff $b_i \leq b_j$ for all $0 \leq i < j \leq k$. This says that $a_1, \ldots, a_k$ and $b_1, \ldots, b_k$ agree on all atomic formulas and hence on all quantifier-free formulas. Using this as basis, an inductive argument using the lemma shows that if $a_1, \ldots, a_k \equiv^m_k b_1, \ldots, b_k$ then

$$
Q_{k+1} x_{k+1} \ldots Q_{k+m} x_{k+m} \ \varphi(a_1, \ldots, a_k, x_{k+1}, \ldots, x_{k+m})
$$
$$
\text{iff}
$$
$$
Q_{k+1} x_{k+1} \ldots Q_{k+m} x_{k+m} \ \varphi(b_1, \ldots, b_k, x_{k+1}, \ldots, x_{k+m}).
$$

This argument is similar to the one given in Lecture 21 for the theory of dense linear order.

The $\equiv^m_k$-equivalence class of a k-tuple $a_1, \ldots, a_k$ can be represented by a permutation giving the order of the a's and the distance between each adjacent pair of a's up to a maximum of 2^m. This information can be represented in polynomial space. Given such a representation, representations of all possible $\equiv^{m-1}_{k+1}$ equivalence classes obtained by adding a new a_{k+1} can be generated by a branching computation in polynomial time. This gives rise to an alternating polynomial-time algorithm to eliminate quantifiers similar to the one given in Lecture 21 for the theory of dense linear order.

2. These games are examples of so-called *Ehrenfeucht–Fraïssé games*.

   (a) Round 1: Sonja plays $0 \in \mathcal{B}$; David plays some $p \in \mathcal{A}$.

   Round 2: Sonja plays $1 \in \mathcal{B}$; David plays some $q \in \mathcal{A}$, $q > p$ (if David plays some $q \leq p$, then it is an immediate loss).

   Round 3: Sonja plays $(p + q)/2 \in \mathcal{A}$. David cannot play between 0 and 1 in $\mathcal{B}$, so Sonja wins.

   Note that Sonja wins by taking advantage of the fact that one order is dense and the other is not.

   (b) Both structures are dense linear orders without endpoints, so no matter where Sonja plays, it is always possible for David to play on the other structure so as to preserve the order of the pebbles.

   (c) For simplicity, we transform formulas so that negations are applied to atomic formulas only. Every first-order formula can be transformed to an equivalent formula in this form using the following rules.

   $$\neg(\varphi \vee \psi) \quad \Rightarrow \quad (\neg\varphi) \wedge (\neg\psi)$$
   $$\neg(\varphi \wedge \psi) \quad \Rightarrow \quad (\neg\varphi) \vee (\neg\psi)$$
   $$\neg(\exists x \ \varphi) \quad \Rightarrow \quad \forall x \ \neg\varphi$$
   $$\neg(\forall x \ \varphi) \quad \Rightarrow \quad \exists x \ \neg\varphi$$
   $$\neg\neg\varphi \quad \Rightarrow \quad \varphi.$$

   Let φ' be the result of performing this transformation to $\neg\varphi$.

   In one direction, assume that

   $$\mathcal{A} \models \varphi \quad \text{and} \quad \mathcal{B} \models \varphi',$$

   where φ is a sentence of quantifier depth at most n. We want to give a winning strategy for Sonja. We show that Sonja can play so as to maintain the invariant that after k rounds, there is a formula $\psi(\overline{x})$ of quantifier depth at most $n - k$ and free variables $\overline{x} = x_1, \ldots, x_k$ such that

   $$\mathcal{A} \models \psi(\overline{a}) \quad \text{and} \quad \mathcal{B} \models \psi'(\overline{b}),$$

   where $\overline{a} = a_1, \ldots, a_k$ and $\overline{b} = b_1, \ldots, b_k$ are the pebbles played so far on $\mathcal{A}$ and $\mathcal{B}$, respectively. This is true by assumption for $k = 0$. Now suppose it is true for k.

   (i) If

   $$\psi(\overline{x}) \quad = \quad \psi_1(\overline{x}) \vee \psi_2(\overline{x}),$$

   then

   $$\psi'(\overline{x}) \quad = \quad \psi_1'(\overline{x}) \wedge \psi_2'(\overline{x}),$$

hence either

$$\mathcal{A} \models \psi_1(\overline{a}) \quad \text{and} \quad \mathcal{B} \models \psi_1'(\overline{b})$$

or

$$\mathcal{A} \models \psi_2(\overline{a}) \quad \text{and} \quad \mathcal{B} \models \psi_2'(\overline{b}),$$

say the former without loss of generality. Continue the argument with the smaller formula ψ_1 in place of ψ.

(ii) If

$$\psi(\overline{x}) \quad = \quad \psi_1(\overline{x}) \wedge \psi_2(\overline{x}),$$

the argument is similar to case (i).

(iii) If $\psi(\overline{x})$ is an atomic formula $x_i \leq x_j$, then

$$a_i \leq a_j \quad \text{and} \quad b_i \not\leq b_j,$$

which is a win for Sonja.

(iv) If

$$\psi(\overline{x}) \quad = \quad \neg\rho(\overline{x}),$$

then $\rho(\overline{x})$ is an atomic formula of the form $x_i \leq x_j$, thus

$$a_i \not\leq a_j \quad \text{and} \quad b_i \leq b_j,$$

which is again a win for Sonja.

(v) If

$$\psi(\overline{x}) \quad = \quad \exists x_{k+1} \; \rho(\overline{x}, x_{k+1}),$$

then

$$\psi'(\overline{x}) \quad = \quad \forall x_{k+1} \; \rho'(\overline{x}, x_{k+1})$$

and

$$\mathcal{A} \quad \models \quad \exists x_{k+1} \; \rho(\overline{a}, x_{k+1}).$$

Let Sonja play a pebble on a witness $a_{k+1} \in \mathcal{A}$ for the existential quantifier, so that

$$\mathcal{A} \quad \models \quad \rho(\overline{a}, a_{k+1}).$$

Because

$$\mathcal{B} \quad \models \quad \forall x_{k+1} \; \rho'(\overline{b}, x_{k+1}),$$

no matter where David plays, we have

$$\mathcal{B} \quad \models \quad \rho'(\overline{b}, b_{k+1})$$

and the quantifier depth is one less, so the invariant is maintained.

(vi) If

$$\psi(\overline{x}) \quad = \quad \forall x_{k+1} \; \rho(\overline{x}, x_{k+1}),$$

the argument is similar to (v), except Sonja plays on $\mathcal{B}$ instead of $\mathcal{A}$.

Conversely, assume that $\mathcal{A}$ and $\mathcal{B}$ agree on all sentences of quantifier depth n. Define

$$\mathcal{A}, a_1, \dots, a_n \;\equiv_n^0\; \mathcal{B}, b_1, \dots, b_n$$

if $\mathcal{A}, a_1, \dots, a_k$ and $\mathcal{B}, b_1, \dots, b_k$ agree on all atomic formulas, hence on all quantifier-free formulas; that is, for all quantifier-free formulas $\varphi(x_1, \dots, x_n)$,

$$\mathcal{A} \models \varphi(a_1, \dots, a_k) \;\;\Leftrightarrow\;\; \mathcal{B} \models \varphi(b_1, \dots, b_k).$$

For $m > 0$, define

$$\mathcal{A}, a_1, \dots, a_k \;\equiv_k^m\; \mathcal{B}, b_1, \dots, b_k$$

if for all $a_{k+1} \in \mathcal{A}$ there exists $b_{k+1} \in \mathcal{B}$ such that

$$\mathcal{A}, a_1, \dots, a_k, a_{k+1} \;\equiv_{k+1}^{m-1}\; \mathcal{B}, b_1, \dots, b_k, b_{k+1}$$

and for all $b_{k+1} \in \mathcal{B}$ there exists $a_{k+1} \in \mathcal{A}$ such that

$$\mathcal{A}, a_1, \dots, a_k, a_{k+1} \;\equiv_{k+1}^{m-1}\; \mathcal{B}, b_1, \dots, b_k, b_{k+1}.$$

One can show by an inductive argument that if

$$\mathcal{A} \;\equiv_0^n\; \mathcal{B}$$

then David has a winning strategy in the n-pebble game: just place pebbles so as to maintain the invariant

$$\mathcal{A}, a_1, \dots, a_k \;\equiv_k^{n-k}\; \mathcal{B}, b_1, \dots, b_k$$

after k rounds.

We now show that each equivalence class of $\equiv_k^m$ is definable by a formula with k free variables and quantifier depth m. In other words, for each equivalence class E of $\equiv_k^m$, there is a formula $\varphi_E(x_1, \dots, x_k)$ with free variables $x_1, \dots, x_k$ and quantifier depth m such that

$$\mathcal{A}, a_1, \dots, a_k \in E \quad \text{iff} \quad \mathcal{A} \models \varphi_E(a_1, \dots, a_k).$$

Thus if $\mathcal{A}$ and $\mathcal{B}$ agree on all sentences quantifier depth n, then they agree on the sentences defining the equivalence classes of $\equiv_0^n$, so they must be $\equiv_0^n$-equivalent. Therefore David has a winning strategy.

The formulas φ_E are defined inductively. Each equivalence class of $\equiv_n^0$ is defined by a conjunction of atomic formulas or negations of atomic formulas over the variables $x_1, \dots, x_n$. For $m > 0$, the $\equiv_k^m$-equivalence class of $\mathcal{A}, \overline{a}$, where $\overline{a} = a_1, \dots, a_k$, is defined as follows. Let $\mathcal{E}$ be the set of all $\equiv_{k+1}^{m-1}$-equivalence classes of $\mathcal{A}, \overline{a}, a_{k+1}$ for all possible choices of $a_{k+1} \in \mathcal{A}$. Although there may be infinitely many such a_{k+1}, there are only finitely many

$\equiv_{k+1}^{m-1}$-equivalence classes (this fact follows inductively from this construction as well). By the induction hypothesis, we have formulas $\varphi_E(\overline{x}, x_{k+1})$ of quantifier depth $m-1$ defining E for each $E \in \mathcal{E}$. The formula defining the $\equiv_k^m$-equivalence class of $\mathcal{A}, \overline{a}$ is then

$$\bigwedge_{E \in \mathcal{E}} \exists x_{k+1}\, \varphi_E(\overline{x}, x_{k+1}) \quad \wedge \quad \forall x_{k+1} \bigvee_{E \in \mathcal{E}} \varphi_E(\overline{x}, x_{k+1})$$

which is of quantifier depth m. This formula describes the set of possible $\equiv_{k+1}^{m-1}$ equivalence classes obtainable by eliminating a quantifier.

Homework 8 Solutions

1. The following nondeterministic Büchi automaton accepts all and only (characteristic functions of) finite subsets of ω. It guesses when it has seen the last 1 in the input string, and then enters a final state, from which it must see only 0 thereafter; otherwise it goes to a dead state.

   There is no deterministic Büchi automaton accepting this set. We prove this by contradiction. Assume that there were such an automaton M with n states. Consider the action of M on the infinite string $(10^{n+1})^\omega$. When scanning the kth substring of $n+1$ consecutive 0's, M must repeat a state, and one of the states in the loop between the two occurrences of the repeated state must be an accept state, because M accepts the string $(10^{n+1})^k 0^\omega$. Therefore M is in some accept state infinitely often on input $(10^{n+1})^\omega$, so it erroneously accepts.

2. (a) If addition were definable in S1S, there would be a nondeterminstic Büchi automaton accepting the set of strings over the alphabet $\{0,1\}^3$ representing a, b, c such that $a+b=c$. For example, $7+4 = 11$ would be represented by the string

   $$00000001000000$$
   $$00001000000000 \cdots .$$
   $$00000000000100$$

   We now use a pumping argument to get a contradiction. Suppose M has n states. Consider the input string corresponding to the addition problem $(n+1) + (n+1) = 2n + 2$, which M accepts. The machine must repeat a state q while scanning the substring $(0,0,0)^n$ in the input string between positions $n+1$ and $2n+2$. The nonnull substring between the two occurrences of q can be deleted and M erroneously accepts.

   (b) We showed in Lecture 25 how to say $y \leq x$ and A is finite. To add the bit vectors represented by A and B, we simulate binary addition. The low-order bit is leftmost. The carry is given by another finite set U. For example,

   $$U = 000000001110100001010000000000000\cdots$$
   $$A = 001010011101010010100000000000000\cdots$$
   $$B = 010100010101000110100000000000000\cdots$$
   $$C = 011110000110110101010000000000000\cdots .$$

   To assert that U is the carry string, we assert that the low-order bit of U is 0, and for all i, the $i + 1$st bit of U is 1 if at least two of the ith bits of A, B, and U are 1:

$$
\begin{aligned}
\kappa(A,B,U) \;=\; & 0 \notin U \;\wedge\; \forall x \; \mathsf{s}x \in U \;\leftrightarrow\; ((x \in A \wedge x \in B)\;\vee \\
& (x \in A \wedge x \in U)\;\vee \\
& (x \in B \wedge x \in U)).
\end{aligned}
$$

The sum C is given by the exclusive-or (mod 2 sum) of the bit strings A, B, and the carry:

$$
\begin{aligned}
\varphi(A,B,C) \;=\; & \exists U \; \kappa(A,B,U) \;\wedge \\
& \forall x \;(x \in C \;\leftrightarrow\; x \in A \leftrightarrow x \in B \leftrightarrow x \in U).
\end{aligned}
$$

3. (a) Because $B_1 = 0101010101\ldots$, we can take

$$
\begin{aligned}
\varphi_1(x,y) \;&=\; 1 \\
\psi_1(B) \;&=\; 0 \notin B \wedge \forall x \; x \in B \leftrightarrow \mathsf{s}x \notin B \\
\varphi_2(x,y) \;&=\; \exists B \; \psi_1(B) \wedge (x \in B \leftrightarrow y \in B).
\end{aligned}
$$

Suppose now that we have constructed $\varphi_n(x,y)$ and $\psi_n(B)$. Consider B_n as an infinite binary string partitioned into substrings of n bits as suggested by the definition in the problem description. Call these n-bit substrings n-blocks. The position of the first bit of each n-block is a multiple of n. First we construct some auxiliary formulas.

$$
\begin{aligned}
\rho_n(x,y) \;&=\; \varphi_n(y,0) \wedge y \le x \wedge \forall w \;(\varphi_n(w,0) \wedge w \le x) \\
& \qquad\qquad\qquad\qquad \to w \le y \\
\;&=\; \text{``y is the largest multiple of n less than or equal} \\
& \quad\;\; \text{to x''} \\
\xi_n(x,y) \;&=\; \varphi_n(x,y) \wedge \exists z \; y = \mathsf{s}z \wedge \rho_n(z,x) \\
\;&=\; \text{``x and y are successive multiples of n''} \\
A = \{0\} \;&=\; \forall x \; x \in A \leftrightarrow x = 0 \\
A = \{n\} \;&=\; \forall x \; x \in A \leftrightarrow \xi_n(0,x) \\
\chi_n(A) \;&=\; \forall z \; z \in A \to \rho_n(z,0) \\
\;&=\; A \subseteq \{0,1,\ldots,n-1\}
\end{aligned}
$$

$$\omega_n(A,B) \;=\; \chi_n(A) \wedge \chi_n(B)$$
$$\wedge \, (\varphi(A,\{0\},B) \vee \varphi(A,\{0\},B\cup\{n\}))$$
$$=\; \text{``}A,B \text{ represent binary numbers } 0 \le n(A),$$
$$n(B) \le 2^n - 1 \text{ such that } n(B) = n(A) + $$
$$1 \;(\mathrm{mod}\; 2^n)\text{''} \text{ (here } \varphi(A,B,C) \text{ is the for-}$$
$$\text{mula defined in Exercise 2(b))}$$

$$\sigma_n(A,z,B,w) \;=\; \forall x \, \forall y \, (\rho(x,z) \wedge \rho(y,w) \wedge \varphi_n(x,y))$$
$$\rightarrow \; (x \in A \leftrightarrow y \in B)$$
$$=\; \text{``the } n\text{-blocks of } A \text{ and } B \text{ starting at } y \text{ and}$$
$$z \text{ are the same''}$$

$$\upsilon_n(A,B,y) \;=\; \sigma_n(A,0,B,y) \wedge \chi_n(A)$$
$$=\; \text{``}A \subseteq \{0,1,\dots,n-1\} \text{ and the } n\text{-blocks of}$$
$$A \text{ starting at } 0 \text{ and } B \text{ starting at } y \text{ are the}$$
$$\text{same''}$$

$$\tau_n(x,y,B) \;=\; \forall z \, \forall w \, (\rho_n(x,z) \wedge \rho_n(y,w)) \rightarrow \sigma_n(B,z,B,w)$$
$$=\; \text{``the } n\text{-blocks containing } x \text{ and } y \text{ in } B \text{ are}$$
$$\text{the same.''}$$

Note that these formulas depend on $\varphi_n(B)$ but not on $\psi_n(B)$. We can now define

$$\varphi_{n2^n}(x,y) \;=\; \varphi_n(x,y) \wedge \exists B \; \psi_n(B) \wedge \tau_n(x,y,B)$$
$$=\; \text{``the } n\text{-blocks of } x \text{ and } y \text{ in } B_n \text{ are the same,}$$
$$\text{and } x \text{ and } y \text{ are in the same position in the}$$
$$\text{block; that is, } x \equiv y \;(\mathrm{mod}\; n)\text{''}$$
$$=\; x \equiv y \;(\mathrm{mod}\; n2^n).$$

Once we have $\varphi_{n2^n}(x,y)$, we can construct the auxiliary formulas $\rho_{n2^n}(x,y)$, $\xi_{n2^n}(x,y)$, and so on, as above. Then

$$\psi_{n2^n}(B) \;=\; \forall y \, \forall z \, \forall C \, \forall D \, (\xi_{n2^n}(y,z) \wedge \upsilon_{n2^n}(C,B,y)$$
$$\wedge \, \upsilon_{n2^n}(D,B,z) \;\rightarrow\; \omega_{n2^n}(C,D))$$
$$\wedge \, \forall y \, \rho_{n2^n}(y,0) \rightarrow y \notin B$$
$$=\; \text{``for all pairs of successive multiples of } n2^n, \text{ the}$$
$$n2^n\text{-blocks starting in those two positions repre-}$$
$$\text{sent numbers in the range } \{0,\dots,2^{n2^n}-1\} \text{ that}$$
$$\text{differ by 1 mod } 2^{n2^n}, \text{ and the first block consists}$$
$$\text{only of 0's''}$$
$$=\; B = B_{n2^n}.$$

(b) Using the construction of part (a), one can build formulas of length n describing strings of length at least $2^{2^{2^{\cdot^{\cdot^{\cdot^n}}}}}$. As in the lower bound proof for the theory of real addition given in Lecture 23, these can be used as "yardsticks" to describe the set of accepting computation histories of a Turing machine whose running time is $2^{2^{2^{\cdot^{\cdot^{\cdot^n}}}}}$.

Homework 9 Solutions

1. The problem is *PSPACE*-complete. Here is an alternating polynomial-time algorithm. To test $\mathbb{Z}_n \models \exists x\ \varphi(x)$, we branch existentially, guessing $a \in \mathbb{Z}_n$. Each such a can be represented by a number $\{0, 1, \ldots, n-1\}$ in binary, so the depth of the computation tree is linear in the binary representation of n. Each process at a leaf then tests $\mathbb{Z}_n \models \varphi(a)$ for its guessed value of a. To test $\mathbb{Z}_n \models \forall x\ \varphi(x)$, the procedure is the same, except universal branching is used. The Boolean connectives $\vee$ and $\wedge$ can be handled with binary existential branches and binary universal branches, respectively. We assume negations $\neg$ have already been pushed down to the atomic formulas by the De Morgan laws and the rules $\neg \exists x\ \varphi(x) \mapsto \forall x\ \neg\varphi(x)$ and $\neg \forall x\ \varphi(x) \mapsto \exists x\ \neg\varphi(x)$. We are left with atomic formulas of the form $s = t$ or $s \neq t$, where s and t are ground terms over constants in $\mathbb{Z}_n$ and arithmetic operators $\cdot$ and $+$. These can be checked in polynomial time using ordinary arithmetic modulo n.

   We show that the problem is hard for *PSPACE* by a reduction from QBF. Given a quantified Boolean formula

   $$Q_1 x_1\ Q_2 x_2\ \cdots\ Q_n x_n\ \mathcal{B}(x_1, \ldots, x_n)$$

   of QBF, replace each Boolean variable x_i with the atomic formula $x_i = 0$. This gives a sentence of the language of number theory that is true in $\mathbb{Z}_2$, or in any $\mathbb{Z}_n$ for $n \geq 2$, iff the original quantified Boolean formula was true in the two-element Boolean algebra $\{0, 1\}$.

   A similar reduction works for all nontrivial first-order theories T. All we need is the existence of a relation R such that $T \models \exists \overline{x}\ R(\overline{x})$ and $T \models \exists \overline{x}\ \neg R(\overline{x})$ (this is what is meant by nontrivial). In the application at hand, we can take $R(x)$ to be $x = 0$, which is nontrivial provided the structure has at least two elements.

2. Let

   $$M\ =\ (Q, \{0, 1\}, \delta, s, \mathcal{F})$$

   be the given Muller automaton. Let Y_q be a set variable corresponding to state q and let X be a set variable corresponding to the input. We first write down an S1S formula $\mathsf{run}(X, \overline{Y})$ describing the runs of M on input X. The variable Y_q gives the times at which the machine is in state q.

$$\mathsf{run}(X, \overline{Y}) \quad = \quad 0 \in Y_s \tag{8}$$

$$\wedge \; \forall n \bigwedge_q (n \in Y_q \wedge n \notin X \rightarrow \bigvee_{p \in \delta(q,0)} s(n) \in Y_p) \tag{9}$$

$$\wedge \; \forall n \bigwedge_q (n \in Y_q \wedge n \in X \rightarrow \bigvee_{p \in \delta(q,1)} s(n) \in Y_p) \tag{10}$$

$$\wedge \; \forall n \bigwedge_{p \neq q} \neg(n \in Y_p \wedge n \in Y_q) \tag{11}$$

The subformula (8) says that the machine starts at time 0 in its start state. The subformula (9) says that transitions on input symbol 0 are correct. Similarly, (10) says that transitions on input symbol 1 are correct. Finally, the subformula (11) says that the machine is in at most one state at any time. For any sets A and B_q, $q \in Q$, the S1S formula $\mathsf{run}(A, \overline{B})$ is true if the B_q describe a run of M on input A in the sense that for all n, $n \in B_q$ iff the machine is in state q at time n.

We now describe acceptance. Define

$$\mathsf{finite}(Y) \quad = \quad \exists x \; \forall y \; y \in Y \rightarrow y \leq x.$$

Then $\neg\mathsf{finite}(Y_q)$ says that M is infinitely often in state q. For $F \subseteq Q$, define

$$\mathsf{io}_F(\overline{Y}) \quad = \quad \bigwedge_{q \in F} \neg\mathsf{finite}(Y_q) \wedge \bigwedge_{q \notin F} \mathsf{finite}(Y_q).$$

This says that F is the IO set of the run described by $\overline{Y}$. Finally, define

$$\mathsf{accept}(\overline{Y}) \quad = \quad \bigvee_{F \in \mathcal{F}} \mathsf{io}_F(\overline{Y}).$$

This says that M accepts according to the Muller acceptance condition. Now take

$$\varphi_M(X) \quad = \quad \exists \overline{Y} \; \mathsf{run}(X, \overline{Y}) \wedge \mathsf{accept}(\overline{Y}).$$

3. We first show how to simulate polynomial-size circuits $B_0, B_1, \ldots$ by a polynomial-time oracle machine M^C with a sparse oracle C. We encode the circuits B_n in the oracle C.

Suppose $A \in \{0,1\}^*$ is the set accepted by the circuits. Thus for $x \in \{0,1\}^n$, $x \in A$ iff $B_n(x) = 1$. Because the circuits are of polynomial size, there is a constant d and an encoding of circuits B_n as strings $b_n \in \{0,1\}^{n^d}$ such that a Turing machine, given b_n and $x \in \{0,1\}^n$, can compute $B_n(x)$ in polynomial time.

Now we save the string b_n in the oracle as the characteristic function of strings of length n. That is, for n so large that $n^d \leq 2^n$, we put the ith string of length n in C iff the ith bit of b_n is 1. Thus B_n can be determined by querying C on at most n^d strings of length n. For the finitely many values of n for which $n^d > 2^n$, the circuit B_n is just encoded in the finite control of M.

Because $|b_n| \leq n^d$, the oracle is sparse. The machine M^C on input $x \in \{0,1\}^n$ first queries C on the first n^d strings of length n to determine B_n, then computes $B_n(x)$ and accepts if the value is 1.

For the other direction, suppose we are given a polynomial-time oracle machine M^C with sparse oracle C, $|C \cap \{0,1\}^n| \leq n^d$, accepting a set A. We wish to construct (nonuniform) polynomial-size circuits $B_0, B_1, \ldots$ equivalent to M^C. We must somehow encode the oracle information in the circuits. We do this in two steps. First we show that for each n there exists a string y_n of length polynomial in n encoding all the oracle information needed by M on inputs of length n. The string y_n is essentially a list of all elements in C up to the maximum length that could be queried by M on inputs of length n. Because these strings are of polynomial length in n and C is sparse, the string y_n is of polynomial length. More accurately, let n^k be the time bound of M. On inputs of length n, M can only query the oracle on strings of length at most n^k, because it must write them down. As C is n^d-sparse, there are at most

$$|C \cap \{0,1\}^{\leq n^k}| \;=\; \sum_{m=0}^{n^k} |C \cap \{0,1\}^m| \;\leq\; \sum_{m=0}^{n^k} m^d \;\leq\; n^{k(d+1)}$$

nonzero elements of C in the range that could possibly be queried by M on an input of length n, and all of them are of length at most n^k, so they can be written down end-to-end separated by 2's in a string $z_n \in \{0,1,2\}^*$ of length at most $O(n^{k(d+2)})$. Now convert the ternary string z_n to binary to get y_n. Then

$$|y_n| \;\leq\; \log_2 3 \cdot |z_n| \;=\; O(n^{k(d+2)}).$$

The binary string y_n contains all the oracle information necessary to process input strings of length n. That is, there is a deterministic polynomial-time TM N such that for any string x of length n, M^C accepts x iff N accepts $x \# y_n$. The machine N first converts y_n to z_n, then simulates M on x; whenever M would consult its oracle C, N searches the list z_n.

Let $C_0, C_1, \ldots$ be the circuits obtained from Ladner's construction for the machine N (Theorem 6.1). Then for any n, $C_{n+|y_n|}$ has $n + |y_n|$ Boolean inputs and is of size polynomial in $n + |y_n|$, which is polynomial in n, and for any x of length n, $C_{n+|y_n|}(x, y_n) = 1$ iff N accepts $x \# y_n$ iff M^C accepts x. The circuit B_n is obtained by specializing the inputs of $C_{n+|y_n|}$ corresponding to y_n to the Boolean values in y_n.

Homework 10 Solutions

1. To construct const, let k be an index for the projection π_1^2 and let ℓ be an index for s_1^1. Then

$$
\begin{aligned}
i &= \pi_1^2(i, x) \\
&= \varphi_k(i, x) \\
&= \varphi_{s_1^1(k, i)}(x) \\
&= \varphi_{\varphi_\ell(k, i)}(x) \\
&= \varphi_{\varphi_{s_1^1(\ell, k)}(i)}(x),
\end{aligned}
$$

so we can take $\mathsf{const} = \varphi_{s_1^1(\ell, k)}$.

To construct pair, let n be an index for s_2^1 and let m be an index for the partial recursive function

$$
<U \circ <\pi_1^3, \pi_3^3>, U \circ <\pi_2^3, \pi_3^3>>.
$$

Then

$$
\begin{aligned}
<\varphi_i, \varphi_j>(x) &= <\varphi_i(x), \varphi_j(x)> \\
&= <U(i, x), U(j, x)> \\
&= <U \circ <\pi_1^3, \pi_3^3>, U \circ <\pi_2^3, \pi_3^3>>(i, j, x) \\
&= \varphi_m(i, j, x) \\
&= \varphi_{s_2^1(m, i, j)}(x) \\
&= \varphi_{\varphi_n(m, i, j)}(x) \\
&= \varphi_{\varphi_{s_1^2(n, m)}(i, j)}(x),
\end{aligned}
$$

so we can take $\mathsf{pair} = \varphi_{s_1^2(n, m)}$.

2. Let h be a total recursive function that on input $<v, j>$ produces the index of a function that on input x

   (i)   computes $\varphi_v(v, j)$;

   (ii) if $\varphi_v(v, j)\downarrow$, applies φ_j to $\varphi_v(v, j)$; and

   (iii) if $\varphi_j(\varphi_v(v, j))\downarrow$, interprets the result as an index and applies the function with that index to x.

Thus

$$\varphi_{h(v,j)}(x) \quad = \quad \varphi_{\varphi_j(\varphi_v(v,j))}(x)$$

if $\varphi_v(v,j)$ and $\varphi_j(\varphi_v(v,j))$ are defined, undefined otherwise. Note that h itself is a total recursive function: it does not do any of the steps (i)–(iii) above, it only computes the index of a function that does them.

Now let u be an index for h. If φ_j is total, then $h(u,j)$ is a fixpoint of φ_j:

$$\varphi_{h(u,j)} \quad = \quad \varphi_{\varphi_j(\varphi_u(u,j))} \quad = \quad \varphi_{\varphi_j(h(u,j))}.$$

Thus we can define $\tau = \lambda j.h(u,j)$.

More formally, letting ℓ and m be indices for the functions

$$U \circ <U \circ <\pi_2^3, U \circ <\pi_1^3, \pi_1^3, \pi_2^3>>, \pi_3^3>$$

and s_2^1, respectively, we have

$$
\begin{aligned}
\varphi_{\varphi_j(\varphi_v(v,j))}(x) \quad &= \quad U(U(j, U(v,v,j)), x) \\
&= \quad U \circ <U \circ <\pi_2^3, U \circ <\pi_1^3, \pi_1^3, \pi_2^3>>, \pi_3^3>(v,j,x) \\
&= \quad \varphi_\ell(v,j,x) \\
&= \quad \varphi_{s_2^1(\ell,v,j)}(x),
\end{aligned}
$$

so we can take $h = \lambda<v,j>.s_2^1(\ell,v,j)$. The function h is total because s_2^1 is. An index for h is then given by

$$u \quad \overset{\text{def}}{=} \quad s_1^2(m,\ell),$$

and we can take

$$\tau \quad = \quad \lambda j.h(u,j) \quad = \quad \lambda j.\varphi_u(u,j) \quad = \quad \varphi_{s_1^1(u,u)}.$$

3. Suppose we have already constructed a finite set A of fixpoints of f. We show how to obtain another fixpoint not in A effectively. Modify f to get f' such that

$$
f'(i) \quad = \quad
\begin{cases}
\mathsf{const}(0), & \text{if } i \in A \\
f(i), & \text{otherwise,}
\end{cases}
$$

where $\mathsf{const}(0)$ is an index of the constant function $\lambda x.0$. Using the recursion theorem, find a fixpoint j of f'. If $j \notin A$, we are done. Otherwise, we know that $\varphi_j = \lambda x.0$. In this case redefine $f'(j) := \mathsf{const}(1)$. Now we are guaranteed that j cannot be a fixpoint of f'. Repeat the process with the new f'. Whenever we get a fixpoint k in A, redefine $f'(k) := \mathsf{const}(1)$ and repeat. This can happen at most $|A|$ times before no element of A can be a fixpoint of f'. The next application of the recursion theorem gives a fixpoint of f' outside A, which is also a fixpoint of f, because f and f' agree outside A.

Homework 11 Solutions

1. It was argued in Lecture 37 that $K^\varnothing = K$ is Σ_1^0-complete. Proceeding by induction, it suffices to show that K^A is Σ_{n+1}^0-complete whenever A is Σ_n^0-complete. Writing

$$K^A = \{x \mid \varphi_x^A(x)\downarrow\} = \{x \mid \exists t \; \varphi_x^A(x)\downarrow^t\},$$

because the predicate $\varphi_x^A(x)\downarrow^t$ is recursive in $A \in \Sigma_n^0$, K^A is r.e. in A, thus $K^A \in \Sigma_{n+1}^0$ by the definition given in Lecture 35.

Now we show that K^A is Σ_{n+1}^0-hard whenever A is Σ_n^0-hard. Let B be an arbitrary element of Σ_{n+1}^0. By Theorem 35.1, B can be expressed as

$$B = \{x \mid \exists y \; x\#y \in C\},$$

where $C \in \Pi_n^0$. Because A is Σ_n^0-hard, its complement $\sim A$ is Π_n^0-hard, therefore there exists a total recursive map σ such that

$$\sigma(x\#y) \in {\sim}A \quad \Leftrightarrow \quad x\#y \in C.$$

Now define the total recursive map τ that on input x gives the index of a machine with oracle A that on any input

- enumerates $y = 0, 1, 2, \ldots$ in order,
- calculates $\sigma(x\#y)$ for each one,
- consults its oracle to determine whether $\sigma(x\#y) \notin A$, and
- halts if it ever finds one.

Then

$$
\begin{aligned}
x \in B \quad &\Leftrightarrow \quad \exists y \; x\#y \in C \\
&\Leftrightarrow \quad \exists y \; \sigma(x\#y) \notin A \\
&\Leftrightarrow \quad \varphi_{\tau(x)}^A(\tau(x))\downarrow \\
&\Leftrightarrow \quad \tau(x) \in K^A,
\end{aligned}
$$

thus τ constitutes a reduction from B to K^A. Because B was arbitrary, K^A is Σ_{n+1}^0-hard.

2. (a) Suppose for a contradiction that $K^A \in \Sigma_n^0$. Because A is $\leq_m$-complete for Σ_n^0, there is a total recursive map σ such that for all x,

$$x \in K^A \quad \Leftrightarrow \quad \sigma(x) \in A.$$

Diagonalizing, let m be the index of an oracle machine with oracle A that halts on input y iff $\sigma(y) \notin A$. Then

$$\sigma(m) \in A \quad \Leftrightarrow \quad m \in K^A$$
$$\Leftrightarrow \quad \varphi_m^A(m)\!\downarrow$$
$$\Leftrightarrow \quad \sigma(m) \notin A,$$

a contradiction. Therefore $K^A \notin \Sigma_n^0$.

(b) Let $n \geq 1$. It suffices to show $\Pi_n^0 \not\subseteq \Sigma_n^0$. By Theorem 35.1, any Σ_{n+1}^0 set B can be expressed as

$$B \;=\; \{x \mid \exists y\; x\#y \in A\},$$

where $A \in \Pi_n^0$. If $\Pi_n^0 \subseteq \Sigma_n^0$, then $A \in \Sigma_n^0$, therefore $B \in \Sigma_n^0$ as well by combining the initial existential quantifiers in the representation of Theorem 35.1. Because B was arbitrary, $\Sigma_{n+1}^0 \subseteq \Sigma_n^0$. This contradicts the conclusion of part (a).

3. The true ones are (a) and (c). The proof of the recursion theorem (Theorem 33.1) goes through verbatim with the decoration A on φ and with or without the decoration A on σ.

To refute (b), we can construct a total σ with an oracle for the halting problem that has no fixpoint. Let $K = \{x \mid \varphi_x(x)\!\downarrow\}$, and let σ^K be the map

$$\sigma^K(x) \;=\; \begin{cases} \mathsf{const}(\varphi_x(x) + 1), & \text{if } \varphi_x(x)\!\downarrow, \\ \mathsf{const}(0), & \text{if } \varphi_x(x)\!\uparrow. \end{cases}$$

The function σ^K is total. On input x, it consults its oracle to determine which of the two cases applies. If $\varphi_x(x)\!\uparrow$, it just outputs $\mathsf{const}(0)$. If $\varphi_x(x)\!\downarrow$, it computes $\varphi_x(x)$ directly (it knows that it must halt), then adds 1 and applies const to that value. Thus for all x,

$$\varphi_{\sigma^K(x)}(x) \;=\; \begin{cases} \varphi_x(x) + 1, & \text{if } \varphi_x(x)\!\downarrow, \\ 0, & \text{if } \varphi_x(x)\!\uparrow \end{cases}$$
$$\neq \;\varphi_x(x),$$

so σ^K has no fixpoint.

4. Say a recursive graph (ω, E) is represented by a Turing machine accepting the set of strings $x\#y$ such that x, y are binary strings and

$(x, y) \in E$. Denote the graph represented by Turing machine M by G_M. We wish to show that the set

$$\mathrm{SC} \;=\; \{M \mid G_M \text{ is strongly connected}\}$$

is Π_2^0-complete. The set SC is in Π_2^0, because it can be defined by a Π_2^0 predicate:

$$\mathrm{SC} \;=\; \{M \mid \forall x \, \forall y \, \exists \sigma \, \exists \tau \; \mathsf{path}(M, x, y, \sigma, \tau)\},$$

where $\mathsf{path}(M, x, y, \sigma, \tau)$ says that σ and τ are natural numbers encoding sequences $x_0, x_1, \ldots, x_n$ and $t_0, t_1, \ldots, t_{n-1}$ respectively, such that $x = x_0$, $x_n = y$, and M accepts $x_i \# x_{i+1}$ within t_i steps, $0 \le i \le n - 1$.

To show that SC is hard for Π_2^0, we reduce an arbitrary Π_2^0 set

$$\{x \mid \forall y \, \exists z \; R(x, y, z)\}$$

to SC. Given x, consider the graph

$$\{(2y, 2z + 1) \mid R(x, y, z)\} \cup \{(2z + 1, w) \mid w, z \in \omega\}.$$

This graph is strongly connected iff $\forall y \, \exists z \, R(x, y, z)$. We can easily build a machine accepting this set of edges from a given x and a machine for the recursive relation R.

Homework 12 Solutions

1. (a) Get rid of each occurrence of $y := \exists$ by replacing it with the following code.

    $$y := 0$$
    ℓ_1: $\ell_2 \lor \ell_3$
    ℓ_2: $y := y + 1$
        `goto` ℓ_1
    ℓ_3: ... .

    This changes each countable existential branch

    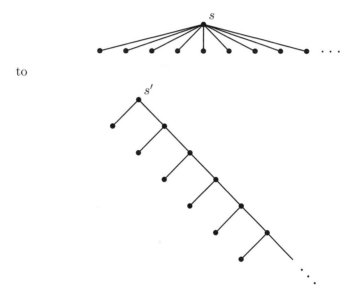

    to

    but node s is labeled 1 iff node s' is labeled 1. This would not work for $\forall$-branching because of the infinite path below s'. Call this new program p.

    Now get rid of the $\ell_i \lor \ell_j$ construct in p the same way you would simulate a nondeterministic Turing machine with a deterministic one, the only difference being that there are some $y := \forall$ steps thrown in from time to time. The simulating program p' keeps a finite list of configurations $(\ell_1, \beta_1), \dots, (\ell_n, \beta_n)$ of p that it is currently simulating, where the ℓ_i are statement labels of p and the $\beta_i : \{y_1, \dots, y_k\} \to \omega$ are valuations of the variables of p. The program p' goes through the list in a round-robin fashion, simulating one step of p from each configuration (ℓ, β) on the list, and updating (ℓ, β) accordingly. The program p' makes a $\forall$-branch whenever p would. Whenever p' is simulating p on (ℓ, β) and ℓ is a statement of the form $\ell : \ell_i \vee \ell_j$, p' just deletes (ℓ, β) from the list and replaces it with (ℓ_i, β) and (ℓ_j, β). The equivalence of these two computations might be called *iterated distributivity*. The program p' has only simple assignments $y := e(\overline{y})$ and universal assignments $y := \forall$. Finally, p' halts and accepts if any one of the configurations it is simulating is an `accept` statement.

(b) To show that the problem of deciding whether a given recursive relation is well-founded is Π_1^1-hard, observe that the computation tree of p' on any input is a recursive tree, and is well-founded iff p' accepts. But p' accepts iff the original program p accepts, and acceptance of IND programs is Π_1^1-hard, because by Kleene's theorem (Theorem 40.1) IND programs accept exactly the Π_1^1 relations over $\mathbb{N}$.

The problem is also in Π_1^1, because it is accepted by an IND program, as shown in Lecture 39.

2. Our plan is to use the lazy conditional test

$$\varphi_{\mathsf{cond}(i,j)}(x, y) \quad = \quad \begin{cases} \varphi_i(y), & \text{if } x = 0, \\ \varphi_j(y), & \text{if } x \neq 0 \end{cases}$$

of Miscellaneous Exercise 111 and the recursion theorem (Theorem 33.1) to construct a function h such that

$$h(x, y) \quad = \quad \begin{cases} y, & \text{if } f(x, y) = 0, \\ h(x, y + 1), & \text{if } f(x, y) \neq 0, \end{cases} \tag{12}$$

then take

$$g(x) \quad \overset{\text{def}}{=} \quad h(x, 0).$$

Let $j, a, b, s,$ and i be indices for f, π_1^2, π_2^2, the successor function, and the identity function, respectively, and let

$$\sigma(x) \quad = \quad \mathsf{pair}(\mathsf{cond}(b, \mathsf{comp}(j, \mathsf{pair}(a, \mathsf{comp}(s, b)))), \mathsf{pair}(x, i)). \tag{13}$$

The function σ is total, therefore has a fixpoint $h = \varphi_x = \varphi_{\sigma(x)}$ by the recursion theorem, thus

$$h \quad = \quad \varphi_x \quad = \quad \varphi_{\mathsf{pair}(\mathsf{cond}(b,\mathsf{comp}(j,\mathsf{pair}(a,\mathsf{comp}(s,b)))),\mathsf{pair}(x,i))}.$$

To make a long story short, unwinding the definitions gives exactly (12).

An index for σ can be obtained effectively from j, because by (13), σ is just a combination of j and some constants using composition and pairing, and an index for h can be obtained effectively from an index for σ by the effective version of the recursion theorem (Homework 10, Exercise 2).

Although you may think otherwise, this exercise was not meant as an endurance test. Its purpose was to gain an appreciation of the programming difficulties that Gödel and his colleagues faced in the 1930s, before the invention of modern programming languages. Cumbersome as these constructs were, one can see in them the seeds of the more versatile programming constructs we use today.

3. By Theorem 2.5,

$$DTIME(T(n)) \quad \subseteq \quad DSPACE(T(n)) \quad \subseteq \quad DTIME(T(n)^{T(n)})$$

for any $T(n) \geq \log n$. By the gap theorem (Theorem 32.1), there exists a T such that

$$DTIME(T(n)) \quad = \quad DSPACE(T(n)) \quad = \quad DTIME(T(n)^{T(n)}).$$

Hints for Selected Miscellaneous Exercises

8. (b) Consider $\{x\#x \mid x \in \Sigma^*\}$.

14. Use Miscellaneous Exercise 11.

15. Finite fingers.

27. Pad the input.

31. Convert to finite automata and use Homework 6, Exercise 2 and Miscellaneous Exercise 15. For information on regular expressions, see [76, Lectures 7–9].

36. Use Miscellaneous Exercise 33.

38. (a) Induction.

40. Reduce to integer sorting.

42. Show that if $n \neq 0$ and q, r are the quotient and remainder, respectively, obtained when dividing m by n using ordinary integer division, that is, if $m = nq + r$ where $0 \leq r < n$, then $\gcd(m, n) = \gcd(n, r)$.

43. Use Miscellaneous Exercise 42, along with the fact that all integer combinations of a and n are multiples of $\gcd(a, n)$.

45. (b) Show using the definition of conditional expectation that if the F_i are disjoint events such that $\Pr(F_i) \neq 0$ and $F = \bigcup_i F_i$, and if X is any random variable, then

$$\mathcal{E}(X \mid F) \quad = \quad \sum_i \mathcal{E}(X \mid F_i) \cdot \Pr(F_i \mid F).$$

46. This hint applies to both (a) and (b). Let s, t be truth assignments to the variables of a given 3CNF formula. Consider the statement, "There exists a truth assignment u such that $s \leq u \leq t$ in lexicographic order and for all truth assignments v, the number of clauses of φ satisfied by v is no more than the number of clauses satisfied by u." Use the fact that if $P = NP$, then $\Sigma_2^{\mathrm{p}} = P$. Do binary search.

47. (b) Use (a) to show that the size of the maximum clique is strongly related to the probability of acceptance.

49. (a) Encode QBF.

50. Take the board positions to be the equivalence classes of the equivalence relations $\equiv_{n,k}^m$ defined inductively as follows.

   - $\mathcal{A}, a_1, \ldots, a_k \equiv_{0,k}^m \mathcal{B}, b_1, \ldots, b_k$ iff $\mathcal{A}, a_1, \ldots, a_k$ and $\mathcal{B}, b_1, \ldots, b_k$ agree on all quantifier-free formulas of length at most m with free variables among $x_1, \ldots, x_k$; that is, if for all such formulas φ,

     $\mathcal{A}, a_1, \ldots, a_k \models \varphi$ iff $\mathcal{B}, b_1, \ldots, b_k \models \varphi$.

   - $\mathcal{A}, a_1, \ldots, a_k \equiv_{n+1,k}^m \mathcal{B}, b_1, \ldots, b_k$ iff both of the following conditions hold.
     (a) For all $a_{k+1} \in \mathcal{A}$, there exists $b_{k+1} \in \mathcal{B}$ such that
        $\mathcal{A}, a_1, \ldots, a_k, a_{k+1} \quad \equiv_{n,k+1}^m \quad \mathcal{B}, b_1, \ldots, b_k, b_{k+1}$.
     (b) For all $b_{k+1} \in \mathcal{B}$, there exists $a_{k+1} \in \mathcal{A}$ such that
        $\mathcal{A}, a_1, \ldots, a_k, a_{k+1} \quad \equiv_{n,k+1}^m \quad \mathcal{B}, b_1, \ldots, b_k, b_{k+1}$.

53. (b)   Show that if there is a string not accepted, then there is one of the form xy^ω for $|x|$ and $|y|$ at most $2^{O(n \log n)}$, where n is the number of states.

60. (b)   Generate a random $n \times n$ matrix A such that for all j there exists $i \geq j$ such that $A_{ij} \neq 0$. There are exactly $\prod_{i=0}^{n-1}(q^n - q^i)$ such matrices, the same as the number of nonsingular matrices. Construct a linearly independent sequence of vectors $x_1, \ldots, x_n$ starting with the standard basis (columns of the identity matrix), using the columns of A as coefficients of linear combinations of previously generated x_i and standard basis elements.

66. (iii)   Use amplification (Lemma 14.1) and the law of sum (Lecture 13).

    (iv)   Use amplification (Lemma 14.1).

    (v)   Use Miscellaneous Exercise 63.

    (vi)   For BP, use amplification (Lemma 14.1).

68. (a)   Use Miscellaneous Exercise 6.

69. (c)   Simulate M with another Σ_k oracle machine N that does not make the oracle queries when M would, but just records the query strings and guesses the response of the oracle, then verifies the guesses at the end of the computation. Use (a) and (b) to show that the guessing can be done either universally or existentially so as not to increase the number of alternations along any computation path.

70.   Induction on the structure of φ. Use the stronger induction hypothesis: for any Boolean formula $\varphi(x_1, \ldots, x_n)$ with Boolean variables $x_1, \ldots, x_n$ and possibly other variables,

$$\varphi(\operatorname{excl}(z, \sigma_1, \tau_1), \ldots, \operatorname{excl}(z, \sigma_n, \tau_n))$$
$$\equiv \operatorname{excl}(z, \varphi(\sigma_1, \ldots, \sigma_n), \varphi(\tau_1, \ldots, \tau_n)).$$

89.   Approximate $\ln(1 - \delta)$ with the first few terms of its Taylor expansion. Recall that

$$\ln(1 + x) = \sum_{n=1}^{\infty} \frac{(-1)^{n+1}}{n} x^n$$

for $|x| < 1$.

93. Use Miscellaneous Exercise 92.

94. Writing $A(m, n)$ as $A_m(n)$, show that $A_{m+1}(n) = A_m^{n+1}(1)$, where f^n denotes the n-fold composition of f with itself:

$$f^0(x) \stackrel{\text{def}}{=} x$$
$$f^{n+1}(x) \stackrel{\text{def}}{=} f(f^n(x)).$$

Use the second characterization of Miscellaneous Exercise 91.

96. Build an instance of PCP such that any nonnull solution is an accepting computation history of a given TM M on a given input x. Let $\Sigma = \{0, 1, \ldots, k - 1\}$ for some k and $\Gamma = \{0, 1, \#\}$. Let $f(0) = \#\alpha_0\#$ and $g(0) = \#$, where α_0 is the start configuration of M on x.

98. Use the recursion theorem.

99. Enumerate some subset in increasing order.

100. (b) Let $M_{\sigma(i)}$ be a machine that on input x simulates M_i on inputs $0, 1, \ldots, x$ and accepts iff M_i halts on all $0, 1, \ldots, x$ and accepts x.

107. Show that $(R \circ S)^{-1} = S^{-1} \circ R^{-1}$ and $(R^{-1})^{-1} = R$. Using these facts, show that $R \circ R^{-1} \circ R \subseteq R$ implies $R^{-1} \circ R \circ R^{-1} \subseteq R^{-1}$. Now use a back-and-forth argument as in Lecture 34 to produce a chain $h_0, h_1, \ldots$ of approximations to $h : \omega \to \omega$, each with finite domain, maintaining the invariant that each h_n is one-to-one on its domain and graph $h_n \subseteq R$.

108. Use effective padding (Lemma 34.2) to assume without loss of generality that σ and τ are one-to-one, then apply Miscellaneous Exercise 107.

109. Apply Miscellaneous Exercise 107.

113. Use Miscellaneous Exercise 112.

114. (b) Consider the sets $\{x \mid \varphi_x(x) \text{ is even}\}$ and $\{x \mid \varphi_x(x) \text{ is odd}\}$.

115. Use the recursion theorem.

116. Use Homework 12, Exercise 2.

118. (a) Show that there exist arbitrarily complex 0,1-valued functions.

121. This has a two-line proof using the recursion theorem.

124. Diagonalize, then use Miscellaneous Exercise 123(b).

125. (b) Let Φ be arbitrary, and define $\Psi_i \overset{\text{def}}{=} \Phi_i + \varphi_i + 1$.

127. (c) An enumeration of the constant functions is given by $\varphi_{\text{const}(k)}$, $k \geq 0$. Let Φ be an arbitrary abstract complexity measure. Define a new complexity measure Ψ by

$$\Psi_i(n) \overset{\text{def}}{=} \begin{cases} 0, & \text{if } \exists k \; i = \text{const}(k) \text{ and } M_k(k)\!\uparrow^n \\ \Phi_i(n) + 1, & \text{otherwise.} \end{cases}$$

132. Encode COF. Use the set A from Miscellaneous Exercise 110(b), except include complements of sets in $\mathcal{L}$ before constructing A. Let $L(M_{\sigma(i)}) = \{f(x) \mid x \in L(M_i)\} \cup A$, where $f(x)$ is the xth element of $\sim A$.

133. Kill three birds with one stone.

135. Enumerate proofs. Build a total recursive function that grows asymptotically faster than any provably total recursive function.

137. Use effective padding.

138. The answers may not be what you think.

139. This is quite tricky. Check your solution to make sure the following is not a counterexample.

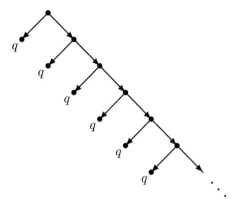

Solutions to Selected Miscellaneous Exercises

6. Let $A \subseteq DSPACE(G(x))$, say by a machine M running in space $G(x)$ on input x. Let

$$A' \stackrel{\text{def}}{=} \{x\#^k \mid M \text{ halts on input } x \text{ in space } k + |x|\}$$

$$A'' \stackrel{\text{def}}{=} \{x\#^k \mid M \text{ halts and accepts } x \text{ in space } k + |x|\}.$$

Then both $A', A'' \in DSPACE(n)$. By assumption, both $A', A'' \in DTIME(T(n))$, say by machines M' and M'', respectively, running in time $T(n)$ on inputs of length n. Now build a machine N that, given x, runs M' on input $x\#^i$ for $i = 0, 1, 2, \ldots$ until it accepts, which it must by the time $i = G(x) - |x|$ at the latest. For the maximum value of i attained, N runs M'' on $x\#^i$ to determine whether $x \in A$, and accepts or rejects accordingly. Thus $L(N) = A$. For each xa^i, the simulation takes time $T(i + |x|)$, which by monotonicity of T is at most $T(G(x))$. Thus the total time is at most $(G(x) - |x| + 2)T(G(x))$.

17. See [63, pp. 377ff.]. Here is a simpler proof using alternating TMs. By the relationship between deterministic time and alternating space (Corollary 7.5), it suffices to show that nondeterministic and deterministic $S(n)$-space-bounded APDAs are equivalent to $S(n)$-space-bounded alternating TMs.

Let M be a nondeterministic $S(n)$-space-bounded APDA. Assume without loss of generality that M empties its stack before accepting. A *configuration* consists of worktape contents, state of the finite control, worktape head position, and the symbol on the top of the stack or a special flag indicating that the stack is empty. It does not include the stack contents below the top symbol. Configurations can be represented in $S(n)$ space.

For configurations α, β and stack σ, write $\alpha \to \beta$ if there is a computation starting in configuration α with stack σ and ending in configuration β with stack σ that does not pop the top element of σ at any time. The computation may push items on the stack above σ and pop them off as much as it wants, but it may not touch σ. Note that the question of whether $\alpha \to \beta$ is independent of σ, because the contents of σ are invisible to the computation except for the top symbol, which is represented in α and β.

Now we describe a recursive procedure to determine whether $\alpha \to \beta$. This can be used to determine whether M accepts x by checking whether **start** $\to$ **accept**, where **start** and **accept** are the start and accept configurations of M, respectively, which we can assume without loss of generality are unique. The procedure can be implemented on an alternating TM in $S(n)$ space.

The procedure works as follows. Given α and β, it first checks whether $\alpha = \beta$ or whether α derives β in one step without pushing or popping the stack, and accepts immediately if so. If not, it nondeterministically guesses whether there exists an intermediate configuration γ such that $\alpha \to \gamma$ and $\gamma \to \beta$. If so, it guesses γ using $\vee$-branching, then checks in parallel using $\wedge$-branching that $\alpha \to \gamma$ and $\gamma \to \beta$. Otherwise, in order for $\alpha \to \beta$, there must exist α' and β' such that α derives α' in one step while pushing a symbol on the stack, β' derives β in one step while popping a symbol off the stack, and $\alpha' \to \beta'$ by a shorter computation. The procedure guesses α' and β', checks that α derives α' in one step and β' derives β in one step, then calls itself tail recursively to check whether $\alpha' \to \beta'$. There is no need to remember α and β, so at most $S(n)$ space is needed.

Conversely, an $S(n)$-space-bounded alternating TM N can be simulated by a deterministic $S(n)$-space-bounded APDA. The APDA simply performs a depth-first search of the computation tree of N, constructing the tree on the fly and calculating the accept/reject values recursively. It uses its stack in the depth-first search to remember where it is in the tree.

20. Finitary set operators are chain-continuous: if τ is finitary and $\mathcal{C}$ is a chain,

$$
\begin{aligned}
\tau(\bigcup \mathcal{C}) &= \bigcup\{\tau(B) \mid B \subseteq \bigcup \mathcal{C},\ B \text{ finite}\} \\
&= \bigcup\{\tau(B) \mid \exists C \in \mathcal{C}\ B \subseteq C,\ B \text{ finite}\} \\
&= \bigcup_{C \in \mathcal{C}} \bigcup \{\tau(B) \mid B \subseteq C,\ B \text{ finite}\} \\
&= \bigcup_{C \in \mathcal{C}} \bigcup\{\tau(B) \mid B \subseteq C,\ B \text{ finite}\} \\
&= \bigcup_{C \in \mathcal{C}} \tau(C).
\end{aligned}
$$

That chain-continuous operators are finitary can be shown by transfinite induction as follows. Recall $A \equiv B$ if there exists $f : A \xrightarrow[\text{onto}]{1\text{-}1} B$. The *cardinality* of A is the least ordinal α such that $\alpha \equiv A$. The cardinality of A is either finite or a limit ordinal, because $\alpha + 1 \equiv \alpha$ for infinite α (map $\beta \mapsto \beta$ for $\omega \leq \beta < \alpha$, $n \mapsto n+1$ for $n < \omega$, and $\alpha \mapsto 0$).

Now suppose τ is a set operator on X and $\tau(\bigcup \mathcal{C}) = \bigcup_{C \in \mathcal{C}} \tau(C)$ for any chain $\mathcal{C}$. For any $A \subseteq X$, let α be its cardinality and $f : \alpha \xrightarrow[\text{onto}]{1\text{-}1} A$. If A is finite, there is nothing to prove. Otherwise, α is a limit ordinal. For any $\beta < \alpha$, define $A_\beta = \{f(\gamma) \mid \gamma < \beta\}$. Then the A_β form a chain and $\bigcup_{\beta < \alpha} A_\beta = A$. Moreover, all the A_β are of smaller cardinality because $A_\beta \equiv \beta < \alpha$, so by the induction hypothesis, $\tau(A_\beta) = \bigcup\{\tau(B) \mid B \subseteq A_\beta,\ B \text{ finite}\}$. Then

$$
\begin{aligned}
\tau(A) &= \tau\left(\bigcup_{\beta < \alpha} A_\beta\right) \\
&= \bigcup_{\beta < \alpha} \tau(A_\beta) \quad \text{by continuity} \\
&= \bigcup_{\substack{\beta < \alpha}} \bigcup_{\substack{B \subseteq A_\beta \\ B \text{ finite}}} \tau(B) \\
&= \bigcup_{\substack{B \subseteq A \\ B \text{ finite}}} \tau(B).
\end{aligned}
$$

The last equation follows from the fact that for finite B, $B \subseteq A$ iff $B \subseteq A_\beta$ for some $\beta < \alpha$.

21. (b)   The set operator on subsets of ω given by

$$A \;\mapsto\; \begin{cases} A, & \text{if } A \text{ is finite} \\ \omega, & \text{otherwise} \end{cases}$$

is monotone but not chain-continuous.

23. If τ is chain-continuous, then

$$
\begin{aligned}
\tau^{\omega+1}(\varnothing) &= \tau(\tau^{\omega}(\varnothing)) \\
&= \tau\Big(\bigcup_{n<\omega} \tau^{n}(\varnothing)\Big) \\
&= \bigcup_{n<\omega} \tau(\tau^{n}(\varnothing)) \\
&= \varnothing \cup \bigcup_{n<\omega} \tau^{n+1}(\varnothing) \\
&= \bigcup_{n<\omega} \tau^{n}(\varnothing) \\
&= \tau^{\omega}(\varnothing).
\end{aligned}
$$

For an example of a τ whose closure ordinal is $\omega + 1$, let X be the set of nodes of the tree pictured in Lecture 40, p. 264, and let

$$\tau(A) \;=\; \{x \mid \text{all successors of } x \text{ are in } A\}.$$

Then $\tau(\varnothing)$ are the leaves, $\tau^2(\varnothing)$ are the leaves and the nodes above the leaves, and so on; $\tau^{\omega}(\varnothing)$ is the set of all nodes except the root; and $\tau^{\omega+1}(\varnothing)$ is the set of all nodes. This is the least fixpoint.

28. This is a refinement of the proof of Savitch's theorem given in Lecture 2. Let M be a nondeterministic TM running in space $S(n)$ and time $T(n)$ on inputs of length n (thus no computation of M on input x uses more than $S(n)$ space or $T(n)$ time). Let start be the start configuration of M on input x, and assume that M has a unique accept configuration accept and a unique reject configuration reject (thus no other configuration halts). Assume also that M erases its worktape and moves all the way to the left before accepting or rejecting. Let the configurations of M be encoded as strings over a finite alphabet Δ.

If $S(n)$ and $T(n)$ are constructible in space $S(n) \log T(n)$, we can first compute $T(n)$ and $S(n)$, then call $\mathrm{SAV}(\text{start}, \text{accept}, T(n), S(n))$, where $\mathrm{SAV}(\alpha, \beta, t, s)$ is the recursive procedure described in Lecture 2 that attempts to find a computation path from α to β of length at most t

through configurations of length at most s. Each recursive instantiation of SAV requires $S(n)$ space, and the depth of the recursion is $\log T(n)$, giving a space bound of $S(n) \log T(n)$.

In case $S(n)$ and $T(n)$ are not constructible, we modify the SAV procedure slightly:

```
boolean EXACTSAV(α, β, t, s) {
    if t = 0 then return α = β;
    if t = 1 then return α →¹ β;
    for γ ∈ Δˢ {
        if EXACTSAV(α, γ, ⌈t/2⌉, s) ∧ EXACTSAV(γ, β, ⌊t/2⌋, s)
            then return 1;
    }
    return 0;
}


boolean SAV(α, β, t, s) {
    for t' := 1 to t {
        if EXACTSAV(α, β, t', s) then return 1;
    }
    return 0;
}
```

The difference here is that $\text{EXACTSAV}(\alpha, \beta, t, s)$ attempts to find a computation path of length *exactly* t from α to β through configurations of length at most s. Neither SAV nor EXACTSAV uses more than $s \log t$ space.

We first determine $S(n)$ and $T(n)$. We start with $S = T = 1$ and alternately check whether T or S is too small, and if so increment it by 1 and check again. To check whether T is too small, we call $\text{EXACTSAV}(\texttt{start}, \alpha, T, S)$ for all nonhalting configurations $\alpha \in \Delta^S$ in turn. If this procedure ever returns successfully, then T is too small. To check whether S is too small, we call $\text{SAV}(\texttt{start}, \alpha, T, S)$ for all configurations $\alpha \in \Delta^S$ in which the head is scanning the Sth worktape cell and wants to move its worktape head right, thereby using $S + 1$ space. If this procedure ever returns successfully, then S is too small. Eventually we find values of S and T just large enough that the entire computation remains within these bounds. At that point we call $\text{SAV}(\texttt{start}, \texttt{accept}, T, S)$.

29.   To show that the problem is in *PSPACE*, we guess a string that is accepted by all the M_i and verify that it is accepted. We start with a

pebble on the start state of each M_i, then guess an input string symbol by symbol, moving pebbles on the states of the M_i according to their transition functions. We do not have to remember the guessed string, just the states that the M_i are currently in. We accept if we ever get to a situation in which all pebbles occupy accept states of their respective automata. Because the pebble configuration can be represented in polynomial space, this is a nondeterministic $PSPACE$ computation. It can be made deterministic by Savitch's theorem.

To show that the problem is hard for $PSPACE$, let N be an arbitrary deterministic n^k-space bounded TM. Assume without loss of generality that N has a unique accept configuration. Given an input x of length n, we build a family of n^k deterministic finite automata M_i with $O(n^k)$ states each whose intersection is the set of accepting computation histories of N on input x. Then $\bigcap_i L(M_i) = \varnothing$ iff N does not accept x.

Recall that an accepting computation history of N on input x of length n is a string of the form

$$\#\alpha_0\#\alpha_1\#\alpha_2\#\ \cdots\ \#\alpha_{m-1}\#\alpha_m\#, \tag{14}$$

where each α_i is a string of length n^k over some finite alphabet Δ encoding a configuration of N on input x, such that

(i) α_0 is the start configuration of N on x,

(ii) α_m is the accept configuration of N on x, and

(iii) each α_{i+1} follows from α_i according to the transition rules of N.

If a string is an accepting computation history, then it must be of the correct format (14) and must satisfy (i), (ii), and (iii). Checking that the input string is of the form (14) requires checking that the input string is in the regular set $(\#\Delta^{n^k})^*\#$, plus some other simple format checks (exactly one state of N per configuration, each configuration begins and ends with endmarkers, etc.). This requires an automaton with $O(n^k)$ states. Checking (i) or (ii) involves just checking whether the input begins or ends with a certain fixed string of length n^k. Again, each of these conditions can be checked by an automaton with $O(n^k)$ states. Finally, to check (iii), recall from the proof of the Cook–Levin theorem that there is a finite set of local conditions involving the $j-1$st, jth, and $j+1$st symbols of α_i and the jth symbol of α_{i+1} such that (iii) holds iff these local conditions are satisfied for all j, $1 \leq j \leq n^k$. The local conditions depend only on the description of N. To check that (iii) holds, we use n^k automata with $O(n^k)$ states. The jth of these automata scans across and checks for all i that the local condition involving the $j-1$st, jth, and $j+1$st symbols of α_i and the jth symbol of α_{i+1} is satisfied. This only involves counting the distance from each # out to a distance of j in each configuration, remembering three symbols in the finite control, then moving to the next #, counting a distance j from there, and comparing symbols. Each machine requires only $O(n^k)$ states to do the counting.

30. First we show that if $P = NP$, then every deterministic polynomial-time-computable length-preserving map $f : \Sigma^* \to \Sigma^*$ is invertible.

Please note that the following solution is incorrect.

> In NP, we can guess x and verify that $f(x) = y$. Because $P = NP$, we can do the same thing deterministically.

This is incorrect because you have not shown how to produce x deterministically when it exists.

Assume the alphabet is binary, say $\{0, 1\}$. Let $\leq$ denote the prefix order on strings; thus $u \leq x$ iff there exists v such that $x = uv$. The set

$$\{(x, y, u) \mid |x| = |y|, \ f(x) = y, \ \text{and} \ u \leq x\}$$

is in P, because all three conditions can be checked deterministically in polynomial time; therefore the set

$$B \quad = \quad \{(y, u) \mid \exists x \ |x| = |y|, \ f(x) = y, \ \text{and} \ u \leq x\}$$

is in NP. By the assumption $P = NP$, the set B is in P. Using this fact, given y of length n we can do a binary search on strings of length n to find x such that $f(x) = y$. First ask whether $(y, \varepsilon) \in B$. If not, then no such x exists; halt and report failure. If so, ask whether $(y, 0) \in B$. If yes, there is an x with $f(x) = y$ whose first bit is 0, and if no, all such x have first bit 1. Now depending on the previous answer, ask whether $(y, 00) \in B$ or $(y, 10) \in B$ as appropriate. The answer determines the second bit of x. Continue in this fashion until all the bits of some x with $f(x) = y$ have been determined.

For the other direction, let φ denote a Boolean formula, say with m variables, and let t be a bit string of length m denoting a truth assignment to the variables of φ. Consider the function

$$f(\varphi \# t) \quad = \quad \begin{cases} \varphi \# 1^{|t|}, & \text{if } \varphi(t) = 1, \\ \varphi \# 0^{|t|}, & \text{if } \varphi(t) = 0. \end{cases}$$

Let $f(y) = y$ for y not of the form $\varphi \# t$. Then f is length-preserving and computable in polynomial time: first determine whether the input is of the form $\varphi \# t$, and if so, evaluate φ on t. If f is invertible, then $P = NP$, because φ is satisfiable iff there exists x such that $f(x) = \varphi \# 1^{|t|}$.

32. (a) Given $A \in NP$, let M be a nondeterministic n^c-time-bounded non-deterministic TM accepting A. We can express membership of x in A by

    > There is a sequence of configurations of M of length at most n^c describing an accepting computation history of M on input x.

    Suitably formalized in first-order logic, this formula is of the required form.

    Conversely, given a condition

    $$\exists y \, |y| \leq |x|^c \wedge R(x, y)$$

    for membership in A, where R is a deterministic polynomial-time predicate, we can build a nondeterministic polynomial-time machine N for A that on input x guesses a witness y of length at most $|x|^c$ and verifies deterministically that $R(x, y)$.

41. (a) Let σ be a deterministic logspace-computable function. There is a constant c such that for all n, $|\sigma(x)| \leq |x|^c$. We construct a family of logspace-uniform, polylog-depth, polynomial-size circuits B_n with n^c output ports computing σ. Because σ is computable in deterministic logspace, the function σ' defined by

    $$\sigma'(x, i) \overset{\text{def}}{=} \quad \text{the } i\text{th bit of } \sigma(x)$$

    is also computable in deterministic logspace (say with i presented in binary), and σ' is Boolean-valued. As such it is the characteristic function of a set in $LOGSPACE$. Because $LOGSPACE \subseteq NC$, there is a family of NC circuits C_m computing σ'; that is, for $|x| = n$, $C_m(x, i) = \sigma'(x, i)$, where $m = n + c \log n$. The first n input ports of C_m are for x and the last $c \log n$ are for i in binary. Then $\sigma(x)$ is given by $C_m(x, i)$, $1 \leq i \leq n^c$. We construct B_n from n^c disjoint copies of C_m. We feed x into the first n input ports of each copy of C_m, then feed the binary constant i into the last $c \log n$ input ports of the ith copy of C_m.

    This construction is logspace uniform because the family C_m is, and to create the multiple copies we only need an outer loop that counts to n^c.

    (b) We know that CVP $\in P$, therefore if $P = NC$ then CVP $\in NC$. Conversely, suppose CVP $\in NC$. Because CVP is $\leq_{\mathrm{m}}^{\log}$-complete for P, for any $A \in P$, $A \leq_{\mathrm{m}}^{\log}$ CVP. By (a), there is a logspace-uniform family of polylog-depth, polynomial-size circuits B_n such that $x \in A$ iff $B_{|x|}(x) \in$ CVP. Plugging the outputs of these circuits into the inputs of the NC circuits for CVP, we get a family of NC circuits for A.

49. (a) Call a structure $\mathcal{A} = (A, R, \dots)$ *nontrivial* if it has at least one distinguished k-ary relation R, $k \geq 1$, such that $R(\overline{a})$ is true for some k-tuple $\overline{a} \in A^k$ and $R(\overline{b})$ is false for some k-tuple $\overline{b} \in A^k$. Note that the identity relation $=$ need not be a distinguished relation of the structure, and there need not be constant symbols for $\overline{a}$ and $\overline{b}$. To show $PSPACE$-hardness, we encode QBF. Given a QBF formula

$$Q_1 x_1 \ \cdots \ Q_n x_n \ B(\overline{x}),$$

replace each quantifier $Q_i x_i$ with k quantifiers $Q_i x_i^1 \ \cdots \ Q_i x_i^k$, and replace each occurrence of x_i in $B(\overline{x})$ with $R(x_i^1, \dots, x_i^k)$.

If in addition $\mathcal{A}$ is finite, and the functions and relations of the structure are given by tables, then sentences in the language of $\mathcal{A}$ can be decided by the following $APTIME$ algorithm. First put φ in prenex form. We then have

$$Q_1 x_1 \ \cdots \ Q_n x_n \ B(\overline{x})$$

with $B(\overline{x})$ quantifier-free. Guess $x_1 \in \{0, 1, 2\}$ existentially if $Q_1 = \exists$, universally if $Q_1 = \forall$, and repeat for x_2, x_3, and so on. When done with the quantifiers, evaluate $B(\overline{x})$ on the guessed values of $\overline{x}$ by table lookup.

50. Define $\equiv_{n,k}^m$ as in the hint for this exercise on p. 362. Let $[\mathcal{A}, \overline{a}]_{n+1,k}^m$ denote the $\equiv_{n+1,k}^m$-equivalence class of $\mathcal{A}, \overline{a}$, where $\overline{a} = a_1, \dots, a_k$. Then $\equiv_{n,k}^m$ has only finitely many classes, because there are only finitely many formulas of length m. Define MOVE so that $([\mathcal{A}, \overline{a}]_{n+1,k}^m, \alpha) \in$ MOVE if and only if $\alpha = [\mathcal{A}, \overline{a}, a_{k+1}]$ for some a_{k+1}.

This construction does not make all theories decidable because it is not effective in general.

57. (a) For the inequality, by the strict monotonicity of the natural logarithm on the interval in question, it suffices to show $z \ln(1 - \frac{1}{z}) \leq -1$ for all $z > 1$, or equivalently $\ln x \leq x - 1$ for all $0 < x < 1$. In fact, this inequality holds for all positive real x. The curves $y = \ln x$ and $y = x - 1$ are tangent at the point $(1, 0)$, as both have slope 1 there; and $y = \ln x$ has strictly decreasing slope everywhere on the interval $x > 0$, because its second derivitive is negative, whereas $y = x - 1$ is flat, so the curve $y = \ln x$ lies below the curve $y = x - 1$.

For the limiting behavior (which we do not need for the rest of this exercise), by the continuity of the exponential on the interval in question, it suffices to show $\lim_{z \to \infty} z \ln(1 - \frac{1}{z}) = -1$. Expanding $\ln(1 - \frac{1}{z})$ in a Taylor series gives

$$
\begin{aligned}
z \ln(1 - \frac{1}{z}) &= z \left(\frac{-1}{z-1} + \frac{1}{2} \left(\frac{-1}{z-1} \right)^2 + \frac{1}{3} \left(\frac{-1}{z-1} \right)^3 + \cdots \right) \\
&= \frac{-z}{z-1} + z \left(\frac{1}{2} \left(\frac{1}{z-1} \right)^2 - \frac{1}{3} \left(\frac{1}{z-1} \right)^3 + \cdots \right).
\end{aligned}
$$

The first term tends to -1 as $z \to \infty$. The remaining expression is bounded in absolute value by

$$
z \left(\left(\frac{1}{z-1} \right)^2 + \left(\frac{1}{z-1} \right)^3 + \cdots \right) = \frac{z}{(z-1)(z-2)},
$$

which tends to 0 as $z \to \infty$.

66. (ii) If $L \in \Pi^{\log} \cdot \oplus \cdot \mathcal{C}$, then there exist $A \in \oplus \cdot \mathcal{C}$ and $k \geq 0$ such that for all x,

$$
x \in L \quad \Leftrightarrow \quad \forall w \; |w| = k \log |x| \Rightarrow x \# w \in A.
$$

Likewise, there exist $B \in \mathcal{C}$ and $m \geq 0$ such that for all $x \# w$,

$$
\begin{aligned}
x \# w \in A \quad &\Leftrightarrow \quad |\{z \mid |z| = |x \# w|^m \wedge x \# w \# z \in B\}| \text{ is odd} \\
&\Leftrightarrow \quad |W(n^m, B, x \# w)| \text{ is odd}.
\end{aligned}
$$

Assume for simplicity of notation that $|x|$ is a power of 2. Let $\{w_0, w_1, \ldots, w_{N-1}\}$ be the set of all binary strings of length $k \log |x|$, where $N = |x|^k$. Let

$$
B' \stackrel{\text{def}}{=} \{ x \# z_0 z_1 \cdots z_{N-1} \mid |z_i| = |x \# w_i|^m \wedge x \# w_i \# z_i \in B, \\ 0 \leq i \leq N - 1 \}.
$$

Then $B' \in \mathcal{C}$ because $\mathcal{C}$ is closed downward under $\leq^{\mathrm{P}}_{\mathrm{T}}$. Let

$$
p(n) \stackrel{\text{def}}{=} n^k (n + k \log n)^m.
$$

Then

$$
\begin{aligned}
W(p, B', x) &= \{z \mid |z| = p(|x|) \wedge x \# z \in B'\} \\
&= \prod_{i=0}^{N-1} \{z \mid |z| = |x \# w_i|^m \wedge x \# w_i \# z \in B\} \\
&= \prod_{i=0}^{N-1} W(n^m, B, x \# w_i),
\end{aligned}
$$

where the product is with respect to the set-theoretic concatenation operation

$$UV \stackrel{\text{def}}{=} \{uv \mid u \in U, \ v \in V\}.$$

Then

$$
\begin{aligned}
x \in L \quad &\Leftrightarrow \quad \forall i < N \ \ x\#w_i \in A \\
&\Leftrightarrow \quad \forall i < N \ \ |W(n^m, B, x\#w_i)| \text{ is odd} \\
&\Leftrightarrow \quad \Big| \prod_{i=0}^{N-1} W(n^m, B, x\#w_i) \Big| \text{ is odd} \\
&\Leftrightarrow \quad |W(p, B', x)| \text{ is odd},
\end{aligned}
$$

therefore $L \in \oplus \cdot \mathcal{C}$.

(iii) If $A \in BP \cdot \mathcal{C}$, then there exist $B \in \mathcal{C}$ and $m \geq 0$ such that for all y,

$$
\begin{aligned}
y \in A \quad &\Rightarrow \quad \mathrm{Pr}_w(y\#w \in B) \geq \tfrac{3}{4}, \\
y \notin A \quad &\Rightarrow \quad \mathrm{Pr}_w(y\#w \in B) \leq \tfrac{1}{4},
\end{aligned}
$$

where the w are chosen uniformly at random among all binary strings of length $|y|^m$. By the amplification lemma (Lemma 14.1), we can make the probability of error vanish exponentially by repeated trials. In particular, we can assume that B and m have been chosen so that

$$
\begin{aligned}
y \in A \quad &\Rightarrow \quad \mathrm{Pr}_w(y\#w \notin B) \leq 2^{-(|y|+2)}, \\
y \notin A \quad &\Rightarrow \quad \mathrm{Pr}_w(y\#w \in B) \leq 2^{-(|y|+2)}.
\end{aligned}
\tag{15}
$$

(Lemma 14.1 was proved for $BPP = BP \cdot P$, but a quick check of that proof reveals that the only property needed of P was closure under polynomial-time Turing reduction, which we have explicitly assumed of $\mathcal{C}$.) By the law of sum, for any n,

$$
\begin{aligned}
\mathrm{Pr}_w(\exists y \in \{0,1\}^n \ & y\#w \in B \Leftrightarrow y \notin A) \\
&\leq \ \textstyle\sum_{|y|=n} \mathrm{Pr}_w(y\#w \in B \Leftrightarrow y \notin A) \\
&\leq \ 2^n \cdot 2^{-(n+2)} \\
&= \ \tfrac{1}{4}.
\end{aligned}
\tag{16}
$$

Now suppose $L \in \oplus \cdot BP \cdot \mathcal{C}$. Then there exist $A \in BP \cdot \mathcal{C}$ and $k \geq 0$ such that for all x,

$$x \in L \quad \Leftrightarrow \quad |\{z \mid |z| = |x|^k \wedge x\#z \in A\}| \text{ is odd}.\tag{17}$$

Choosing $B \in \mathcal{C}$ and $m \geq 0$ satisfying (15), we have by (16) that for all x,

$$\mathrm{Pr}_w(\exists z \in \{0,1\}^{|x|^k} \; x\#z\#w \in B \Leftrightarrow x\#z \notin A) \;\; \leq \;\; \tfrac{1}{4}. \tag{18}$$

Let

$$B' \stackrel{\mathrm{def}}{=} \{x\#w\#z \mid x\#z\#w \in B\}$$

$$B'' \stackrel{\mathrm{def}}{=} \{x\#w \mid |\{z \mid |z| = |x|^k \wedge x\#w\#z \in B'\}| \text{ is odd}\}$$

$$= \{x\#w \mid |\{z \mid |z| = |x|^k \wedge x\#z\#w \in B\}| \text{ is odd}\}.$$

Then $B' \in \mathcal{C}$ and $B'' \in \oplus \cdot \mathcal{C}$. Now combining (17) with (18),

$$x \in L \;\; \Leftrightarrow \;\; |\{z \mid |z| = |x|^k \wedge x\#z \in A\}| \text{ is odd}$$
$$\Rightarrow \;\; \mathrm{Pr}_w(|\{z \mid |z| = |x|^k \wedge x\#z\#w \in B\}| \text{ is odd}) \geq \tfrac{3}{4}$$
$$\Leftrightarrow \;\; \mathrm{Pr}_w(x\#w \in B'') \geq \tfrac{3}{4},$$

$$x \notin L \;\; \Leftrightarrow \;\; |\{z \mid |z| = |x|^k \wedge x\#z \in A\}| \text{ is even}$$
$$\Rightarrow \;\; \mathrm{Pr}_w(|\{z \mid |z| = |x|^k \wedge x\#z\#w \in B\}| \text{ is odd}) \leq \tfrac{1}{4}$$
$$\Leftrightarrow \;\; \mathrm{Pr}_w(x\#w \in B'') \leq \tfrac{1}{4}.$$

This proves that $L \in BP \cdot \oplus \cdot \mathcal{C}$.

67. (d) Let $f \in \#P$. Then there is a polynomial-time nondeterministic TM M that has exactly $f(x)$ accepting computation paths on input x. We build a new machine N that has exactly $g(x)(f(x))$ accepting computation paths on input x. The machine N first computes all the coefficients of $g(x) \in \mathbb{N}[z]$. The degree of $g(x)$ is at most $|x|^d$ for some constant d, and the coefficients are polynomial-time computable by assumption. Then N calls a recursive procedure R with input $g(x)$.

The recursive procedure R on input $p \in \mathbb{N}[z]$ works as follows.

- If p consists of more than one term, write p as $q + r$, where q and r are polynomials each with roughly half the terms of p. Make a nondeterministic branch, the two branches calling R recursively with q and r, respectively.

- If p consists of a single term az^i with coefficient $a \geq 2$, where a is represented in binary with polynomially many bits, make a nondeterministic branch, the two branches calling R recursively with $\lceil a/2 \rceil z^i$ and $\lfloor a/2 \rfloor z^i$, respectively.

- If p consists of a single term z^i with $i \geq 1$, run M on input x, branching as M branches. For every computation path of M leading to rejection, just reject. For every computation path of M leading to acceptance, call R recursively with z^{i-1}.

- If p consists of the single term 1, accept.

One can show inductively that the number of accepting computation paths generated by R on input p is exactly $p(f(x))$, therefore the number of accepting computation paths of N on input x is exactly $g(x)(f(x))$. The running time of N is still polynomial, because each of the steps takes polynomial time and the depth of the recursion is bounded by $\log d + c + dn^k$, where d is the degree of $g(x)$, c is a bound on the number of bits needed to represent any coefficient of $g(x)$, and n^k is the running time of M.

69. (c) Let M be a Σ_k oracle machine with oracle A. We simulate M with another Σ_k oracle machine N as described in the hint for this exercise on p. 363.

At any point in the simulation, the simulating machine N has recorded on its tape the current configuration α of M, the oracle queries $y_1, \ldots, y_m$ that M has made up to that point in the computation, and a Boolean formula $\psi(x_1, \ldots, x_m, z)$ with a single positive occurrence of z encoding an acceptance condition

$$\psi(A(y_1), \ldots, A(y_m), \mathrm{acc}(\alpha, A)),$$

where the $A(y_i)$ are the (as yet undetermined) responses of the oracle to the queries y_i, and $\mathrm{acc}(\alpha, A)$ represents the assertion that M accepts when started in configuration α with oracle A. Initially, the machine N starts with $\psi = z$, the null list of queries, and the start configuration of M. If α is a $\vee$-configuration with successors α_0 and α_1, it branches existentially, each branch taking one of the successors. The list of queries and ψ do not change. This is correct, because by (b),

$$\psi(A(y_1), \ldots, A(y_m), \mathrm{acc}(\alpha, A))$$
$$\equiv \quad \psi(A(y_1), \ldots, A(y_m), \mathrm{acc}(\alpha_0, A) \vee \mathrm{acc}(\alpha_1, A))$$
$$\equiv \quad \psi(A(y_1), \ldots, A(y_m), \mathrm{acc}(\alpha_0, A))$$
$$\vee\, \psi(A(y_1), \ldots, A(y_m), \mathrm{acc}(\alpha_1, A)).$$

The procedure for simulating $\wedge$-configurations of M is similar.

For an oracle query y_{m+1} with "yes" successor α_1 and "no" successor α_0, N branches existentially if the previous branch was existential and universally if the previous branch was universal. If there have been no branches yet and $k \geq 1$, N branches existentially. (If $k = 0$, so M is deterministic, there is nothing to do.) If the branch is existential, the successor taking the "yes" guess proceeds with α_1, $y_1, \ldots, y_{m+1}$, and $\psi(x_1, \ldots, x_m, x_{m+1} \wedge z)$, and the successor taking the "no" guess proceeds with α_0, $y_1, \ldots, y_{m+1}$, and

$\psi(x_1, \dots, x_m, x_{m+1} \wedge \neg z)$. This is correct, because by (b),

$$
\begin{aligned}
\psi&(A(y_1), \dots, A(y_m), \mathrm{acc}(\alpha, A)) \\
&\equiv \psi(A(y_1), \dots, A(y_m), (A(y_{m+1}) \wedge \mathrm{acc}(\alpha_1, A)) \\
&\qquad\qquad\qquad\qquad \vee (\neg A(y_{m+1}) \wedge \mathrm{acc}(\alpha_0, A))) \\
&\equiv \psi(A(y_1), \dots, A(y_m), A(y_{m+1}) \wedge \mathrm{acc}(\alpha_1, A)) \\
&\qquad\qquad \vee \psi(A(y_1), \dots, A(y_m), \neg A(y_{m+1}) \wedge \mathrm{acc}(\alpha_0, A)).
\end{aligned}
$$

If the branch is universal, the argument is similar, except that we use (a) to make the branch in the condition universal:

$$
\begin{aligned}
\psi&(A(y_1), \dots, A(y_m), \mathrm{acc}(\alpha, A)) \\
&\equiv \psi(A(y_1), \dots, A(y_m), (A(y_{m+1}) \rightarrow \mathrm{acc}(\alpha_1, A)) \\
&\qquad\qquad\qquad\qquad \wedge (\neg A(y_{m+1}) \rightarrow \mathrm{acc}(\alpha_0, A))) \\
&\equiv \psi(A(y_1), \dots, A(y_m), A(y_{m+1}) \rightarrow \mathrm{acc}(\alpha_1, A)) \\
&\qquad\qquad \wedge \psi(A(y_1), \dots, A(y_m), \neg A(y_{m+1}) \rightarrow \mathrm{acc}(\alpha_0, A)).
\end{aligned}
$$

80. Write φ as $\varphi_0 \wedge \varphi_1$, where φ_1 consists of all clauses of φ containing a literal ℓ such that $\rho(\ell) = 1$, and φ_0 consists of the remaining clauses of φ. Then $\rho(\varphi_1) = 1$ and $\rho(\varphi_0) \leq \varphi_0$, where $\leq$ is the natural order in the free Boolean algebra on generators X. Thus $\rho(\varphi) \leq \varphi_0$.

Let K be the conjunction of all literals ℓ such that $\rho(\ell) = 1$. Then K has a literal in common with every clause of φ_1, so $K \leq \varphi_1$. Also, $N \leq \rho(\varphi_0) \leq \varphi_0$, and N and K are over disjoint sets of variables. Then $NK \leq \varphi_0 \varphi_1 = \varphi$, therefore $NK \leq M$ for some minterm M of φ. Moreover, M is of the form $N'K'$, where $N \leq N'$ and $K \leq K'$. Then $\rho(M) = \rho(N')\rho(K') = N' \leq \rho(\varphi)$; but because N is a minterm of $\rho(\varphi)$, $N = N'$.

81. Take $\varphi = x_1, \dots, x_n$, $\psi = x_2, \dots, x_n$, and $s = 1$. Then

$$
\begin{aligned}
\Pr(\exists M \in \mathrm{m}(\rho(\varphi)) \, |M| \geq 1) &= 2^{-n}((1+p)^n - (1-p)^n) \\
\Pr(\exists M \in \mathrm{m}(\rho(\varphi)) \, |M| \geq 1 \mid \rho(\psi) = 1) &= (1-p)/2.
\end{aligned}
$$

90. We show inductively that every program is equivalent to one of the form

$p; \texttt{while } b \texttt{ do } q$

where p and q are *while*-free. This can be done using the following transformations.

(a) Replace

```
while b do q; p
```

with

```
c := b;
while c {
    q;
    c := b;
    if ¬c then p
}
```

where c is a new Boolean variable not occurring in p or q. (In reality we do not have Boolean variables, but we can simulate them.)

(b) Replace

```
while b₁ do p₁;
while b₂ do p₂;
⋮
while bₖ do pₖ;
```

with

```
i := 1;
while i ≤ k {
    case i of {
        1 : if b₁ then p₁ else i := i + 1;
        2 : if b₂ then p₂ else i := i + 1;
        ⋮
        k : if bₖ then pₖ else i := i + 1;
    }
}
```

where i is a new integer variable and k is a constant. We can simulate the `case` statement with `if-then-else`'s.

(c) Replace

```
if b
    then p₁; while c₁ do q₁
    else p₂; while c₂ do q₂
```

with

```
c := b;
if c then p₁ else p₂;
while (c ∧ c₁) ∨ (¬c ∧ c₂) {
    if c then q₁ else q₂
}
```

where c is a new Boolean variable.

(d) Replace

```
while b {
    p;
    while c do q
}
```

with

```
if b {
    p;
    while b ∨ c {
        if c then q else p
    }
}
```

and then use (c).

91. We show first that $\mathcal{P} \subseteq \mathcal{C}$ by induction on the definition of $f \in \mathcal{P}$. All cases are straightforward except when f is defined by primitive recursion. Suppose $f : \omega^{m+1} \to \omega^n$ is defined by primitive recursion from $h : \omega^m \to \omega^n$ and $g : \omega^{m+n+1} \to \omega^n$ defined previously:

$$
\begin{aligned}
f(0, \overline{x}) &= h(\overline{x}) \\
f(s(y), \overline{x}) &= g(y, \overline{x}, f(y, \overline{x})).
\end{aligned}
$$

We wish to show that $f \in \mathcal{C}$. By the induction hypothesis, h and g are in $\mathcal{C}$. Define

$$
\begin{aligned}
\widehat{g}(y, \overline{x}, \overline{z}) &= (s(y), \overline{x}, g(y, \overline{x}, \overline{z})) \\
\widehat{f}(n, y, \overline{x}, \overline{z}) &= \widehat{g}^n(y, \overline{x}, \overline{z}).
\end{aligned}
$$

Both $\widehat{f}$ and $\widehat{g}$ are in $\mathcal{C}$. Arguing by induction on n,

$$
\begin{aligned}
\widehat{g}^0(0, \overline{x}, h(\overline{x})) &= (0, \overline{x}, h(\overline{x})) \\
&= (0, \overline{x}, f(0, \overline{x})) \\
\widehat{g}^{s(n)}(0, \overline{x}, h(\overline{x})) &= \widehat{g}(\widehat{g}^n(0, \overline{x}, h(\overline{x}))) \\
&= \widehat{g}(n, \overline{x}, f(n, \overline{x})) \quad \text{(induction hypothesis)} \\
&= (s(n), \overline{x}, g(n, \overline{x}, f(n, \overline{x}))) \\
&= (s(n), \overline{x}, f(s(n), \overline{x})).
\end{aligned}
$$

Using projections, tupling, and composition, we can define

$$
\begin{aligned}
e(n, \overline{x}) &= \widehat{f}(n, z(n), \overline{x}, h(\overline{x})) \\
&= \widehat{f}(n, 0, \overline{x}, h(\overline{x})) \\
&= \widehat{g}^{n}(0, \overline{x}, h(\overline{x})) \\
&= (n, \overline{x}, f(n, \overline{x})),
\end{aligned}
$$

so f can be obtained from e by projecting.

We now show conversely that $\mathcal{C} \subseteq \mathcal{P}$ by induction on the definition of $f \in \mathcal{C}$. Again, all cases are straightforward except when f is defined from g using the n-fold composition rule given in the statement of the problem:

$$
f(n, \overline{y}) = g^{n}(\overline{y}).
$$

By the induction hypothesis, $g \in \mathcal{P}$. Then so is

$$
\widehat{g}(x, \overline{y}, \overline{z}) = g(\overline{z}),
$$

and f can be defined from $\widehat{g}$ by primitive recursion:

$$
\begin{aligned}
f(0, \overline{y}) &= \overline{y} \\
&= g^{0}(\overline{y}) \\
f(s(n), \overline{y}) &= \widehat{g}(n, \overline{y}, f(n, \overline{y})) \\
&= \widehat{g}(n, \overline{y}, g^{n}(\overline{y})) \\
&= g(g^{n}(\overline{y})) \\
&= g^{s(n)}(\overline{y}).
\end{aligned}
$$

93. Let p be a *for* program. Let t be a new variable not occurring in p. Insert $t := t+1$ before every statement, and insert $t := 0$ at the beginning of the program. The resulting program is a *for* program p' such that the final value of t on an input x is the number of steps taken by the original program p on input x. This function is certainly primitive recursive, because it is computed by the *for* program p'.

Conversely, let p be a *while* program with a primitive recursive time bound. Then there is a *for* program q computing the running time of p on input x and leaving the value in a variable t. We assume without loss of generality that q has no variables in common with p except x, and does not change the value of x. Because the body of any *while* loop in p is executed at most t times on input x, the original p is equivalent to the *for* program $q; p'$, where p' is obtained from p by replacing each *while* loop

```
while b do r
```

with the *for* loop

```
for t {
    if b then r
}
```

98. It is certainly infinite, because there are infinitely many distinct partial recursive functions. Suppose it contained an infinite r.e. subset enumerated by enumeration machine M. Define $f(i)$ to be the first element greater than i enumerated by M. Then f is a total recursive function without a fixpoint, contradicting the recursion theorem.

107. Construct a sequence of finite approximations to $h : \omega \to \omega$ using a back-and-forth argument. Start with h completely undefined. Now suppose that after n stages we have constructed a map h with finite domain such that h is one-to-one and $\mathbf{graph}\, h \subseteq R$. If n is even, let x be the least element not in the domain of h. Execute the following program.

```
y := f(x);
while (y ∈ range(h)) {
    z := h⁻¹(y);
    y := f(z);
}
h(x) := y;
```

After the initial assignment, $(x, y) \in R$. Because $\mathbf{graph}\, f \subseteq R$, $\mathbf{graph}\, h \subseteq R$, and $R \circ R^{-1} \circ R \subseteq R$, the loop maintains the invariant $(x, y) \in R$. Thus if the loop terminates, then the final value of y is not in the range of h, so we have increased the domain of definition of h by one element while maintaining the invariant that h is one-to-one on its domain and $\mathbf{graph}\, h \subseteq R$.

To argue that the loop terminates, let Y and Z be the sets of elements that the variables y and z, respectively, ever take on during the execution of the loop. Then $Y = f(\{x\} \cup Z)$ and $Z = h^{-1}(Y \cap \mathrm{range}\, h)$. If the loop never terminates, then $Y \subseteq \mathrm{range}\, h$, therefore $Z = h^{-1}(Y)$ and $Y = h(Z)$. Because f and h are one-to-one and the domain of h is finite, $|Z| = |Y| = |\{x\} \cup Z|$, therefore $x \in Z = h^{-1}(Y)$. But this contradicts the fact that x is not in the domain of h.

If n is odd, we take x to be the least element not in the range of h and work in the other direction, replacing f by g and R by R^{-1} in the argument above. We need the property $R^{-1} \circ R \circ R^{-1} \subseteq R^{-1}$. As indicated in the hint for this exercise on p. 364, this follows from the assumption $R \circ R^{-1} \circ R \subseteq R$ by elementary set-theoretic reasoning.

110. (a) Let $M_0, M_1, \ldots$ be an enumeration of total TMs equipped with polynomial-time clocks that shut the machine down after n^k steps. The enumeration contains a copy of each TM with an n^k clock for each k. Then

   - every M_i runs in polynomial time, and
   - every set in P is accepted by some M_i.

   Here is a procedure for enumerating A and $\sim A$. Maintain a finite list of indices, initially empty. Each index on the list is either marked or unmarked. At stage x, put x unmarked on the list. Then simulate every machine on the list on input x and pick the smallest i on the list such that M_i accepts x. If no such M_i exists, just put $x \in A$ and go on to stage $x + 1$. If i is unmarked, put $x \in A$ and mark i. If i is already marked, put $x \in \sim A$ and cross i off the list. Every M_i gets on the list eventually. If M_i is on the list and accepts x, then the only way M_i would not be chosen to be marked or deleted in stage x is if there is a machine with a smaller index on the list that is marked or deleted in stage x, and this can happen at most $2i$ times. Thus if M_i accepts an infinite set, it will eventually be the highest priority machine on the list and will eventually be marked and deleted. When M_i is marked, we put the current $x \in A$, ensuring that $L(M_i) \not\subseteq \sim A$. When M_i is deleted, we put the current $x \in \sim A$, ensuring that $L(M_i) \not\subseteq A$.

   Both A and $\sim A$ are r.e., because the construction above enumerates them, so A is recursive. Both A and $\sim A$ are infinite, because infinitely many machines are marked and deleted.

126. We exhibit total recursive space bounds S_0 and S_1 such that

$$DSPACE(S_0) \cup DSPACE(S_1) \neq DSPACE(S)$$

for any S. Let M be a total TM such that every TM accepting $L(M)$ uses at least one tape cell on almost all inputs (see Miscellaneous Exercise 117). Let $T(n)$ be the maximum space usage of M on inputs of length n, and define

$$S_0(n) \overset{\text{def}}{=} \begin{cases} T(n), & \text{if } n \text{ even,} \\ 0, & \text{if } n \text{ odd} \end{cases}$$

$$S_1(n) \overset{\text{def}}{=} \begin{cases} T(n), & \text{if } n \text{ odd,} \\ 0, & \text{if } n \text{ even.} \end{cases}$$

Then

$$A_0 \stackrel{\text{def}}{=} L(M) \cap \{x \mid |x| \text{ is even}\} \in DSPACE(S_0)$$
$$A_1 \stackrel{\text{def}}{=} L(M) \cap \{x \mid |x| \text{ is odd}\} \in DSPACE(S_1).$$

If $DSPACE(S_0) \cup DSPACE(S_1) \subseteq DSPACE(S)$, then both A_0 and A_1 are in $DSPACE(S)$, therefore their union $L(M)$ is. But $L(M) \notin DSPACE(S_0)$ and $L(M) \notin DSPACE(S_1)$, because all machines for $L(M)$ require at least one tape cell a.e.

131. (b)   Let $\varphi_i(x)\downarrow^t = y$ denote the recursive predicate, "The machine computing φ_i halts on input x in t or fewer steps and outputs y." The set EQUAL is in Π_2^0, because it can be expressed as

$$\text{EQUAL} = \{(i,j) \mid \forall x \, \forall t \, \forall y \, \exists s \, (\varphi_i(x)\downarrow^t = y \Rightarrow \varphi_j(x)\downarrow^s = y) \\ \wedge \, (\varphi_j(x)\downarrow^t = y \Rightarrow \varphi_i(x)\downarrow^s = y)\}.$$

To show EQUAL is Π_2^0-hard, we reduce ALL from part (a) to it. (You did part (a), right?) Let $\tau(i)$ be an index of the partial recursive function

$$\varphi_{\tau(i)}(x) = \begin{cases} 0, & \text{if } M_i \text{ accepts } x \\ \text{undefined}, & \text{otherwise.} \end{cases}$$

The index $\tau(i)$ can be obtained from i effectively, and the domain of $\varphi_{\tau(i)}$ is exactly $L(M_i)$. Now define

$$\sigma(i) = (\tau(i), \mathsf{const}(0)),$$

where $\mathsf{const}(0)$ is an index for the constant function $\lambda x.0$. Then

$$i \in \text{ALL} \iff \varphi_{\tau(i)} = \lambda x.0 \iff \sigma(i) \in \text{EQUAL}.$$

134. (a)   Yes. Let

$$A \stackrel{\text{def}}{=} \{M \mid M \text{ runs in polynomial time}\}$$
$$B \stackrel{\text{def}}{=} \{M \mid \text{"}M \text{ runs in polynomial time" is provable in PA}\}.$$

Certainly $B \subseteq A$, because PA is sound. We show that $B \in \Sigma_1^0$ and A is Σ_2^0-complete, therefore the two sets cannot be equal. Certainly $B \in \Sigma_1^0$, because we can enumerate proofs in PA. The set A is in Σ_2^0, because M runs in polynomial time iff $\exists k \, \forall x \, M(x)\downarrow^{|x|^k}$, which is a Σ_2^0 predicate. Also, A is hard for Σ_2^0, because we can reduce FIN $= \{M \mid L(M) \text{ is finite}\}$ to it, which we know is Σ_2^0-complete (Lecture 35). Given a machine M, we want to construct a machine M' that runs in polynomial time iff $L(M)$ is finite. On input x, let M' take the following actions.

(i) Simulate M on all inputs y such that $|y| < \log |x|$ for $|x|$ steps.

(ii) If c of these simulations halt and accept, run for $|x|^c$ more steps and halt.

Step (i) takes $|x|^2$ simulation steps. Step (ii) takes $|x|^c$ simulation steps. Now if $L(M)$ is finite, say $|L(M)| = c$, then M' runs in time n^c. If not, then there is no upper bound on c, and M' does not run in polynomial time.

(b) No. Let A be as in part (a). We are asking whether there exists a partial recursive function f defined on all of A (at least) such that if $M \in A$, then M runs in time $n^{f(M)}$. Suppose there were such an f. Let

$$C \stackrel{\text{def}}{=} \{M \mid \exists t \; f(M){\downarrow}^t \wedge \forall x \; \forall t \; (f(M){\downarrow}^t \Rightarrow M(x){\downarrow}^{|x|^{f(M)}})\}.$$

Then $C \subseteq A$, and we are asking whether $A \subseteq C$. As shown in part (a), A is Σ_2^0-complete, but C is the intersection of a Σ_1^0 set and a Π_1^0 set, therefore is contained in Δ_2^0, so it cannot be Σ_2^0-complete.

136. Here is one possible formalization. Let F be any formal deductive system for number theory (for example, Peano arithmetic). There is a formula $\varphi(x)$ of number theory such that the set $\{x \mid \varphi(x)\}$ is a recursive set, but is not $L(M)$ for any TM M such that $F \vdash \forall x \; M(x){\downarrow}$.

To prove this, let N be a TM that enumerates proofs in F of sentences of the form $\forall x \; M(x){\downarrow}$ (that is, M is total). Whenever such a theorem is enumerated, say at stage x, N runs M on x and enumerates x iff M rejects x. Assuming F is sound, this is decidable, because M really is total. Now let $\varphi(x)$ be the formula $x \in L(N)$. The set $\{x \mid \varphi(x)\} = L(N)$ is recursive, because we only need to run N for finitely many steps to determine whether $x \in L(N)$. But it is not $L(M)$ for any M such that $F \vdash \forall x \; M(x){\downarrow}$, because we diagonalized away from all such machines.

139. Problem (a) is Σ_2^0-complete and problem (b) is Π_1^1-complete.

To show that (a) is in Σ_2^0, first assume without loss of generality that M never halts (modify halt states to enter an infinite loop not containing q; this does not change whether M satisfies (a)). We claim that (a) holds iff

(A) there is a finite computation path π from the root such that for all $n > |\pi|$, there is an extension ρ_n of π of length n such that ρ_n contains no occurrences of q outside of the prefix π.

If you believe this claim, then the problem is in Σ_2^0, because (A) can be expressed in the form $\exists \pi \forall n$ followed by a recursive predicate.

To prove the claim, first note that (a) implies (A) easily: given an infinite path σ satisfying (a), let π be a finite prefix containing all occurrences of q. Then for all $n > |\pi|$, the prefix of σ of length n is an extension ρ_n of π satisfying the condition of (A).

The other direction requires König's lemma. Suppose (A) holds. Consider the subtree consisting of π and all the extensions ρ_n for each $n > |\pi|$ given by (A). This is an infinite finitely branching tree, so by König's lemma contains an infinite path, which must satisfy the condition of (a).

To show that (a) is Σ_2^0-hard, we reduce FIN to it. Given a machine M, we wish to produce a nondeterministic machine N with a state q such that N has a computation path with only finitely many occurrences of q iff M accepts a finite set. Assume without loss of generality that M never rejects (modify reject states to enter an infinite loop instead). Let N first guess a number n nondeterministically. It does this by entering a loop that repeatedly adds one to a counter, then nondeterministically chooses whether to exit the loop or keep going. Let q be the first state of this loop. For every branch that exits the loop with a guessed number n, simulate M on all inputs of length greater than n in a timesharing fashion. The state q is never entered in these simulations. If any one of these simulations accepts, erase the tape and restart the entire computation from scratch.

Now if M accepts a finite set, then there will be an n such that M accepts no string of length greater than n, and for the computation path of N corresponding to the guessed number n (or any larger number), q will never occur again, because N will be stuck simulating M forever. On the other hand, if M accepts an infinite set, then every computation path of N corresponding to a guessed number n will discover an $x \in L(M)$ of length greater than n and will restart the computation from scratch, thereby reentering state q. The infinite computation path corresponding to N remaining in the n-guessing loop forever is not an issue, because it enters q infinitely often.

To show that (b) is in Π_1^1, note that the condition (b) can be expressed

$$\forall \pi \ \exists n \ \forall m \geq n \ \text{state}(\pi(m)) \neq q$$

consisting of a universal second-order quantifier ranging over (possibly infinite) paths π in the computation tree, followed by a first-order predicate expressing that π contains only finitely many occurrences of q.

To show that (b) is Π_1^1-hard, we reduce the fair termination problem to it. Consider the fairness condition $(\text{true}, \text{last}(0))$ of Lecture 41 for the computation tree of a given binary-branching nondeterministic machine M. Thus an infinite path is fair iff it contains infinitely many left branches. Modify M to enter state q momentarily whenever it takes a left branch. Then the modified machine is fairly terminating iff there are no infinite fair paths iff every path contains only finitely many occurrences of q.

140. (b) This is an example of a *promise problem*. We are promised that we will only ever be given instances of the problem that satisfy a certain property—in this case, that the given machine is total—that may not be decidable. However, we do not have to decide it; we may assume that it always holds of the inputs we are given.

Formally, a *promise problem* is a pair $(A, P) \in \Sigma^* \times \Sigma^*$, where A is the decision problem and P is the promise. A promise problem (A, P) is in a complexity class $\mathcal{C}$ defined by some resource bound on machines if there is a machine that on inputs in P operates within that resource bound and accepts exactly the strings in $A \cap P$. It is not required to accept strings in A or respect the resource bound on input strings not in P. For example, a promise problem (A, P) is *decidable* if there exists a Turing machine that halts on all elements of P, and of those inputs, accepts exactly the elements of $A \cap P$. (Note that this is not the same as saying that there exists a recursive set B such that $A \cap P = B \cap P$.) A promise problem (A, P) is said to be $\leq_m$-*hard* for $\mathcal{C}$ if every B in $\mathcal{C}$ $\leq_m$-reduces to A via a total recursive function σ that fulfills the promise; that is, for all x, $\sigma(x) \in P$.

In our case the promise problem consists of

$$
\begin{aligned}
A \;&=\; \{M \mid M \text{ accepts a transitive binary relation}\} \\
&=\; \{M \mid \forall x\, \forall y\, \forall z \;\; (x, y) \in L(M) \wedge (y, z) \in L(M) \\
&\qquad\qquad\qquad \rightarrow (x, z) \in L(M)\}, \\
P \;&=\; \{M \mid M \text{ is total}\}.
\end{aligned}
$$

This is in Π_1^0, because there is an IND program with $\forall$-branches only that always halts on inputs in P and accepts exactly $A \cap P$. To show Π_1^0-hardness, we can reduce the complement of the halting problem to A. We need to construct effectively from a given $N\#x$ a total machine M accepting a binary relation that is transitive iff N does not halt on x. Let M accept input (s, t) iff either (i) $s \neq t$, or (ii) $s = t$ and N does not halt on x in t steps. Surely M can be made total. If N does not halt on x, then $L(M)$ contains all pairs, therefore is a transitive relation. If N halts on x in t steps, then $(t, t+1), (t+1, t) \in L(M)$, but $(t, t) \notin L(M)$, so $L(M)$ is not transitive.

References

[1] L. ADLEMAN AND M. HUANG, Recognizing primes in random polynomial time, in *Proc. 19th Symp. Theory of Computing*, New York: ACM, 1987, pp. 462–469.

[2] M. AGRAWAL AND S. BISWAS, Primality and identity testing via Chinese remaindering, in *Proc. 40th Symp. Foundations of Computer Science*, Los Alamitos, CA: IEEE, 1999, pp. 202–208.

[3] M. AGRAWAL, N. KAYAL, AND N. SAXENA, PRIMES is in P, *Ann. Math.*, 160 (2004), pp. 781–793.

[4] A.V. AHO, J.E. HOPCROFT, AND J.D. ULLMAN, *The Design and Analysis of Computer Algorithms*, Reading, MA: Addison-Wesley, 1975.

[5] M. AJTAI, Σ_1^1 formulae on finite structures, *Ann. Pure Appl. Logic*, 24 (1983), pp. 1–48.

[6] S. ARORA, C. LUND, R. MOTWANI, M. SUDAN, AND M. SZEGEDY, Proof verication and hardness of approximation problems, *J. Assoc. Comput. Mach.*, 45 (1998), pp. 501–555.

[7] S. ARORA AND S. SAFRA, Probabilistic checking of proofs: A new characterization of *NP*, *J. Assoc. Comput. Mach.*, 45 (1998), pp. 780–112.

[8] L. BABAI, Trading group theory for randomness, in *Proc. 17th Symp. Theory of Computing*, New York: ACM, April 1985, pp. 421–429.

[9] L. BABAI, L. FORTNOW, AND C. LUND, Nondeterministic exponential time has two-prover interactive protocols, *Comput. Complex.*, 1 (1991), pp. 3–40.

[10] T. BAKER, J. GILL, AND R. SOLOVAY, Relativizations of the $P \overset{?}{=} NP$ question, *SIAM J. Comput.*, 4 (1975), pp. 431–442.

[11] M. BELLARE, D. COPPERSMITH, J. HÅSTAD, M. KIWI, AND M. SU-DAN, Linearity testing in characteristic two, *IEEE Trans. Inf. Theory*, 42 (1996), pp. 1781–1795.

[12] C. BENNETT AND R. GILL, Relative to a random oracle $P \neq NP \neq$ co-NP with probability 1, *SIAM J. Comput.*, 10 (1981), pp. 96–103.

[13] E.R. BERLEKAMP, Factoring polynomials over large finite fields, *Math. Comp.*, 24 (1970), pp. 713–735.

[14] L. BERMAN, The complexity of logical theories, *Theor. Comput. Sci.*, 11 (1980), pp. 71–77.

[15] P. BERMAN, Relationship between the density and deterministic complexity of NP-complete languages, in *Proc. 5th Int. Colloq. Automata, Languages, and Programming*, vol. 62 of *Lect. Notes in Comput. Sci.*, New York: Springer-Verlag, 1978, pp. 63–71.

[16] M. BLUM, A machine-independent theory of the complexity of recursive functions, *J. Assoc. Comput. Mach.*, 14 (1967), pp. 322–336.

[17] ——, On effective procedures for speeding up algorithms, *J. Assoc. Comput. Mach.*, 18 (1971), pp. 290–305.

[18] M. BLUM AND D. KOZEN, On the power of the compass, in *Proc. 19th Symp. Found. Comput. Sci.*, IEEE, October 1978, pp. 132–142.

[19] M. BLUM, M. LUBY, AND R. RUBINFELD, Self-testing/correcting with applications to numerical problems, *J. Comput. Syst. Sci.*, 47 (1993), pp. 549–595.

[20] R.B. BOPPANA AND M. SIPSER, The complexity of finite functions, in *Handbook of Theoretical Computer Science (vol. A): Algorithms and Complexity*, Cambridge, MA: MIT Press, 1990, pp. 757–804.

[21] A.B. BORODIN, Computational complexity and the existence of complexity gaps, *J. Assoc. Comput. Mach.*, 19 (1972), pp. 158–174.

[22] ——, On relating time and space to size and depth, *SIAM J. Comput.*, 6 (1977), pp. 733–744.

[23] J.R. BÜCHI, Weak second order arithmetic and finite automata, *Zeitschrift für Math. Logik und Grundlagen Math.*, 6 (1960), pp. 66–92.

[24] ——, On a decision method in restricted second order arithmetics, in *Proc. Int. Congr. Logic, Methodology and Philosophy of Science*, Stanford, CA: Stanford University Press, 1962, pp. 1–12.

[25] J.Y. CAI AND M. OGIHARA, Sparse hard sets, in *Complexity Theory Retrospective*, L.A. Hemaspaandra and A.L. Selman, eds., New York: Springer-Verlag, 1997, pp. 53–80.

[26] A. CHANDRA, D. KOZEN, AND L. STOCKMEYER, Alternation, *J. Assoc. Comput. Mach.*, 28 (1981), pp. 114–133.

[27] R. CHANG, B. CHOR, O. GOLDREICH, J. HARTMANIS, J. HÅSTAD, D. RANJAN, AND P. ROHATGI, The random oracle hypothesis is false, *J. Comput. Syst. Sci.*, 49 (1994), pp. 24–39.

[28] A. CHURCH, A set of postulates for the foundation of logic, *Ann. Math.*, 33–34 (1933), pp. 346–366, 839–864.

[29] ———, An unsolvable problem of elementary number theory, *Amer. J. Math.*, 58 (1936), pp. 345–363.

[30] A. COBHAM, The intrinsic computational difficulty of functions, in *Proc. 1964 Cong. for Logic, Methodology and Philosophy of Science II*, Y. Bar-Hillel, ed., Amsterdam: North-Holland, 1964, pp. 24–30.

[31] S.A. COOK, The complexity of theorem proving procedures, in *Proc. 3rd Symp. Theory of Computing*, New York: ACM, 1971, pp. 151–158.

[32] ———, Deterministic CFL's are accepted simultaneously in polynomial time and log squared space, in *Proc. 11th Symp. Theory of Computing*, New York: ACM, April 1979, pp. 338–345.

[33] D.C. COOPER, Theorem proving in arithmetic without multiplication, *Mach. Intell.*, 7 (1972), pp. 91–100.

[34] L. CSANKY, Fast parallel matrix inversion algorithms, *SIAM J. Comput.*, 5 (1976), pp. 618–623.

[35] H.B. CURRY, An analysis of logical substitution, *Amer. J. Math.*, 51 (1929), pp. 363–384.

[36] R.A. DEMILLO AND R.J. LIPTON, A probabilistic remark on algebraic program testing, *Inf. Proc. Lett.*, 7 (1978), pp. 193–195.

[37] J. EDMONDS, Paths, trees, and flowers, *Canadian J. Math.*, 17 (1965), pp. 449–467.

[38] U. FEIGE, S. GOLDWASSER, L. LOVASZ, S. SAFRA, AND M. SZEGEDY, Interactive proofs and the hardness of approximating cliques, *J. Assoc. Comput. Mach.*, 43 (1996), pp. 268–292.

[39] W. FELLER, *An Introduction to Probability Theory and Its Applications*, vol. 1, New York: Wiley, 1950.

[40] J. FERRANTE AND C. RACKOFF, A decision procedure for the first order theory of real addition with order, *SIAM J. Comput.*, 4 (1975), pp. 69–76.

[41] ———, *The Computational Complexity of Logical Theories*, vol. 718 of *Lecture Notes in Mathematics*, New York: Springer-Verlag, 1979.

[42] M.J. FISCHER AND M.O. RABIN, Superexponential complexity of Presburger arithmetic, in *Complexity of Computation, SIAM–AMS Proceedings*, vol. 7, Amer. Math. Soc., 1974, pp. 27–41.

[43] S. FORTUNE, A note on sparse complete sets, *SIAM J. Comput.*, 8 (1979), pp. 431–433.

[44] N. FRANCEZ, *Fairness*, New York: Springer-Verlag, 1986.

[45] R.M. FRIEDBERG, Two recursively enumerable sets of incomparable degrees of unsolvability, *Proc. Nat. Acad. Sci.*, 43 (1957), pp. 236–238.

[46] M. Furst, J. Saxe, and M. Sipser, Parity, circuits, and the polynomial time hierarchy, *Math. Syst. Theory*, 17 (1984), pp. 13–27.

[47] K. Gödel, On undecidable propositions of formal mathematical systems, in *The Undecidable*, M. Davis, ed., Hewlitt, NY: Raven Press, 1965, pp. 5–38.

[48] O. Goldreich, S. Micali, and A. Wigderson, Proofs that yield nothing but their validity, and a methodology of cryptographic protocol design, in *Proc. 27th Symp. Foundations of Computer Science*, IEEE, October 1986, pp. 174–187.

[49] S. Goldwasser, S. Micali, and C. Rackoff, The knowledge complexity of interactive proof systems, *SIAM J. Comput.*, 18 (1989), pp. 186–208.

[50] S. Goldwasser and M. Sipser, Private coins vs. public coins in interactive proof systems, in *Proc. 18th Symp. Theory of Computing*, ACM, May 1986, pp. 59–68.

[51] G.H. Hardy and E.M. Wright, *An Introduction to the Theory of Numbers*, 5th ed., Oxford, UK: Oxford University, 1979.

[52] D. Harel, Effective transformations on infinite trees, with applications to high undecidability, dominoes, and fairness, *J. Assoc. Comput. Mach.*, 33 (1986), pp. 224–248.

[53] D. Harel and D. Kozen, A programming language for the inductive sets, and applications, *Inf. Control*, 63 (1984), pp. 118–139.

[54] D. Harel, D. Kozen, and J. Tiuryn, *Dynamic Logic*, Cambridge, MA: MIT Press, 2000.

[55] J. Hartmanis and R.E. Stearns, On the computational complexity of algorithms, *Trans. Amer. Math. Soc.*, 117 (1965), pp. 285–306.

[56] J. Håstad, Almost optimal lower bounds for small depth circuits, in *Proc. 18th Symp. Theory of Computing*, New York: ACM, 1986, pp. 6–20.

[57] ———, Clique is hard to approximate within $n^{1-\varepsilon}$, *Acta Math.*, 182 (1999), pp. 105–142.

[58] ———, Some optimal inapproximability results, *J. Assoc. Comput. Mach.*, 48 (2001), pp. 798–859.

[59] F.C. Hennie, One-tape off-line Turing machine computations, *Inf. Control*, 8 (1965), pp. 553–578.

[60] F.C. Hennie and R.E. Stearns, Two-tape simulation of multitape turing machines, *J. Assoc. Comput. Mach.*, 13 (1966), pp. 533–546.

[61] J. Hopcroft, R. Motwani, and J. Ullman, *Introduction to Automata Theory, Languages, and Computation*, 2nd ed., Reading, MA: Addison-Wesley, 2001.

[62] J.E. Hopcroft and R.M. Karp, An $n^{5/2}$ algorithm for maximum matching in bipartite graphs, *SIAM J. Comput.*, 2 (1973), pp. 225–231.

[63] J.E. HOPCROFT AND J.D. ULLMAN, *Introduction to Automata Theory, Languages and Computation*, Reading, MA: Addison-Wesley, 1979.

[64] D. IERARDI AND D. KOZEN, Parallel resultant computation, in *Synthesis of Parallel Algorithms*, J. Reif, ed., San Francisco: Morgan Kaufmann, 1993, pp. 679–720.

[65] N. IMMERMAN, Nondeterministic space is closed under complement, *SIAM J. Comput.*, 17 (1988), pp. 935–938.

[66] D. JOHNSON, Approximation algorithms for combinatorial problems, *J. Comput. Syst. Sci.*, 9 (1974), pp. 256–278.

[67] N.D. JONES, Space-bounded reducibility among combinatorial problems, *J. Comput. Syst. Sci.*, 11 (1975), pp. 68–85.

[68] N.D. JONES, E. LIEN, AND W. LAASER, New problems complete for nondeterministic logspace, *Math. Syst. Theory*, 10 (1976), pp. 1–17.

[69] H. KARLOFF AND U. ZWICK, A 7/8 approximation algorithm for MAX3SAT?, in *Proc. 38th Symp. Foundations of Computer Science*, Los Alamitos, CA: IEEE, 1997, pp. 406–415.

[70] R.M. KARP, Reducibility among combinatorial problems, in *Complexity of Computer Computations*, R.E. Miller and J.W. Thatcher, eds., New York: Plenum, 1972, pp. 85–103.

[71] S.C. KLEENE, A theory of positive integers in formal logic, *Amer. J. Math.*, 57 (1935), pp. 153–173, 219–244.

[72] ——, On notation for ordinal numbers, *J. Symbolic Logic*, 42 (1938), pp. 150–155.

[73] ——, *Introduction to Metamathematics*, Princeton, NJ: D. van Nostrand, 1952.

[74] ——, On the forms of the predicates in the theory of constructive ordinals (second paper), *Amer. J. Math.*, 77 (1955), pp. 405–428.

[75] D. KOZEN, *The Design and Analysis of Algorithms*, New York: Springer-Verlag, 1991.

[76] ——, *Automata and Computability*, New York: Springer-Verlag, 1997.

[77] ——, Computational inductive definability, *Ann. Pure Appl. Logic*, 126 (2004), pp. 139–148.

[78] R.A. LADNER, The circuit value problem is logspace complete for P, *SIGACT News*, 7 (1975), pp. 18–20.

[79] L.A. LEVIN, Universal sorting problems, *Prob. Inf. Transmission*, 9 (1973), pp. 265–266.

[80] L. LOVÁSZ, On determinants, matchings, and random algorithms, in *Proc. Symp. on Fundamentals of Computing Theory*, L. Budach, ed., Berlin: Akademia-Verlag, 1979, pp. 565–574.

[81] C. LUND, L. FORTNOW, H. KARLOFF, AND N. NISAN, Algebraic methods for interactive proof systems, *J. Assoc. Comput. Mach.*, 39 (1992), pp. 859–868.

[82] S.R. MAHANEY, Sparse complete sets for *NP*: solution of a conjecture by Berman and Hartmanis, *J. Comput. Syst. Sci.*, 25 (1982), pp. 130–143.

[83] E.M. McCREIGHT AND A.R. MEYER, Classes of computable functions defined by bounds on computation: Preliminary report, in *Proc. 1st ACM Symp. Theory of Computing (STOC'69)*, New York: ACM, 1969, pp. 79–88.

[84] R. McNAUGHTON, Testing and generating infinite sequences by a finite automaton, *Inf. Control*, 9 (1966), pp. 521–530.

[85] A.R. MEYER AND D.M. RITCHIE, The complexity of loop programs, in *Proc. ACM Natl. Meeting*, 1967, pp. 465–469.

[86] G.L. MILLER, Riemann's hypothesis and tests for primality, *J. Comput. Syst. Sci.*, 13 (1976), pp. 300–317.

[87] Y.N. MOSCHOVAKIS, *Elementary Induction on Abstract Structures*, Amsterdam: North-Holland, 1974.

[88] A.A. MUCHNIK, On the unsolvability of the problem of reducibility in the theory of algorithms, *Doklady Akademii Nauk SSSR*, 108 (1956), pp. 194–197.

[89] D. MULLER, Infinite sequences and finite machines, in *Proc. 4th Symp. Switching Circuit Theory and Logical Design*, Los Alamitos, CA: IEEE, 1963, pp. 3–16.

[90] J. MYHILL, Creative sets, *Zeitschrift für mathematische Logik und Grundlagen der Mathematik*, 1 (1955), pp. 97–108.

[91] D.C. OPPEN, An upper bound on the complexity of Presburger arithmetic, *J. Comput. Syst. Sci.*, 16 (1978), pp. 323–332.

[92] C.H. PAPADIMITRIOU, *NP*-completeness: A retrospective, in *24th Int. Colloq. Automata, Languages and Programming (ICALP'97)*, P. Degano, R. Gorrieri, and A. Marchetti-Spaccamela, eds., vol. 1256 of *Lecture Notes in Computer Science*, Bologna, Italy, New York: Springer-Verlag, July 1997, pp. 2–6.

[93] C.H. PAPADIMITRIOU AND S. ZACHOS, Two remarks on the power of counting, in *Proc. 6th GI Conf. Theoretical Computer Science*, New York: Springer-Verlag, 1982, pp. 269–276.

[94] E.L. POST, Finite combinatory processes-formulation, I, *J. Symbolic Logic*, 1 (1936), pp. 103–105.

[95] ——, Formal reductions of the general combinatorial decision problem, *Amer. J. Math.*, 65 (1943), pp. 197–215.

[96] ——, Recursively enumerable sets of positive integers and their decision problems, *Bull. Amer. Math. Soc.*, 50 (1944), pp. 284–316.

[97] V.R. PRATT, Every prime has a succinct certificate, *SIAM J. Comput.*, 4 (1975), pp. 214–220.

[98] M. PRESBURGER, Über die Vollständigkeit eines gewissen Systems der Arithmetik ganzer Zahlen, in welchen die Addition als einzige Operation hervortritt, *Comptes Rendus, I. Congrès des Math. des Pays Slaves*, (1929), pp. 192–201.

[99] M.O. RABIN, Decidability of second order theories and automata on infinite trees, *Trans. Amer. Math. Soc.*, 141 (1969), pp. 1–35.

[100] ——, Probabilistic algorithms for testing primality, *J. Num. Theory*, 12 (1980), pp. 128–138.

[101] O. REINGOLD, Undirected ST-connectivity in log-space, in *Proc. 37th Symp. Theory of Computing*, New York: ACM, May 2005, pp. 376–385.

[102] H.G. RICE, Classes of recursively enumerable sets and their decision problems, *Trans. Amer. Math. Soc.*, 89 (1953), pp. 25–59.

[103] ——, On completely recursively enumerable classes and their key arrays, *J. Symbolic Logic*, 21 (1956), pp. 304–341.

[104] H. ROGERS, *Theory of Recursive Functions and Effective Computability*, New York: McGraw-Hill, 1967.

[105] R. RUBINFELD AND M. SUDAN, Robust characterizations of polynomials with applications to program testing, *SIAM J. Comput.*, 25 (1996), pp. 252–271.

[106] W. RUZZO, On uniform circuit complexity, *J. Comput. Syst. Sci.*, 22 (1981), pp. 365–383.

[107] S. SAFRA, On the complexity of ω-automata, in *Proc. 29th Symp. Foundations of Comput. Sci*, Los Alamitos, CA: IEEE, October 1988, pp. 319–327.

[108] W. SAVITCH, Relationship between nondeterministic and deterministic tape complexities, *J. Comput. Syst. Sci.*, 4 (1970), pp. 177–192.

[109] M. SCHÖNFINKEL, Über die Bausteine der mathematischen Logik, *Math. Annalen*, 92 (1924), pp. 305–316.

[110] J.T. SCHWARTZ, Fast probabilistic algorithms for verification of polynomial identities, *J. Assoc. Comput. Mach.*, 27 (1980), pp. 701–717.

[111] A. SHAMIR, *IP = PSPACE*, in *Proc. 31st Symp. Foundations of Computer Science*, Los Alamitos, CA: IEEE, 1990, pp. 11–15.

[112] M. SIPSER, A complexity theoretic approach to randomness, in *Proc. 15th Symp. Theory of Computing*, New York: ACM, 1983, pp. 330–335.

[113] ——, *Introduction to the Theory of Computation*, Pacific Grove, CA: Brooks Cole, 1996.

[114] R.I. SOARE, *Recursively Enumerable Sets and Degrees*, Berlin: Springer-Verlag, 1987.

[115] R. STEARNS, J. HARTMANIS, AND R. LEWIS, Hierarchies of memory limited computations, in *Proc. IEEE Conf. Switching Circuit Theory and Logical Design*, 1965, pp. 179–190.

[116] L. STOCKMEYER AND A. CHANDRA, Provably difficult combinatorial games, *SIAM J. Comput.*, 8 (1979), pp. 151–174.

[117] L.J. STOCKMEYER, The polynomial-time hierarchy, *Theor. Comput. Sci.*, 3 (1976), pp. 1–22.

[118] L.J. STOCKMEYER AND A.R. MEYER, Word problems requiring exponential time, in *Proc. 5th Symp. Theory of Computing*, New York: ACM, 1973, pp. 1–9.

[119] R. SZELEPCSÉNYI, The method of forcing for nondeterministic automata, *Bull. EATCS*, 33 (1987), pp. 96–100.

[120] S. TODA, On the computational power of PP and $\oplus P$, in *Proc. 30th Symp. Foundations of Computer Science (FOCS'89)*, Los Alamitos, CA: IEEE, 1989, pp. 514–519.

[121] ——, PP is as hard as the polynomial-time hierarchy, *SIAM J. Comput.*, 20 (1991), pp. 865–877.

[122] B.A. TRAKHTENBROT, Turing computations with logarithmic delay, *Algebra i Logika*, 3 (1964), pp. 33–48.

[123] A.M. TURING, On computable numbers with an application to the Entscheidungsproblem, *Proc. London Math. Soc.*, 42 (1936), pp. 230–265. Erratum: Ibid., 43 (1937), pp. 544–546.

[124] L.G. VALIANT, The complexity of computing the permanent, *Theor. Comput. Sci.*, 8 (1979), pp. 189–201.

[125] L.G. VALIANT AND V.V. VAZIRANI, NP is as easy as detecting unique solutions, in *Proc. 17th ACM Symp. Theory of Computing (STOC'85)*, New York: ACM, 1985, pp. 458–463.

[126] A.C. YAO, Separating the polynomial-time hierarchy by oracles, in *Proc. 26th Symp. Foundations of Computer Science (FOCS'85)*, Los Alamitos, CA: IEEE, 1985, pp. 1–10.

[127] R.E. ZIPPEL, Probabilistic algorithms for sparse polynomials, in *Proc. EUROSAM 79*, Ng, ed., vol. 72 of *Lect. Notes in Comput. Sci.*, New York: Springer-Verlag, 1979, pp. 216–226.

Notation and Abbreviations

Index

TEXTS IN COMPUTER SCIENCE   *(continued from page ii)*